Emerging Johannesburg

PERSPECTIVES ON THE POSTAPARTHEID CITY

EDITED

ROBERT A. BEAUREGARD,
LINDSAY BREMNER, AND
XOLELA MANGCU

Routledge
New York • London

Published in 2003 by
Routledge
29 West 35th Street
New York, NY 10001
www.routledge-ny.com

Published in Great Britain by
Routledge
11 New Fetter Lane
London EC4P 4EE
www.routledge.co.uk

Routledge is an imprint of the Taylor & Francis Group.
Printed in the United States of America on acid-free paper.

10 9 8 7 6 5 4 3 2 1

Library of Congress Cataloging-in-Publication Data

Emerging Johannesburg / edited by Richard Tomlinson . . . [et al.]
        p. cm.
Includes bibliographical references and index.
    ISBN 0-415-93558-X (hardback) — ISBN 0-415-93559-8 (pbk.)
    1. Johannesburg (South Africa)—Social conditions. 2. Johannesburg (South Africa)—Social conditions.
3. Johannesburg (South Africa)—Politics and government. I. Tomlinson, Richard, 1952-

    HN801.J64 E44 2003
    306'.096822'1—dc21

2002010902

Contents

Acknowledgments vii

Introduction ix
Richard Tomlinson, Robert A. Beauregard, Lindsay Bremner, and Xolela Mangcu

Section I REORGANIZING SPACE

1 The Postapartheid Struggle for an Integrated Johannesburg 3
Richard Tomlinson, Robert A. Beauregard, Lindsay Bremner, and Xolela Mangcu

2 Villas of the Highveld: A Cultural Perspective on Johannesburg
and Its "Northern Suburbs" 21
André P. Czeglédy

3 The Race, Class, and Space of Shopping 43
Richard Tomlinson and Pauline Larsen

4 New Forms of Class and Racial Segregation: Ghettos or Ethnic Enclaves? 56
Ulrich Jürgens, Martin Gnad, and Jürgen Bähr

5 Property Investors and Decentralization: A Case of False Competition? 71
Soraya Goga

iii

Section II EXPERIENCING CHANGE

6 Making a Living in the City: The Case of Clothing Manufacturers 85
Anna Kesper

7 Violent Crime in Johannesburg 101
Ingrid Palmary, Janine Rauch, and Graeme Simpson

8 On Belonging and Becoming in African Cities 123
Graeme Gotz and AbdouMaliq Simone

Photographic Essay: Rodney Place and ZAR Works, Johannesburg:
RETREKS, Post-CARDS (1999) 148

Section III GOVERNING AND INSTITUTION BUILDING

9 Reclaiming Democratic Spaces: Civics and Politics in
Posttransition Johannesburg 155
Patrick Heller

10 HIV/AIDS: Implications for Local Governance, Housing,
and Delivery of Services 185
Elizabeth Thomas

11 Social Differentiation and Urban Governance in Greater Soweto:
A Case Study of Postapartheid Meadowlands 197
Jo Beall, Owen Crankshaw, and Susan Parnell

12 The Limits of Law: Social Rights and Urban Development 215
Erica Emdon

13 Johannesburg Art Gallery and the Urban Future 231
Jillian Carman

Section IV REREPRESENTING

14 Johannesburg's Futures: Beyond Developmentalism and Global Success 259
Jennifer Robinson

15 Johannesburg in Flight from Itself: Political Culture Shapes Urban Discourse ... 281
Xolela Mangcu

About the Editors 293

Contributing Authors 295

Index 299

Acknowledgments

This book is an outcome of an international, multidisciplinary conference—Urban Futures—held in Johannesburg from July 10–14, 2000. Professor Colin Bundy, then Vice Chancellor of the University of the Witwatersrand and now Head of the School of African and Oriental Studies at the University of London, initiated and supported the conference. Without Professor Bundy's vision of a new partnership between the University and the Greater Johannesburg Metropolitan Council (now the City of Johannesburg), neither the conference nor this book would have been possible.

Cohosted by the University of the Witwatersrand and the City of Johannesburg, the conference brought together academics, policymakers, local government administrators, and community activists from around the world to discuss and debate understandings and themes relating to urban life at the turn of the millenium. Its sessions included those normally appearing on an academic conference program—plenaries, paper and poster presentations—as well as a series of "Urban Policy Lekgotlas" (a gathering or meeting, in Sesotho) where key policy issues related to local government and its transformation were discussed. The conference was complemented by a rich and varied cultural program, highlighting the role of culture in urban life.

The sections of this book were constructed around key themes that emerged at the conference: race, democracy, citizenship, urban demographics, space, and identity. Its chapters were selected from conference papers, presentations, and exhibitions that considered how these themes were framed in the conference's host city, Johannesburg. In doing so, the editors trust that the volume makes a contribution to developing an understanding of the postapartheid city, as well as to broader international debates.

We would like to express our appreciation to the Ford Foundation for funding Robert Beauregard's trip to South Africa in early 2001 for the purpose of developing the structure of the book and mapping its content. In addition, our gratitude goes to Werner Fourie of Global Image for providing the maps that appear in the book and Melisa Olivero at the New School University for her clerical assistance. David McBride of Routledge recognized the importance of the book, found two supportive reviewers, and offered numerous helpful suggestions. We also acknowledge the production staff—particularly Nicole Ellis—for making the process of producing a high-quality book go smoothly. Finally we thank our contributors not only for their insightful papers but also for their patience in the face of editorial demands.

Introduction

RICHARD TOMLINSON, ROBERT A. BEAUREGARD,
LINDSAY BREMNER, AND XOLELA MANGCU

In the long decades of grand apartheid, there was a single Johannesburg. Its white population understood it to be theirs and the black population could neither escape nor significantly change the city's role in their lives. A small minority dissented; the dominant view was always a mixture of fact and fiction. Love crossed racial divides, blacks made homes in white areas, and white liberals ("disloyal" to their race) criticized the apartheid regime, among other resistances. Nonetheless, only a few people managed to elude the control of an authoritarian state. The spaces of South Africa (and those spaces "not" of South Africa, such as the infamous Bantustans) had to be clearly and precisely defined. At the commanding heights of apartheid, the truth was widely known if not universally shared. There was one official Johannesburg, all others were hidden or suppressed.

With the election in 1994 of Nelson Mandela as the country's President, the final dissolution of grand apartheid began in earnest and the nation embraced a nonracial democracy. The truth about Johannesburg disappeared. With the extension of civil, political, and social rights to those who had been apartheid's main victims, the bindings that had held multi-racialism, tribal affiliations, and class divisions in check were removed. A multiplicity of Johannesburgs came into being, each with a different imaginative moment.

The process of rethinking Johannesburg, of course, was underway prior to 1994. As far back as the early 1980s, with reforms weakening the apartheid state, it seemed possible to remake the city. With nonracial democracy came the extension of rights and a greater sense of belonging for the majority black population. Expectations soared as the first national election neared; all adults, regardless of race, could finally vote. Since then, a cold reality has drifted across the social landscape. Poverty, stubbornly high unemployment, widespread lack of access to decent housing, and a host of other durable inequalities have not been radically transformed. More and more, the country and its cities are faced with the problems—realizing neoliberal "economic fundamentals" viewed (by some) as essential for attracting foreign capital, redistributing income, expanding the economy, balancing local government budgets, or counteracting the epidemic of HIV/AIDS—that must be solved before justice can be achieved. While solutions await implementation, Johannesburg changes so rapidly as to defy comparison.

The fall of the apartheid state was part of a breakdown of those modernist institutions by which the state had maintained its dominance. Wide-ranging restrictions on travel and behavior (such as the pass laws), the bureaucratization of racial categories, the rigid application of city planning principles (specifically zoning) to produce separate races, and the fascination with a "pure" national identity all collapsed. Their absence reverberated through local government, housing markets, transit systems, labor–business relations, and schools. From the mid-1980s to the mid-1990s, the vacuum created by the collapse of modernist institutions was unfilled.

During this time, South Africa hoped to benefit from a return to the good graces of the international community. With sanctions lifted and its pariah status removed, foreign direct investment and involvement in global markets would bolster the private economy and the public sector. The nation would take a new role in southern Africa and the world. The replacement of tainted modernist institutions, though, particularly in the face of pressure to solidify political power and remove racial inequalities, was an even more formidable task. Models of urban policy, especially those represented in the Habitat Agenda and World Bank documents, essentially suppressed more indigenous alternatives.

Transformation from above, however, cannot proceed independently of the legitimacy engendered by building from below. Nation building without city building is a hollow exercise. To remove the barriers to non-racial justice, the cities, townships, suburbs, rural communities, and peri-urban places must also be transformed.

The central government's rebuilding of places has had to serve at least three masters. Previously oppressed blacks view it as a test of the country's commitment to social justice and democracy and as a measure of their ability to govern. Once-protected whites need reassurance that they are still important. And international investors and corporate leaders view the dismantling of apartheid in the light of the government's commitment to a neoliberal global agenda. The visions are as different as the means to achieve them.

Within this nexus of city building and nation building, Johannesburg occupies a unique position. As the country's premier industrial city, center of finance, and site of many of the most important struggles against apartheid, the success or failure of new institutional and social relations will be widely noticed within and without the country. How Johannesburg fares will also be seen, unfairly or not, as indicative of the success or failure of South Africa.

Just as a nation requires those who live within its boundaries to imagine themselves as citizens, a successful city requires that its residents also identify with it and feel a moral attachment to its fortunes. In the post-apartheid era, institutions are in disarray and cultural identifies flourish. There are now as many Johannesburgs as there are cultural identities. Each group experiences the city differently and all of these differences are valid. Less confined by the straightjacket of apartheid, group relations are more problematic. They are also more central to the city's governance. What institutions will enable the citizens of Johannesburg to live "together in difference" (Young, 2000:221–8)?

The forging of these institutions is more than a matter of allocating resources and organizing tasks, it also requires a shift in the way we think about cities. Discourses and representations have to change with material conditions, if not before. The city has to be reimagined in ways that both remember the past and resist the modernist logics of states and capitalist markets.

From a policy perspective, the perceptions of international governmental leaders, CEOs of transnational corporations, international consultants, and heads of global finance agencies are crucial. Their ability to regulate the flow of investment and trade will influence employment, the distribution of incomes, and even the location of real estate investment and reinvestment.

One sees this clearly in the expectations attendant to the dispensation of 1994. Political and corporate leaders assumed that the dismantling of apartheid would remove the stigma from the country and lead to the lifting of international sanctions. As a result, businesses and investors would take advantage of the country's skilled labor and relatively low wages and tourists would visit to experience its coastlines and game parks. The election of

Nelson Mandela to the presidency would dispell the apartheid legacy and change how South Africa was perceived across the world.

As the center of finance for South Africa, Johannesburg would thrive. It would become the gateway for South Africa's entry into the global economy. Consequently, publicists for the city touted its ostensible inclusivity and rapid transition to equal opportunity. The city adopted a strategic plan that would promote Johannesburg as a world city, an international metropolis, and even bid to become the South African nominee for the 2004 Olympic Games (Rogerson, 1996). Without the burden of apartheid, Johannesburg could be imagined as a global city.

Despite the city's efforts, the perception of the city as dangerous and the economy as unstable persisted. Investors did not rush in and tourism has remained minimal. Although Johannesburg discarded one off-putting image, it seemed to attract another, an image of a crime-ridden and deteriorating city, an inner city in decline.

Among international aid agencies and consultants, Johannesburg was reimagined as comparable to other large cities in developing countries (Tomlinson, 2002). They viewed the end of apartheid as an invitation to require the city to make the transition to a neoliberal regime of downsized government, privatized public services, wage restraints, and "open" markets. International "best practices" could now be deployed and Johannesburg could restructure itself to be attractive to a relatively unfettered capitalism. Policymakers and consultants imagined Johannesburg as capable of making itself market friendly. Only then could its developmental problems—lingering poverty and a large redundant population, among others—be addressed. This perspective was at odds with that of the majority of the black population; they still believed in the efficacy of the government.

Johannesburg has also changed in the minds of those who live there. Once a white, European (predominantly English) city in Africa, its purity eroded in the 1980s as a mixture of European and African life-styles appeared. No longer unrelentingly white, the city became dappled with gray areas, black spots, and illegal residences. By 2001, parts of the inner city had become almost wholly black and African, with a recent population of "foreigners" providing a new arena of racial and ethnic tensions.

From the perspective of whites, particularly those who have moved to the northern suburbs, Johannesburg has gone from the citadel of white dominance to the declining inner city of crime and grime. The city is imagined almost like a declining industrial city in the United States (Beauregard, 2003): as obsolete, so deteriorated as to be beyond redemption, or as taken

over by (or left behind for) Africans. It has become the "other" to the affluent northern suburbs and white, working class communities alike. From a
white suburban perspective, Johannesburg, the symbol, has become the city
left behind; Sandton is the new center. The old Johannesburg exists in nostalgia; the new Johannesburg exists in *absentia*.

Inner city Johannesburg has occupied a much different place in the collective psyche of black South Africans. For most of the country's history, it
was a source of employment, a place of opportunity: first under compulsion
and later (as the rural black economy was destroyed) due to necessity. The
city also symbolized the oppression of the black population. Influx laws
made the city a forbidden place even though black labor—the men confined
to hostels and the mines and the women serving as domestic servants—
made it livable. Blacks moved there to minimize the costs of having to work
in the white, formal economy. Even before the removals in the 1930s and
1950s, nonwhite groups were barely tolerated in the city, even if they managed to make a place for themselves there.

Many blacks believed that once state apartheid was no more, Johannesburg would thrive—the new government would ensure this—and they could
move there and live well. The 1994 dispensation that brought the African
National Congress (ANC) to power would yield a government committed
to providing new housing, jobs, and improved and more accessible services
quickly and to everyone. The government has delivered housing and services,
but many blacks view them as inferior and their locations worse than they
had received under apartheid. In Johannesburg, housing and services might
be better than in the townships to the south, and commuting less onerous,
but the city's hospitality toward blacks leaves much to be desired.

Just when blacks were able to imagine Johannesburg as also belonging
to them, and thus to imagine themselves as city dwellers, two events
ensued. First, whites fled and took with them their retail businesses, insurance companies, and stock exchange. The "shell" they left to the poorer
black population. A multiracial Johannesburg was quickly taken off the
agenda. Second, Johannesburg witnessed an influx of blacks from other
African countries who moved aggressively into street trading and illegal
activities such as prostitution and drugs. Black South Africans have been
forced to reimagine Johannesburg not as belonging to all South Africans
or to them alone but shared with others in a new African cosmopolitanism
(Tomlinson, 1999).

Numerous other stories can be told. Those of white suburbanites and
former township residents represent only the dominant ones, and even

these, as we have presented them, are simplifications. What matters in the restructuring of local government and the ebb and flow of investment, of course, are the understandings and imaginings of investors and policymakers, civic leaders, and consultants. These are the groups who will shape local policy.

On the policy side, the challenge is to incorporate the egalitarian and social justice impulses of the antiapartheid struggle while recognizing the imperatives of global capitalism and its neoliberal mode of governance. If we imagine Johannesburg as a global city, our gaze turns to one set of policies involving local government, residential choice, and business regulations and inducements (Robinson, this volume). If we imagine Johannesburg as a Third World city, another set of policy options comes into view (Mabin, 1999). Of utmost importance is to imagine Johannesburg in a way that captures its current transformations and possibilities. Johannesburg is not yet a global city. But neither is it a Third World city in the classic sense. What is the new postapartheid Johannesburg?

On the theory side, Johannesburg is an opportunity. The rapidity of its transformation, the simultaneous destabilization of government structures, economic relations, social understandings, cultural images, and spatial orderings, and the avowed goal to recenter South Africa and its cities in global networks make Johannesburg emblematic of the forces operating in cities around the world. Unique in its own ways, Johannesburg is also similar to many other places. Cities in the United States and Brazil, to take two examples, also struggle with concentrated poverty, racial segregation, and inner city slums.

The authors in this book confront the policy biases and dominant theoretical inclinations about how best to think about Johannesburg. In its place, they emphasize the symbolic transformation of Johannesburg from a city of apartheid to one representative of a nonracial, multicultural, democratic, and just nation. In this complex city, to create a cultural space for the Johannesburg Art Gallery, sangomas from different cultures proposed sacrificing animals and conducting vigils to bring together the spirits of the departed. To cope with the extraordinary depredations of HIV/AIDS, policymakers must devise new ways of delivering housing and services for which no recipe exists in international best practice. To revive the central business district, formal retailing and informal street trading must coexist. The contributors to this volume write in the midst of this uncertainty, knowing the transformation's origins but at a loss to name its destination.

The chapters are organized around broad themes. We begin with the reorganization of space in which market forces only partially replace apartheid's urban form and racism continues to burden the poor. We next turn to the experience of change in informal enterprises, in the face of crime, and in the search for new identities. The third grouping of chapters focuses on drafting legislation and building new institutions for local governance in ways that recognize apartheid's heritage and the wrenching disruptions emanating from HIV/AIDS. The book ends with two chapters that investigate how we should think about the transformation of Johannesburg and its future in the global economy.

REFERENCES

Beauregard, Robert A. 2003. *Voices of Decline: The Postwar Fate of U.S. Cities.* New York: Routledge.

Mabin, Alan. 1999. "The Urban World Through a South African Prism," in R. A. Beauregard and S. Body-Gendrot, editors, *The Urban Moment: Cosmopolitan Essays on the Late 20th Century City.* Thousand Oaks, CA: Sage Publications, pp. 141–52.

Rogerson, Chris. 1996. "Image Enhancement and Local Economic Development in Johannesburg." *Urban Forum* 7:139–58.

Tomlinson, Richard. 1999. "From Exclusion to Inclusion: Rethinking Johannesburg's Central City." *Environment & Planning A* 31:1655–78.

———. 2002. "International Best Practices, Enabling Frameworks and the Policy Process: A South African Case Study." *International Journal of Urban and Regional Research* 26:377–388.

Young, Iris M. 2002. *Inclusion and Democracy.* Oxford: Oxford University Press.

Section I

REORGANIZING SPACE

South African apartheid was built on spatial divisions; separateness was taken literally to mean that whites and blacks were to live apart from each other. In postapartheid Johannesburg, spatial divisions persist, though no longer solely based on racial differences. In turn, these spatial divisions—as they did under apartheid—reinforce existing structures of privilege and make it difficult to create a just, democratic, and egalitarian society.

This section begins—Chapter 1—with an overview of these issues, tracing spatial divisions back in time and into the present. Czeglédy and Goga explore the most visibly obvious of them in their reflections, respectively, on the emergence of the wealthy, northern suburbs and the shift of office investment from the city to the areas surrounding the new edge city of Sandton. Czeglédy notes how even the architecture in the northern suburbs reinforces divisions between the private and the public (particularly through gated communities) and the individual and the state. Goga, later in this section, reveals the decision making that went into the northern migration of prime office space as large corporate investors responded to an excess of capital in a time of increasing uncertainty about the inner city.

Tomlinson and Larsen investigate these themes from a retail perspective and further deepen our understanding of the shift of wealth to the north. The rise of informal trading within the inner city, the flight of upscale retail, and the emergence of shopping malls on the outskirts changed the shopping behavior of whites and blacks, poor and wealthy alike. These behaviors are now based as much on class as they are on race.

1

Implicated in these transformations is the rising residential fragmentation of the metropolitan region; the dismantling of apartheid strictures led to greater residential mobility and heightened demands for security. Jürgens, Gnad, and Bähr document the rise of middle-income gated communities that cater to blacks and Indians and show how inner city neighborhoods retain their allure for migrants yet remain precarious. The end of apartheid has brought a new spatial fragmentation.

1

The Postapartheid Struggle for an Integrated Johannesburg

RICHARD TOMLINSON, ROBERT A. BEAUREGARD,
LINDSAY BREMNER, AND XOLELA MANGCU

Over 230 years after the first white settlers arrived at the tip of Africa, gold was discovered on the high plains, nearly 1,000 kilometers inland from Cape Town. The year was 1886. The discovery led to the founding of Johannesburg, which, throughout most of its history, has since served as the center of the English capital in South Africa.

From the beginning, Johannesburg was viewed as a city of "uitlanders" or foreigners. First applied by the Afrikaners to the English, this label was soon used for blacks also. Until the 1980s, and despite the existence of a large, urban black population, the apartheid philosophy was that blacks were temporarily in the white cities to serve in the mines, work in industries, and provide services. Once their working life was over, they were to go "home" to the rural areas. The philosophy also involved forcibly removing Africans, Indians, and coloreds from Johannesburg and expelling them to the urban periphery. Johannesburg, much like the country of South Africa, was for whites and their descendents.

South Africa's first multiracial national election in 1994 marked the end of state-sanctioned racism. It was also the year in which Johannesburg was reconstituted as a city with white and nonwhite areas brought under the same local government. This meant considerably increasing the city's population while sharing much the same fiscal base.

No longer an international pariah, Johannesburg would reenter the global economy. The promise was that the world's markets would purchase the city's goods and services and foreign corporations would invest in its economy. Once integrated into the global circuits of capital, the city would prosper. Although civic leaders proposed to make Johannesburg a "globally competitive African world-class city," they struggled to give substance to this vision (C. Rogerson, 1996). It seemed that before Johannesburg could join the company of world cities it would have to address apartheid's lingering consequences: enduring poverty, too few jobs, racial divisions, capital flight, low educational levels, and the other social and economic obstacles that made the city seemingly no more egalitarian and no more just than it had been before 1994. In addition, in the 1990s Johannesburg came face to face with the HIV/AIDS epidemic.

The purpose of this chapter is to portray the broad outlines of inner city Johannesburg's transformation from a citadel of white supremacy to a struggling city in Africa. A brief history brings the reader quickly to the times of struggle when apartheid was directly challenged and began to weaken. We then focus on the political integration of the city, its demographic makeover, and its enduring social fragmentation. The chapter concludes with a brief reflection on how policymakers are thinking about the city's future.[1]

HISTORICAL BACKGROUND

Within ten years of its origins, Johannesburg was the biggest city in the country and by 1936 was recognized as the "largest and most densely populated European city in Africa" (cited in Chipkin, 1993:105). A building boom prior to the turn of the century and an economic expansion following the Boer War (1899–1902) between the British and Afrikaners resulted in a rapid increase of population.[2] The latter expansion also produced a consolidation of the gold mining industry and the creation of the city's financial district. During the 1930s depression, the gold standard was abandoned and this led to another surge in investment. Foreign capital flooded into the country and transformed Johannesburg into a little New York or, if not New York, then at least Chicago or Saint Louis (cited in Chipkin, 1993:93). Successive waves of economic activity, corresponding largely to the booms and slumps of gold mining, produced by 1990 a city housing the headquarters of most of the country's largest corporations, the Johannesburg Stock Exchange, and the National Reserve Bank.

The city's most enduring economic upswing emerged out of crisis. On March 21, 1960, in a black township 75 kilometers to the south of the city, police fired on a peaceful crowd protesting the pass laws.[3] The Sharpeville massacre—69 of the protesters were killed—led to massive outflows of capital and a critical fall in gold reserves. The government responded with draconian measures. By 1962, the economic situation had been dramatically reversed. Growth reached 9% per annum in 1963. The South African economic miracle, which was to last for a decade, was underway (Chipkin, 1998:250).

Johannesburg was booming. Foreign capital flooded into the city and the Johannesburg stock market recorded many new listings, acquisitions, and mergers. Tower blocks mushroomed in the downtown area. Nearby, numerous factories were built and between 1951 and 1970 employment in manufacturing rose by over 75% to approximately 230,000 workers (Beavon, 1997:159). *Fortune* magazine quoted an American businessman as saying "This is the only real industrial complex south of Milan" (cited in Chipkin, 1998:250). A sense of confidence and invulnerability prevailed.

By the 1970s, especially after the Soweto uprising in 1976, this image had begun to tarnish. Growing resistance to apartheid revealed the city's racial segregation and political divisions. Its decades of celebrating white dominance and the brushing aside of an alternative black experience of the city were no longer tolerable (C. Rogerson, 1996). By 1986, at the time of the city's centenary celebrations, black opposition to the image of the "city-with-a-golden-heart" reached its peak and rendered the celebrations meaningless. The city's divisions had cracked wide open.

These divisions had been constructed by successive white governments. After the Boer War, an interventionist British government had taken measures to stabilize the white proletariat and the separation of the white from the black working classes. Workers had originally lived close to sites of production in the working class suburbs of Jeppe and Fordsburg or in compounds on mining property; they were gradually segregated and relocated. From the first forced removal of the residents of the "Coolie Location" to Klipspruit to the south of the city in 1904, the history of Johannesburg has been, to the recent past, one of gradual and more rigid segregation on the basis of race, class, and space.

Early efforts to this end produced the Urban Areas Act of 1924 that compelled local authorities to set aside land for black occupation and formalized existing segregation policies. This legislation saw the beginning of a planned program of evacuation of black people from the municipal area and their relocation to state-subsidized housing to the south of the city in the area that became known as Soweto.[4] By 1929, when the Native Urban Areas Act was passed, it was illegal for blacks to rent or purchase property in white designated areas. By 1933 the whole of the city of Johannesburg had been proclaimed white and by 1938 much of the black population had been moved south to the new townships of Orlando and Orlando East in Soweto (Carr, 1990). Of a total population of 500,000 people in Johannesburg, 60,000 had been relocated.

Throughout these years, the city's population grew with every surge in the local economy. By 1936, Johannesburg had over 425,000 legal residents and by the mid-1980s the metropolitan area was the home to over 2.5 million people. This increased to 2.7 million by 1996. The whole urban conurbation, comprised of Johannesburg, Tshwane (formerly Pretoria), and Ekhuruleni (formerly the East Rand) metropolitan areas, was estimated at between 6 and 7 million people. With the country's population at approximately 38 million, this combined metropolitan region accounted for 17% of all South Africans.

By the time the National Party government came to power in 1948 on a platform of "separate development," the segregated landscape of the city had largely been determined. However, the Party's ruthless combination of discriminatory legislation, influx control (that is, mechanisms to prevent blacks from establishing residence in the cities), and the creation of the ethnically based "homeland" areas and racially based administrative zones, enshrined separate development in the total system, which became known as grand apartheid.[5] The implications of this are etched in the landscape (see Figure 1).

In Johannesburg this meant that black people living in the racially mixed area of Sophiatown (established in 1905) were forcibly removed south (in the 1950s) to the new black suburb of Meadowlands. Domestic workers living in quarters on the rooftops of inner city apartment buildings were pushed into single-sex hostels in Soweto. Later, in the 1960s, the Indian population of Pageview was moved to Lenasia 30 kilometers to the south, and coloreds were allocated the areas of Eldorado Park and Ennerdale.

The period from 1948 onward saw the development of vast, sprawling suburbs for people of color south of Johannesburg. These townships, as they became known, had no sustainable commercial or industrial base. Their extraordinarily inefficient layout was designed for security reasons to make internal circulation difficult. All major roads and rail, where they were available, led to employment centers and to the retail shops of the city center.

For whites, the city and the rolling hills to the north were theirs, though the inner city slipped slowly from their grasp beginning in the 1980s. As it did, more and more businesses relocated to the northern suburbs, eventually

Figure 1

Johannesburg's racial landscape, circa 1990.

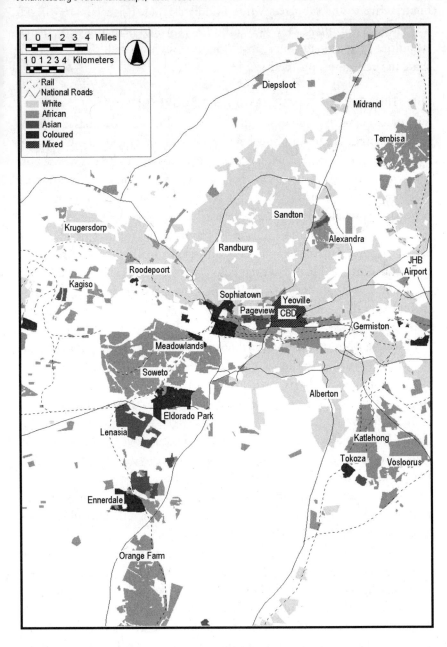

leaving the inner city with unused factories, many abandoned, and empty office buildings, low rentals in the more modern buildings, and retailing unattractive to the suburban white middle class (J. Rogerson, 1996). To leave local government as it was in the 1970s would perpetuate the spatial inequities of apartheid. Reorganization was essential if all people were to share in the region's prosperity.

THE STRUGGLE FOR DEMOCRATIC LOCAL GOVERNMENT

The new, post-1994 Johannesburg was a product of broad-based, democratic negotiations around how best to build an integrated city. Central to the negotiations were the civics, self-organized townships bodies whose purpose was to construct a popular form of grassroots democracy and oppose illegitimate state control (Turok, 1994b). Without the civics, the local negotiations would have proceeded very differently, if at all.[6]

The origins of the civic movement can be traced indirectly to June 16, 1976. On that day Soweto students protested the use of Afrikaans as the medium of instruction in their schools. Two of the protesters were killed by the police. Over the next year, 575 people were killed and 2,389 were wounded in the general strikes that followed the Soweto protest (Lapping, 1987:211–13).

Reeling from the international outcry over the June 16 killings and the subsequent murder by the security police of Steve Biko, the leader of the Black Consciousness movement, the apartheid government attempted a politics of reform under newly elected National Party leader P. W. Botha. In his inauguration speech in 1978 Botha urged whites to "adapt or die." Thereafter he proceeded with a series of constitutional amendments that, *inter alia*, made coloreds and Indians junior partners in a new parliament known as the tricameral parliament. Blacks, however, were still excluded. The only concession, in 1982, was to allow blacks to vote for the newly created black local authorities, while still maintaining that they should exercise their national voting rights within their "homelands."

The creation of black local authorities was the first recognition of the permanence of black residence adjacent to the white cities and a tacit acknowledgment that the old policies of influx control as a way of regulating black urbanization had failed. But the black local authorities were in an impossible situation. As illegitimate political structures and in the absence of a commercial and industrial tax base, they were to collect rents and services payments and to use this revenue base, inadequate as it was, to run the townships. From 1984, numerous campaigns arose to boycott black local authority elections and structures. The civic movement was demanding more than the constricted urban citizenship the government was offering.

The Soweto Rent Boycott began in mid-1986 in response to the increased service charges, deteriorating services, and worsening economic conditions and as a repudiation of the black local authority system. It was eventually joined by 80% of Soweto's formal rent-paying households. The strategy consisted of withholding rental and services payments, the retaliatory refusal to provide services, and consumer boycotts of white-owned stores. The boycott was built around five key demands: (1) the arrears owed by people who have supported the boycott must be written off, (2) the houses must be transferred to the people, (3) services must be upgraded, (4) affordable service charges must be established, (5) and a single tax base for Johannesburg and Soweto must be introduced. "One city, one tax base" became a national rallying cry.

By the early 1990s, fifty-two of eighty-four local authorities in the Transvaal province (in which Johannesburg is located) were hit by rent and service charge boycotts (Murray, 1994). Financial duress and pressure from business people caused local authorities to enter into negotiations with the civic movement and these negotiations preceded negotiations about the broader political dispensation.

The national government initially opposed local-level forums, a situation that changed with the signing of the Soweto Accord in 1990. In the Soweto Accord, government essentially conceded all the Soweto People's Delegation's demands, writing off the arrears owed by residents and agreeing to establish the Central Witwatersrand Metropolitan Chamber to negotiate the implementation of the remaining demands. For its part, the Soweto Civic Association agreed to broker an end to the rent boycott, which it subsequently failed to do.

The Central Witwatersrand Metropolitan Chamber was established as a policymaking body to make decisions by consensus. Its membership eventually expanded to fifty-three organizations: the Transvaal Provincial Administration, all relevant local government structures, many civic associations, white ratepayer and residents associations, and the now unbanned African National Congress (ANC) and other political parties. There were also many observer bodies such as trade unions, educational institutions, and organized business. The Chamber was "a glorious experiment in participatory governance" (Swilling, 2000:E8).

The Chamber represented the "preinterim phase," the first of three phases mapped out by the Local Government Transition Act of 1993. It involved the establishment of local forums to negotiate the appointment of temporary local government councils that would govern until municipal elections in November 1995. The "interim phase" began with the 1995 municipal elections and lasted until a new local government system had been designed and legislated and new local governments were elected. The "final phase" began with local government elections in 2001.

Between 1990, when the Chamber came into being, 1995, when the Greater Johannesburg Metropolitan Council and four Metropolitan Local Councils were established, and 2001, when a single metropolitan government was established, the City of Johannesburg endured an extended period of uncertainty and political dispute, financial difficulties, and cycles of centralization, decentralization, and then centralization again. The City inherited dire financial constraints. The extent of the financial difficulties became apparent in October 1997 when the Provincial Government intervened in the financial affairs of the Greater Johannesburg Metropolitan Council and the local councils because the five councils were heading for an unfunded position of R2 billion by the end of the financial year. The subsequent review led to a provincially imposed, but willingly adopted, structural adjustment program. Subsequently, an ongoing Transformation Lekgotla was appointed in late 1998 to restructure the five councils.[7] It had "delegated powers to decide on key policy issues in relation to the transformation process" (*iGoli 2002*:29).

From the activities of the Transformation Lekgotla in 1999 came the *iGoli 2002* plan for dealing with Johannesburg's financial difficulties and institutional disarray. It "usher[ed] in the birth of the unicity by transforming local government in Greater Johannesburg through changed governance, financial viability, institutional transformation, sustainable development and enhanced service delivery" (*iGoli 2002*:3). Noted in the *iGoli 2002* brochure was that the financial crisis precludes even effective maintenance of the services that exist, let alone the extension of services to overcome backlogs in townships and squatter settlements.

BUT NOT DISMANTLING SEGREGATION

In light of Johannesburg's financial and institutional malaise, and that of most other local governments, central government's Department of Provincial and Local Government promulgated a requirement that all local governments prepare "integrated development plans." The plans were intended to enable local governments to deal with scarcity through aligning their budgets and their service delivery programs and through charting the way for public–private partnerships. Moreover, they would promote compact, integrated urban development (Tomlinson, 2002).

Compact urban development would be achieved through the subsidy of low-income housing and a plan for delivering a million dwelling units within five years. However, most low-income housing projects are located on the outer edges of the townships, heading still further away from jobs, and this has seriously eroded hopes for integrated development (Bremner, 2000; Tomlinson *et al.*, 2000). Residential development in Johannesburg has become increasingly balkanized, with the character of residential change differing according to whether one is referring to the generally low-income population living south of the inner city, the inner-city population, or the generally high-income population living to the north. The differences arise from the still partial shift from race to class in the north, the perpetuation of racial exclusion in the south, and the racial makeover of the inner city. The larger trend is one of settlements proceeding away from the inner city in both the south and north.

The significance of these trends is to be found in Table 1 (also represented in Figure 2) in the form of the eleven new management districts created by the City of Johannesburg. Forty-four percent of Johannesburg's

Table 1

Spatial and Ethnic Distribution of Johannesburg's Population, 1996

Region	African	Asian	Colored	White	Total
Diepslot	22,892	237	216	6,766	30,121
Midrand	102,875	1,246	3,981	24,522	132,624
Sandton	49,526	2,470	1,465	107,904	161,365
Northcliff	38,748	9,473	47,228	104,269	199,717
Roodepoort	41,980	1,723	7,621	124,608	175,933
Soweto	586,314	11,155	11,170	288	608,927
Alexandra	148,868	2,976	1,408	35,052	188,304
Inner City	141,750	13,268	11,506	39,878	206,402
South	60,404	4,897	6,701	77,107	149,109
Diepkloof	513,003	1,405	59,590	1,588	575,586
Orange Farm	196,688	49,106	22,759	611	269,164
Total	1,903,048	97,965	173,646	522,603	2,697,253

Source: Simkins (2000).

Figure 2

Johannesburg's administrative regions.

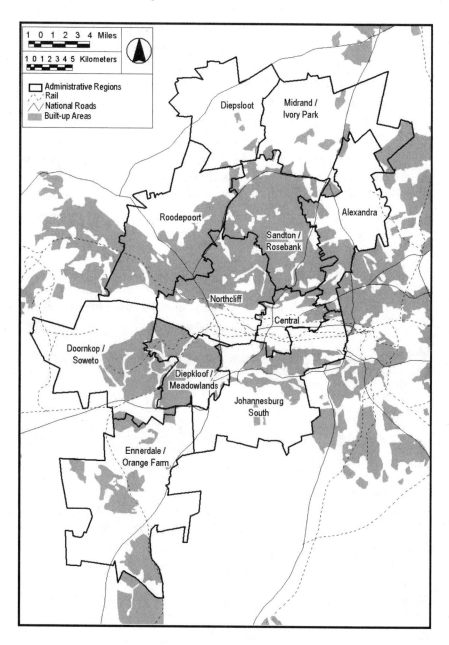

population resides in Soweto and Diepkloof, contiguous areas sited south of the inner city. These districts are far larger than the rest and are almost entirely African. The combined population of the Soweto, Diepkloof, South, and Orange Farm districts is 1.6 million, 59% of the city's total population. As a comparison, only 8% of Johannesburg's residents live in the inner city. The two major population concentrations are those in the south and the north, the latter comprising 33% of the city's total. Blacks comprise 85% of the population living in the south and only 46% of the population living in the north. The latter figure declines to 37% if Alexandra, the township that withstood forced removal, is deleted from the calculations.

In the north, it had long been anticipated that a shift from race to class and the relocation northward of high-income blacks would bring about desegregation (Saff, 1994). This has not happened. On the one hand, between only 1 and 2% of houses in the former whites-only suburbs are being sold to blacks (Beavon, 2000). Furthermore, better-off black households are moving into just a few of the northern suburbs, mostly to Randburg. This is creating "a form of de facto apartheid" with "no real signs of genuine intermingling of races in the former whites-only suburbs being visible by 1999" (Beavon, 2000). On the other hand, Bremner (2002) and Jürgens *et al.* (this volume) have found a much higher percentage of blacks moving into the newer, edge city suburbs, in particular, new walled estates.

In the center, blacks began moving to the inner city from the township areas and into inner city suburbs beginning in the early 1980s.[8] Diminished capacity to enforce apartheid legislation and escalating repression and violence in the townships under a general State of Emergency were key factors. At the time of the city's centenary celebrations in 1986, approximately 20,000 black people of a population of 120,000 lived in the inner city suburb of Hillbrow (Morris, 1999). Migration into inner city suburbs accelerated rapidly. By 1993, 85% of the inner city residential population was black (Morris, 1999) and by 1996 only 5% was white (Crankshaw, 1997). The majority, moreover, were more affluent than those who remained in the segregated townships and in the informal settlements (ICHUT, 1996; Crankshaw, 1997; Morris, 1999).[9]

This rapid "graying" of the inner city was accompanied by physical decline and the racial stereotyping of new residents. In many cases, the exodus of white residents was aided and abetted by landlords who saw the illegal status of black tenants in the 1980s as an opportunity for increasing profits through raising rents, overcrowding, and reducing building maintenance.

Additional demographic shifts occurred during the second half of the 1990s as Africans from countries farther north, mainly Nigeria and the Democratic Republic of Congo, arrived in the inner city. In 1995, 23,000 Congolese were estimated to be residing in Johannesburg (Kadima and Kalombo, 1995), while Morris (1999) estimated that 3,000 Nigerians were living in the inner city. These migrant groups have been subjected to high levels of persecution from South Africans (Dhlomo, 1997). They are blamed for the overcrowded informal trading sector, the growth of the narcotics trade, and the deterioration of the physical environment (Simone, 1998). Increasing xenophobia, assaults, and conflicts over space and access have led to the construction of a defensive, ethnically defined spatiality.

In the south, new urban development was meant to produce a more compact city. Instead of a "deracialization of space," that is, an expansion of townships or informal settlements toward the boundaries of formerly white parts of town (Saff, 1994:382), the growth went the other way. In the case of low-income housing projects, housing delivery was taken forward by private developers. Predictably, they went in search of cheap land and this led them away from the city. More important to developers was political opposition that could hold up projects in the courts for years and dramatically increase the cost of housing. In one instance there were 20,000 written objections to a low-income housing project. The objections mostly came from middle-class black households (Bremner, 2000).

Rather than promote integrated development, the implementation of housing policy is disempowering the recipients of the subsidy (Gear, 1999). For example, there are frequent reports that households, unable to afford the cost of traveling to work and of taking children to school, are informally selling their subsidized plots for as little as R500 and moving closer to the cities.

In the case of informal settlements, people settling close to townships, but mostly on their outer edges, is a natural consequence of high unemployment and family networks that provide a means of survival. It takes money to break away. It also takes organization to invade land closer to the city, but very many erstwhile struggle leaders have been incorporated into government and are thus unavailable for organizing "spontaneous" mass settlements.

In sum, the fact that most new informal settlements and low-income housing projects are located south of the inner city and almost all new jobs are being located along the M1 between the Johannesburg and Tshwane central business districts (CBDs) to the north means that spatial exclusion, so well known under apartheid, continues.

DEEPENING ECONOMIC INEQUALITY

The economic inequality arising from spatial exclusion has been exacerbated by structural changes in the economy and increasing unemployment. Johannesburg's regional economy has seen striking changes over the past fifteen or so years, with a profound out-migration of mining and heavy manufacturing, and an equally profound, if less job-intensive, move into finance and services. These changes parallel national economic restructuring and reflect Johannesburg's (and the country's) reintegration into the global economy.

National economic restructuring is dominated by the decline of formal sector employment. The trend toward shedding jobs started in the late 1980s, largely due to international sanctions. It increased sharply after 1994 with the advent of a democratic government, South Africa's entry into the World Trade Organization, and the embrace of an export-led growth strategy that resulted in a rapid drop in tariff protections against imports. More imports meant an outflow of capital and a rise in unemployment. *Business Day* (26 September 2000:2) reported that a million mostly unskilled jobs were lost between 1993 and 1997, offset against 60,000 new skilled jobs and about a million informal sector jobs [77% of which pay less than $140 per month (*Business Day,* 3 October 2000)]. One million jobs represents about 12% of the formal sector labor force.

Underlying the lost jobs are broader structural trends in the economy (Bhorat and Hodge, 1999). On the one hand, due to changes in the local and global demand for goods and services, investment has shifted out of agriculture, mining, and manufacturing into the services sector. On the other hand, changes in production methods resulting from capital deepening have led to a demand for more skilled labor. Africans and coloreds are mostly employed in the primary and secondary sectors and the changes in production methods have been less favorable for them than for Asians and whites who are mostly found in the tertiary sector. Between 1970 and 1995, total employment in the country increased by 13.8% while employment opportunities for non-Africans increased by 45% and those for Africans (77% of the population) decreased by 3.8% (Bhorat and Hodge, 1999:371–2). About one-third of the labor force now lacks formal employment and well over one-half lacks access to growing economic sectors. Global changes have marginalized different workers in direct proportion to their skills and ability to find employment in sectors competitive in the global economy.

Between 1991 and 1996, employment for people living in the inner city and the south declined by 4.9% per annum, with a precipitous decline among people employed in manufacturing.[10] Over the same period, employment among people in the northern suburbs declined by 1.6% per annum, again with the most rapid decline being in manufacturing. Formal sector unemployment in Sandton was 4%, 23% in the inner city, 45% in Diepkloof, and 55% in Devland (a squatter settlement south of Soweto). Rogerson (2000a) refers to the creation of new industrial spaces arising from the shift of jobs out of heavy manufacturing into knowledge-based manufacturing and the related shift in the location of employment, with only the area between the north of Johannesburg and south of Tshwane along the M1 showing an increase in employment.

The trends are in contrast to Johannesburg's informal sector, which had an explosive expansion, specifically in areas abandoned by the formal economy. For example, at the end of the 1970s, when informal enterprises were banished from the streets of Johannesburg, there were thought to be about 200 to 250 hawkers in the inner city. Two decades later there were about 15,000 street traders there (Rogerson, 2000b). These are not traders formerly bottled up in the townships who are now surfacing in the inner city. The growth of the informal sector is linked to the structural changes in the city's economy and credited to the lack of employment opportunities. Also contributing to the sector's growth is increasing subcontracting to informal enterprises by formal enterprises that are circumventing labor regulations.

Most persons in the informal sector are blacks engaged in survivalist activities. This is reflected in the distribution of informal activities, with 46% being in retail, 31% in services, and only 23% in manufacturing. As a consequence of the low entry barriers in most service and trading activities, there is considerable oversupply, causing earnings for many participants in the informal sector to drop below that necessary for survival. Only by combining earnings are households able to survive.

The shift into knowledge-based manufacturing and business and financial services and the relocation of formal sector jobs to the northern suburbs have reduced access to these jobs among the mass of the labor force. When coupled with spatial exclusion and the struggle to survive in the informal sector, the potential for meaningful integration is diminished.

CHANGING DEMOGRAPHY AND GROWING UNCERTAINTY

Policies for economic and social integration have to be interpreted in light of Johannesburg's changing demography, especially when these changes arise from the consequences of the HIV/AIDS epidemic.[11]

Johannesburg's population is expected to grow slowly at only 0.9% per annum between 2000 and 2010 (Simkins, 2000). Three factors are important. First, fertility in Johannesburg is low. The fertility of Asians and whites is below replacement and that of Africans and coloreds, at 2.6% and 2.5%, respectively, is close to the replacement fertility rate of 2.1%. Second, mortality is increasing due to an aging population and the effects of HIV/AIDS. Third, migration from South Africa's rural areas has virtually ceased.[12] Whereas there had been a massive transfer of unskilled labor from rural areas to Johannesburg to find jobs in the mines and in manufacturing and services, nowadays there is high and increasing unemployment, especially among those without skills. There is little incentive to move to the city.

The different population growth rates are changing the racial mix. In 1996, Africans comprised 71%, whites 19%, Asians 4%, and coloreds 6% of the city's population. By 2010, projections point to Africans increasing to 75% and whites declining to 15% (Simkins, 2000). A remarkably high 2.5% of the city's white population left in 1996. In contrast, despite equivalent fertility rates, the Asian population will continue to grow due to immigration. Inexorably, Johannesburg's population is becoming more African, but not as rapidly as it would have in the absence of HIV/AIDS.

HIV infections and "AIDS mortality will be light among Whites and Asians, moderate among coloureds and heavy among Africans" (Simkins, 2000:5). The incidence of HIV/AIDS is also correlated with employment status, with a much higher rate of AIDS deaths among unskilled workers (Arndt and Lewis, 2000).

The impact of these HIV/AIDS projections is vividly illustrated by the declining life expectancy of the African population. Whereas life expectancy of Asians is expected to stay much the same from 1995 to 2005, that for Africans and coloreds will drop quite sharply. The life expectancy for African males is expected to fall from 59.5 to 46.7 years. In contrast, the life expectancy of white males is expected to drop from 76.0 to 74.4. However, susceptibility to HIV infection is higher among women. The life expectancy of African women, which previously exceeded African males by 4 years, is expected to drop to two years below their male counterparts (Simkins, 2000).

The combination of HIV/AIDS with high unemployment and low household income levels will greatly exacerbate hardship among Africans, while causing the breakdown of traditional support systems. A desperate situation among many living south of the inner city can be anticipated.

LOOKING TO THE FUTURE[13]

Government's initial response to the difficulties facing Johannesburg did not address such issues. *iGoli 2002* was an introverted effort to deal with financial crisis and institutional malaise. The focus was on a three-year revenue-led budget, credit control, institutional rationalization, and partnerships with the private sector for service delivery. However a longer term vision was needed.

iGoli 2010 represented the initial step in building that vision. It was completed at the end of 2000, a little before the 2001 election of the new, executive mayor. *iGoli 2010* sought to position Johannesburg as "a globally competitive African world-class city." The Mayor's review of the strategy led to the view that it was trying to be "all things to all people" and it was replaced by *iGoli 2030.*

The 2030 strategy drops reference to "a globally competitive African world-class city" and calls for Johannesburg to become "a world-class business location." The housing and services backlogs, in South African terms not particularly severe (Tomlinson *et al.,* 2000), are viewed as less of a priority than sustained economic growth in financial and business services, transport, communications, trade, and accommodation.

> It is the belief of the Johannesburg City Council that, by growing the economy of the city, and by basing our dreams of a better life for all our citizens firmly on economic growth, we can aim to confer on the citizens of the City the economic freedom equivalent to the political freedom they achieved in 1994. (City of Johannesburg, 2002:115)

The city's *iGoli 2030* strategy is intended to be aligned with the Gauteng Provincial Government's "Blue IQ" growth strategy.[14] Blue IQ is an R1.7 billion initiative to invest in 10 megaprojects related to tourism, transport, and high value-added manufacturing. Its goal is to create a "smart province" and to attract R100 billion in foreign direct investment. The emphasis is on skilled employment with most projects sited between Johannesburg and the Tshwane CBDs, thereby by-passing the needs of communities to the south.

Both *iGoli 2030* and Blue IQ are likely to reinforce economic, social, and spatial separation and disparities in and around Johannesburg. If such plans are any indication, postapartheid Johannesburg is likely to be no more integrated than its apartheid predecessor. Johannesburg has and will continue to change in significant ways. Yet, divisions of race, class, and space continue to haunt it.

NOTES

1. For background on the history of Johannesburg, see Beavon (1997), Parnell and Pirie (1991), and Tomlinson (1999a). For more general discussion of the apartheid city, see Christopher (1994), Lemon (1991), Mabin (1992), Mabin and Smit (1997), Smith (1992), and Turok (1994a, 1994b). For general histories of South Africa, see Lapping (1987) and Worden (1994).

2. The Boer War is now referred to as the South African War since blacks fought on both sides and the war was significant for black futures.
3. The pass laws restricted access to public facilities (e.g., beaches) and private services (e.g., movie houses), required identification documents of nonwhites, and blocked access to the city.
4. Soweto is really a geographic cluster of townships, Soweto standing for South Western Townships.
5. So-called homelands were lands designated by the apartheid state to house the black population. They were meant to be "independent" and thus where blacks would establish their rights. Called Bantustans by the government, they were established by the 1959 Promotion of Bantu Self-Government Act (Worden, 1994:110–113).
6. By 1989, there were approximately 2,000 civic associations throughout the country (Murray, 1994).
7. Lekgotla is Sotho and is used to refer to a meeting.
8. The term suburbs refers not to the kind of mass suburbs typical of the United States but what would be termed pre-World War II, inner-ring suburbs in that country.
9. Nonetheless, commuters remain a significant part of Johannesburg's daily life, currently estimated at 700,000 persons per day (Tomlinson, 1999b:1659).
10. These data were provided by David Viljoen of the Development Bank of South Africa. StatsSA estimated that for 1996 the total economically active population in Johannesburg was 326,092, for Soweto 232,964, and for Randburg 173,456. In 1991, it was 842,899 for Johannesburg. Between 1991 and 1996, the average annual growth rate for the total economically active population in Johannesburg and Soweto was −7.7%.
11. HIV infection often results in economically active family members leaving the labor force to be cared for by other members whose income earning potential is thereby reduced by their spending time away from work. Household savings, if there are any, are then used to care for the sick and pay for burials. The earnings of relatives without HIV-infected family members also are siphoned off to provide assistance. Children can be taken out of school to care for the sick and earn what they might. It is even anticipated that overburdened aged relatives charged with caring for the children will send those best able to care for themselves to the cities so that they might scavenge a living (Crewe, 2000).
12. No data exist for migrants from elsewhere in Africa, many of whom, we suspect, are illegal.
13. This section is based on an interview with Rashid Seedat, Director: Corporate Planning, of the City of Johannesburg, and on the draft 2030 vision and strategy document that was available at the time of writing.
14. See http://www.blueiq.co.za/frmain.htm.

REFERENCES

Arndt, C., and J. D. Lewis. 2000. "The Macro Implications of HIV/AIDS in South Africa." African Region Working Paper Series, No. 9. Washington, D.C.: The World Bank.

Beavon, Keith. 1997. "Johannesburg: A City and Metropolitan Area in Transformation," in C. Rakodi, editor, *The Urban Challenge in Africa.* Tokyo: United Nations University Press, pp. 150–191.

———. 2000. "Northern Johannesburg: Part of the Rainbow Nation or Neo-apartheid City?" *Mots Pluriels,* no. 13, http://www.arts.uwa.edu.au/MotsPluriels/MP1300kb.html.

Bhorat, H., and J. Hodge. 1999. "Decomposing Shifts in Labour Demand in South Africa." *The South African Journal of Economics* 67(3):348–380.

Bremner, Lindsay. 2000. "Post-Apartheid Urban Geography: A Case Study of Greater Johannesburg's Rapid Land Development Programme." *Development South Africa* 17(1):87–104.

———. 2002. "Sunday Times Bessie Head Award Research." *Sunday Times,* 10 February 2002.

Carr, W. J. P. 1990. *Soweto: Its Creation, Life & Decline.* Johannesburg: Johannesburg Race Relations.

Chipkin, C. 1993. *Johannesburg Style Architecture and Society: 1880s–1964.* Cape Town: David Philip.

———. 1998. "The Great Apartheid Building Boom: The Transformation of Johannesburg in the 1960s," in H. Judin and I. Vladislovic, editors, *Blank_, Architecture, Apartheid and After.* Rotterdam: NAI Publisher, pp. 248–267.

Christopher, A. J. 1994. *The Atlas of Apartheid.* London: Routledge.

City of Johannesburg. 2002. *iGoli 2030.* Johannesburg: City of Johannesburg.

Crankshaw, Owen. 1997. "Challenging the Myths, Johannesburg: Inner City Residents Housing Usage and Attitudes." Report prepared for Inner City Housing Upgrade Trust.

Crewe, M. 2000. "Other Will Not Save Us: AIDS Orphans: The Urban Impact." Paper given at a workshop on "HIV/AIDS Orphans: Building an Urban Response to Protect Africa's Future," 21 July. Johannesburg: Centre for Policy Studies.

Dhlomo, M. 1997. "Attacking Foreign Hawkers Not the Solution." *Sowetan,* September 18, 1999.

Gear, Sasha. 1999. "Numbers or Neighbourhoods? Are the Beneficiaries of Government Subsidized Housing Provision Being Economically Subsidized?" Issues in Development, No. 18. Johannesburg: Friedrich Ebert Stiftung.

ICHUT. 1996. "Inner City Housing Research Report." Johannesburg: Inner City Housing Upgrade Trust.

iGoli 2002. It cannot be business as usual. 1999. Published by the GJMC and the four MLCs.

Kadima, D., and G. Kalomba. 1995. *The Motivation for Emigration and Problems of Integration of the Zairean Community in South Africa.* Johannesburg: Witwatersrand University Press.

Lapping, Brian. 1987. *Apartheid: A History.* London: Grafton Books.

Lemon, Anthony. 1991. "The Apartheid City," in A. Lemon, editor, *Homes Apart: South Africa's Segregated Cities.* Cape Town: David Philip, pp. 1–25.

Mabin, Alan. 1992. "Dispossession, Exploitation, and Struggle: An Historical Overview of South African Urbanization," in David M. Smith, editor, *The Apartheid City and Beyond.* London: Routledge, pp. 13–24.

Mabin, Alan, and Dan Smit. 1997. "Reconstructing South Africa's Cities? The Making of Urban Planning, 1900–2000." *Planning Perspectives* 12:193–223.

Morris, Alan. 1999. *Bleakness & Light: Inner City Transition in Hillbrow.* Johannesburg: Witwatersrand University Press.

Murray, Martin. 1994. *The Revolution Deferred: The Painful Birth of Post-Apartheid South Africa.* London: Verso.

Parnell, S. M., and G. H. Pirie. 1991. "Johannesburg," in A. Lemon, editor, *Homes Apart: South Africa's Segregated Cities.* Cape Town: David Philip, pp. 129–145.

Rogerson, Chris. 1996. "Image Enhancement and Local Economic Development in Johannesburg." *Urban Forum* 7(2):139–158.

———. 2000a. "Manufacturing Change in Gauteng 1989–1999." *Urban Forum* 11(2):311–340.

———. 2000b. "Emerging from Apartheid's Shadow: South Africa's Informal Economy." *Journal of International Affairs* 53(2):673–695.

Rogerson, Jayne. 1996. "The Geography of Property in Inner-City Johannesburg." *GeoJournal* 39:73–79.

Saff, Grant. 1994. "The Changing Face of the South African City." *International Journal of Urban and Regional Research* 18(3):377–391.

Simkins, Charles. 2000. "Greater Johannesburg: Demographic Report." Prepared for the GJMC's *iGoli 2010* visioning process.

Simone, AbdouMaliq. 1998. "Globalization and the Identity of African Urban Practices," in H. Judin and I. Vladislovic, editors, *blank_Architecture, apartheid and after,* Amsterdam: NAI Publishers, pages 156–174.

Smith, David M., editor. 1992. *The Apartheid City and Beyond.* London: Routledge.

Swilling, Mark. 2000. "Rival Futures: Struggling Visions, Post-Apartheid Choices," in *blank_ Architecture, apartheid and after,* H. Judin and I. Vladislavic, editors Amsterdam: NAI Publishers, p. E8.

Tomlinson, Richard. 1999a. "Ten Years in the Making: A History of Metropolitan Government in Johannesburg." *Urban Forum* 10(1):1–40.

———. 1999b. "From Exclusion to Inclusion: Rethinking Johannesburg's Central City." *Environment and Planning A* 31:1655–1678.

———. Forthcoming. "The Local Economic Development Mirage in South Africa." *Geoforum.*

Tomlinson, Richard, Phil Sinnett, Bala Rajaratnam, and Werner Fourie. 2000. "Service Delivery, Equity and Efficiency in Greater Johannesburg." Report commissioned by The World Bank and undertaken on behalf of and with assistance of the Greater Johannesburg Metropolitan Council.

Turok, Ivan. 1994a. "Urban Planning in the Transition from Apartheid, Part 1." *Town Planning Review* 65(3):243–259.

———. 1994b. "Urban Planning in the Transition from Apartheid, Part 2." *Town Planning Review* 65(4):355–374.

Worden, Nigel. 1994. *The Making of Modern South Africa.* Oxford: Blackwell.

2

Villas of the Highveld

A CULTURAL PERSPECTIVE ON JOHANNESBURG AND ITS "NORTHERN SUBURBS"[1]

ANDRÉ P. CZEGLÉDY

Will Johannesburg always be a potluck city on some journey of chance? A city trapped in hazardous indetermination? Who will structure the journey? Who will signpost the route? The city is fickle, innovative, and malleable. With vigorous direction she will continue to develop in her own mold. (Palestrant, ND:133)

This chapter examines the social and cultural ramifications of the built environment in Johannesburg's northern suburbs, a wide swath of exclusive suburban development that is the residential base for a significant proportion of South Africa's political and economic elite. The main interest here is to describe and analyze the constitution of these suburbs in terms of architecture, space, and the built environment, and to do so with respect to the relevance of a sense of civic community for the city as a whole. What does the end of apartheid mean to urban South Africans today, especially for the residents of its major city? In what way can Johannesburg be considered a postapartheid city? To what extent do its iconoclastic northern suburbs reflect the country's past as much as its future? Does their seemingly exponential growth represent a specific response to contemporary shifts in the urban fabric? These and other questions will be addressed through a discussion that draws together the threads of apartheid-era urban planning, postapartheid suburban development, and the more elusive aspects of locality and identity in contemporary South Africa.

I begin with an historical overview of urban development in Johannesburg. Of special interest is the former status of the city as a model for

21

urban design during the apartheid era, and what this role has meant in terms of the shifting dichotomy between the inner city center and outlying suburban areas. Having established the range of articulation between the urban core and its suburban peripheries, the discussion then focuses on suburban Johannesburg. The historical development of the northern suburbs is considered first, followed by an examination of the structural and decorative features of residential housing peculiar to them and the ways in which these features secure local perceptions of style and necessity. The social resonance of Johannesburg's built environment is paralleled by the spatial configuration of given neighborhoods and the articulation of residence and thoroughfare in conjunction with other public land uses. In terms of the more reflective discussion that follows, ideas of community and association are examined in relation to the prospects of civic identity. In conclusion, it is suggested that the distinctly architectonic features of the postapartheid city increasingly symbolize entrenched divisions between private and public life and between the individual and the state. In so doing, they represent a fundamental urban dislocation as well as a suburban disengagement from the city and the nation as a lived identity.

Conducted mainly from the perspective of urban anthropology and architectural history, the discussion at hand draws on continuing research formally initiated in Johannesburg in October 1998 and presaged by thematically parallel research conducted in Budapest (Czeglédy, 1999, 2000). The chapter's emphasis is not, however, on the processes of *urbanization* as traditionally favored by anthropologists (Foster and Kemper, 1988:95), including those interested in urban African contexts (e.g., Epstein, 1967; Du Toit, 1968; Forde, 1963; Mayer, 1971, Mitchell, 1966, 1987). Nor is it intent on "culturally decoding" a representative residence in the vein of Bourdieu's (1990) structuralist analysis of Kabyle housing or Fiske *et al.*'s (1987) examination of "Rembrandt," an Australian show home. Instead, it focuses attention on the lived experience of *urbanism* itself as framed by the built environment.[2] What has emerged is an understanding of Johannesburg as the quintessential postapartheid city, a city rooted in distinct social and political trajectories affecting the built environment while attempting to unburden itself from the grip of a contentious history.

EARLY JOHANNESBURG AND APARTHEID

With some 3.8 million residents, Johannesburg is the most populous city in South Africa and one of the largest urban conurbations on the African continent.[3] It is, however, a relatively young city, even by southern African standards. In September 1886, the Transvaal Republican government proclaimed as "public diggings" nine farms adjoining the "ridge of white waters," or Witwatersrand (Barry and Law, 1985:12). A year later the triangle-shaped Randjeslaagte farm of 246 hectares was officially designated as the township for the new gold fields (Zeederburg, 1972a:7).[4] The wealth produced by the ore-bearing deposits of the so-called "Main Reef" built the city above to such an extent that its Nguni name, *Egoli* (from the root word for "gold"), is far more apt than the formal version of European derivation: Johannesburg. Nevertheless, the name of the city remains Johannesburg—informally "Jo'burg" or "Jozi"—to this day. Its retention is a symbolic reminder that the city has always followed the spirit of commercial reality in South Africa. Not without reason did Transvaal President Paul Kruger view the hustle and bustle of early Johannesburg (in contrast to staid Pretoria) as the activities of "goddelose mense" (Zeederburg, 1972b:32)

Today, Johannesburg's diversified economy generates as much as 35% of the country's GDP,[5] and the mining industry remains a key player in terms of local employment and investment. Not surprisingly, the mining companies financed the town's first major thoroughfare (Main Reef Road) in order to connect their operations along the major ore vein (Maud, 1937:23). More importantly (in terms of the urban landscape), mining headgear, by-product dumps, and vacant land closed to development for reasons of subsidence still dominate the less prestigious, southern suburbs. Along with their neighbors to the east, west, and north, these suburbs form an integral part of the city's urban fabric. During the apartheid era, they were a major planning component of Johannesburg in its ideological role as the primary testing ground for urban racial segregation—essentially the model apartheid city.

In *The Atlas of Apartheid,* Christopher (1994:105–33) gives perhaps the best account of a model apartheid city. He begins by noting the legal basis for urban segregation, including the three legislative acts that served as the foundation for urban apartheid. In 1950, the Group Areas Act established spatial segregation of the population groups defined by the Population Registration Act of the same year. This was succeeded by the Natives Resettlement Act of 1954, which gave authorities the right to evict residents from designated locales. Third, the Natives (Urban Areas) Amendment Act was introduced in 1955 to "remove such concentrations of Blacks as the servants living in central city blocks of flats" (Christopher, 1994:122). Together, these laws substantially increased the state's level of interference—if not control—in the urban arena. At the same time, they consolidated existing cultural dispositions regarding the spatial separation between classes and races—sensibilities already entrenched in the record of colonial development (see Drakakis-Smith, 1995:12–28 *pace* King, 1976), and inscribed in the earliest history of Johannesburg (Kallaway and Pearson, 1986:30).

More specific to the current discussion is Christopher's Figure 4.2, a diagram that depicts the "model apartheid city" in graphic form. Drawing from a previous source (Davies, 1981), this illustration clearly presents a city along *Johannesburg* lines: with a focal downtown core traversed by a single railroad line running from east to west, and flanged by white group areas to the north and a variety of industrial, black, Indian, and colored group areas to the south (Christopher, 1994:105). Not surprisingly, the white group areas to the north are assigned fully half of the urban land area of the model city—an allocation completely out of ratio to the 13% of the (current) population that they represent as a whole in South Africa.[6]

Christopher (1994:105–106) reminds us that the radial nature of this ideal plan allowed for the growth of both urban and suburban areas, thereby accommodating an expanding metropolis—without substantially affecting residential segregation:

> The [urban planning] guidelines proposed that group areas be drawn on a sectoral pattern with compact blocks of land for each group, capable of extension onwards as the city grew. Group areas were separated by buffer strips of open land at least 30 metres wide, which were to act as barriers to movement and therefore restrict social contact. Accordingly, rivers, ridges, industrial areas, railways, etc. were incorporated into the town plan. Links between the different group areas were to be limited, preferably with no direct roads between the different group areas, but access only to commonly used parts of the city, for example, the industrial or central business districts.

Absolute residential segregation, however, could never be successfully enforced due to the modern, integrated division of labor under apartheid—and not least because of its concomitant traditions of labor-intensive industry and on-site domestic service in the residential context.[7] Nevertheless, from the late 1940s onward, the inherited conventions of colonial segregation were intensified and amplified to an oppressive degree by succeeding National Party governments. As a test-bed for apartheid policies, Johannesburg experienced forced removals from Sophiatown and Pageview, among the best-known examples. Such evictions—and the physical destruction of the houses, shops, etc. in these neighborhoods—never succeeded in wholly erasing the memory of past lives and communities.[8] Nevertheless, they clearly demonstrated the authorities' resolve to create "a special form of urbanization marked by a characteristic spatial system, manifested not only in separation and division at the micro level but in decentralization at the macro level" (Chipkin, 1999).

Although the process of decentralization granted a degree of administrative autonomy along the lines of the apartheid policy of "separate development," it was paralleled by an increasing inability to meet the demands of the regime's critics—which included the majority population. In spite of an impressive array of separate administrations, separate facilities, and physical and bureaucratic access points, the state had clearly accepted the limitations of racial segregation by the time of the Free Settlement Act of 1988. This Act allowed for the existence of so-called "gray areas," racially heterogeneous residential enclaves within the designated group areas of larger cities (Christopher, 1994:139). It recognized in law an inexorable trend: the movement of people—irrespective of racial designation—to urban areas in close proximity to services and employment opportunities.

In Johannesburg, the mainly labor-oriented migration gained increasing momentum from the end of the 1980s onward; it quickly impacted the demographic map of the inner city and radically influenced the urban linkage with suburban regions such as the northern suburbs. The authorities' inability to anticipate the response of the private sector, particularly as to shifts in residential and commercial investment, would eventually spell the practical eclipse of Johannesburg's city center and alter the traditional articulation between the urban core and its outlying suburban areas. This changed articulation presages the new role that wealthy suburban areas such as the northern suburbs are now beginning to play in their role of leading urban development in South Africa (Figure 1).

A DECENTERED CITY

Until the last half of the 1980s, Johannesburg's central business district (CBD) was the main focus of social, cultural, and commercial life for the city and for the wider geographic region. Related to one of the original Randlord (mining magnate) families and living in the posh suburb of Dunkeld, 72-year-old Diana Blake recalls today: "We used to go to town every Saturday then, even if it was only to window-shop. *Everybody* would go."[9]

Indeed, Johannesburg's CBD was previously considered by many people to be the premier commercial hub and consumer showcase on the Africa continent. Pritchard, Eloff, and nearby streets were touted to rank on a retailing par with international peers in both Europe and North America.

Figure 1

Suburban communities.

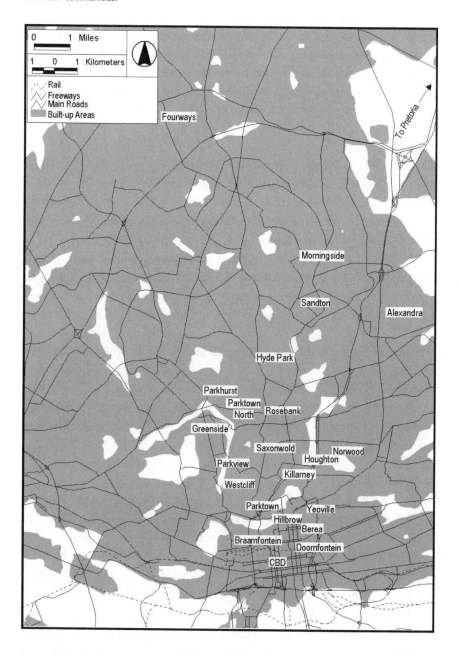

The famous Rand Club on Loveday Street was the meeting place for much of the continent's business fraternity, particularly those involved in the mining industry on which the national economy depended. The main theaters and cinemas, along with many of the country's leading hotels and restaurants, were all located within the boundaries of the CBD's grid-patterned streets originally laid out by the surveyors E. de Villiers and W. A. Pritchard at the end of the nineteenth century (Barry and Law, 1985:20).

With the focal role of the inner city in mind, Patience Malota, a secretary now working in Braamfontein at the University of the Witwatersrand, reminded me that until the 1980s: "All [of] the shops that used to be in Johannesburg . . . all of the shops used to be there [in the CBD]—you remember—shops like Stuttafords, Greatermans, John Orr's. . . . Those used to be *the* shops of the time!"[10]

This prosperity of the CBD lasted until the mid-decade, well beyond the time when many of the large European and North American cities had experienced a downturn in their inner city fortunes. During the late 1980s and the early 1990s, however, a twinned process of urban decline and exponential suburban growth began to occur that would indelibly mark the city to the present day. This process occurred in the longer wake of the international gold standard's demise just as the effects of economic isolation under apartheid increasingly debilitated South Africa's economy. Not least, a series of changes rooted in the combination of increased urban crime and an influx of poorer, nonwhite residents rapidly changed the ethnic and fiscal demographics of the inner city[11] and gave pretext for well-to-do city center residents and businesses to quickly move from the urban core. The majority of downtown businesses began to reestablish their office and retail operations in the well-to-do suburbs to the north, many of them in the commercial nodes of lower Parktown, Rosebank, Hyde Park, and Sandton. This displacement diverted private investment away from inner city regeneration initiatives while simultaneously undercutting the municipality's tax base. In addition, it changed the very nature of urban articulation in Johannesburg, including the manner in which many residents now understand the city *as a city*.

Whereas previously the Johannesburg CBD was a common ground of activity for the monied residents of the city as a whole, by the beginning of the 1990s this had substantially changed. Inner city residents vacated the CBD in favor of suburban certainties removed from the threat of urban crime and infrastructural decay. The once prosperous band of interstitial apartment and office building areas (Berea, Yeoville, Hillbrow, and Braamfontein) quickly lost both commercial and residential tenants at the upper, then middle, range of the market.[12] In the shift to the city's suburbs—especially northward—new and/or enlarged shopping/leisure mall complexes (especially in the Killarney, Rosebank, Hyde Park, and Sandton areas), along with a variety of office building developments, began to anchor a more dispersed form of private sector investment. The difference between this specific shift and its international analogies has been the relative lack of municipal support for the urban core—understandable at a time when new political priorities focused attention on the transfer of societal power rather than on the maintenance of a socially tainted built environment.

Today, most people living outside of the inner city have begun to consider the CBD as a virtual "no-go" zone. Suburban residents readily admit that they "do not even go into town—unless I [absolutely] have to." Their chief motivations tend to be voiced through stigmatic concerns such as crime and a decaying infrastructure. The far subtler variable in their decision making is the way in which the very structure of the city has changed as a result of the most recent spurt in suburban development. Simply said, Johannesburg's city center no longer fulfills the role of *center* in practical terms.

The most significant feature of recent suburban development in Johannesburg has been neither its pace nor its extent (both of which are considerable) but its quality. Specific sites of commercial development in the northern suburbs have captured the commercial heart of the town from the inner city. Although a major portion of this urban substitution has involved the relocation of corporate headquarters—even the Johannesburg Stock Exchange—to the north, much of the commercial development has taken the form of large-scale shopping malls (and lately casinos) constructed in direct proximity to new office and/or residential districts. Not surprisingly, the new (and only) convention center in the wider Witwatersrand area recently opened in the northern suburb of Sandton in 2000. Each of these functional alternatives has replaced the retail, business management, and entertainment/leisure roles that the downtown core once played. In usurping the traditional role of the city center, they have managed a literal *inversion* of the urban structure. Only the municipal bureaucracy remains rooted to the downtown core while the city's private investment is nearly exclusively redirected to a functioning urban fabric that is, literally, less and less a part of the rest of the original city. In this transformation, the periphery *seems* to have become the center just as the center has become the periphery. In many ways, the northern suburbs thereby represent the pattern of "edge cities" introduced by Garreau (1991) and critiqued by Beauregard (1995). However, whereas the edge cities of America seem to compliment their originating urban cores, the suburbs of Johannesburg and South Africa have eclipsed them nearly completely.

The term "seem" is employed above not just because the process of inversion is incomplete, but also because the decentering of Johannesburg is opposed on several levels, most notably the municipal platform of the Greater Johannesburg Metropolitan Council (GJMC) and its *iGoli 2002* and *iGoli 2010* plans. The *iGoli* plans have as a final goal the "creation of a world class city" for Africa by the year 2010. Irrespective of their details with reference to financial, administrative, and management consolidation, these plans signal an attempt to revitalize the inner city, thereby reestablishing the traditional articulation of the urban core to the suburban periphery. Implicitly, this reversal involves more than just urban renewal; it pits public—and ethnic majority—interests against private capital and a minority population. Effectively, it draws together the very confrontations of sociopolitical transformation that are so crucial to the future of South Africa. Such confrontations may be verbalized in the speeches of politicians, but they are enacted on a daily basis in the ways that urban dwellers remain contentedly apart in the ordinary terms of where they live, work, and play.

Although the process of urban decentering and suburban preeminence described above is not new (having occurred in a variety of larger American cities during the 1970s), nowhere has it hollowed out the metropolitan core to the extent that it has in Johannesburg. This process is not simply a question of nostalgia, but one of urban anchorage—an issue that lies at the very heart of traditional visions of the built environment and their attendant questions of community and identity. Before these concerns are addressed, however, it is necessary to consider more closely the actual built environment of Johannesburg's suburbs. How did these suburbs—and specifically the northern suburbs—evolve? What makes them so important to a consideration of South Africa's urban future?

SUBURBAN PANACEA

As the gold mines became concentrated south of Johannesburg around the turn of the nineteenth to the twentieth centuries, and the city center developed primarily through municipal means, the area to the north evolved to become the "northern suburbs." Now, their place in urban history may very well lay claim to extinguishing the very nature of the city's traditional urban structure.

Since their inception, the northern suburbs have become a national metaphor variously representing white wealth, Eurocentricism, capitalist materialism, cultural elitism, political conservatism, and, in spite of all that, social liberalism. In the main, they belong to the Randburg magisterial district, the wealthiest administrative area in the province of Gauteng.[13] Yet, in historic terms, the northern suburbs are really built on wastelands; most were developed as a commercially viable, default option after it was confirmed that mining operations were not sustainable there. The upshot was that these and other suburbs were primarily planned and built by a diversity of property development subsidiaries connected to the separate mining companies. In some cases, a mining connection has been abundantly clear from the beginning. For example, the heavily treed neighborhood now called Saxonwold (but initially named Sachsenwald Forest) was planted to supply timber for the mine tunnels (Rosenthal, 1974:4–5).[14] In other cases, such as Bryanston, the connection to mining was still evident in the corporate name of the original developer—Abe Bailey's South African Townships, Mining and Finance Company (Rosenthal, 1979:115).

Initially, the northern suburbs encompassed the entirety of neighborhoods in the wider Doornfontein, Houghton Estate, and Braamfontein areas (Barry and Law, 1985:6). This status no longer holds true in contemporary social understandings. The neighborhoods of Parktown and Houghton Estate are popularly classified as the beginning of the northern suburbs. Old residential neighbourhoods (Yeoville, Berea, Bellevue, Hillbrow, and Braamfontein) are understood to be economically depressed appendages to the inner city rather than the southern boundaries of the northern suburbs, which they once were in the popular imagination as well as in fact.

In spite of current perceptions, the first exclusive suburb was Doornfontein, an area located in the very band of interstitial suburbs noted above.[15] Its turn-of-the century reputation was such that as far away as London "guests at functions and country homes were often announced as 'Mr. And Mrs. So and So, from Doornfontein, Johannesburg' " (Barry and Law, 1985:32). This sort of social introduction is not without significance for current perspectives on community identity—for it trumpets the status of the local neighbourhood *in precedence* to that of the city in a way that we will see (below) has relevance for current associations of civic identity.

The heyday of Doornfontein was extremely brief, extending only through the first half of the 1890s. By the decade's end, the leading socialite couple of the day (Sir Lionel and Lady Florence Philips) had moved from their Norman House (in Doornfontein) to a new residence located on the Parktown ridge to the north. A main advantage to this new site was its elevation and distance from the mine dumps and the flat, treeless veld plain that together produced high levels of atmospheric dust and made the near rainless Johannesburg winters much more of a trial than newcomers ever expected.[16]

Other wealthy families quickly joined the Philipses to settle in an uneven string of impressive mansions along the Parktown ridge, many of them designed by Herbert Baker, South Africa's most notable architect.[17] These grand homes reaffirmed the race and class-oriented spatial segregation (not least inherited from Britain) that would be refined in its later apartheid forms. In short time, the Parktown mansions provided the fashionable anchor for a wide range of residential development yet to come and marked the historical beginnings of the northern suburbs as a separate zone of early urban sprawl. To this day, they remain the traditional archetype for large-scale villas in South Africa, irrespective of the (just as) fanciful contemporary intrusions in modern and postmodern architectural styles around them.

One of the major problems with the initial creation of the northern suburbs was the inarticulate nature of their independent, commercial development. Dewar sees it as a more general, South African dynamic, one which involves "*explosive* low density sprawl, the direction of which is largely non-managed" (1992:244). Unlike the inner city, which was under relatively tight municipal control, the residential districts of the northern suburbs were developed without central coordination. In this respect, a popular historian (Rosenthal, 1946:154) reminds us that

> Square mile upon mile to the north of the city had been laid out in townships—irregularly joined together in a fashion calculated to break the heart of any town-planner. Because the chessboard served as a universal model to the surveyors, many roadways finished as blind alleys, while others attained gradients so steep that horses could not climb them.

This legacy of separation, of a lack of overall municipal continuity, is still apparent today and being addressed through the variety of strategic plans formulated by the Greater Johannesburg Metropolitan Council.[18] It can be seen in the positioning of the Parkview Golf Course, a meandering line that cuts through the heart of the northern suburbs and reduces east–west access to problematic proportions. It is even more apparent at 1st Avenue West, a north–south street dividing the suburbs of Parkhurst and Parktown North. At this thoroughfare border, the differing plot sizes of the two neighborhoods lead to divergent residential block proportions.[19] As a partial consequence, the adjoining streets never connect with one another across the Avenue and crossing automobiles as well as pedestrians are forced to travel a dogleg path of at least several meters between the ostensibly contiguous streets.

A lack of urban continuity has peculiar advantages. Under the segregationist policies of both colonial and apartheid regimes in South Africa, controlling physical access between areas of the city was a matter of strategic consideration; it restricted the mobility of the general population, particularly the nonwhite majority. It requires no great leap of the imagination to understand the current maintenance of barriers to physical communication as a continuing matter of community "security" rather than an issue of civic efficiency. Indeed, many resident groups in the northern suburbs have taken such ideas of restriction to heart by "gating" their communities, effectively cordoning off the subneighborhood, or even a single street, via fences and a manned boom. In some cases, such actions are carried out with the permission of the municipal authorities; in other cases, official sanction is not solicited. The latter disregard for wider regulation underscores current levels of resentment at the authorities' inability to retard the level of crime against both person and property. They also highlight an independence from civil authority and overarching social norms that frames so much of postapartheid South Africa. This independence is paralleled in the very architecture of the northern suburbs.

FORTRESS ARCHITECTURE IN THE NORTHERN SUBURBS

In terms of architectural style, there is no regional equivalent of the Cape Dutch house with its whitewashed brick and gentle, ornate curves.[20] Instead, a variety of imported styles foreign to the region have taken root in Johannesburg and, especially, in the northern suburbs where wealth has laid open the door to aesthetic appetites.

Beside the transplanted Arts & Crafts style of Herbert Baker's famous Parktown homes, there stand a mixture of Cape Dutch, Spanish Mission, Edwardian, and other "period" architecture.[21] An important addition to the diversity is a curving and straw-thatched *rondeval* style that echoes the first dwellings on the Witwatersrand: the cylindrical huts of the Sotho-Tswana people (Chipkin, 1993:3). This housing style acknowledges an architectural past but is unable to overcome the rising tide of polyform creations that dominate new construction. Apart from modernist fantasies, a variety of mock-Mediterranean flourishes increasingly seem to be preferred of late—particularly in the case of the gated, semidetached housing compounds called "cluster developments" (Thomas, 1997:16) that are a familiar sight along transport corridors.[22] The advantage of Mediterranean style architecture is that it easily slides into the popular imagination as one appropriate to warm climate locations such as the South African highveld. In its more modest (pared-down) versions, it also allows for the minimalist construction designs favored by cost-conscious real estate developers. Most recently, there has been a local resurgence of interest in the use of Georgian facades, each new, white-washed development implicitly suggesting a sense of social stability and quiet prosperity of the kind that seems to physically repudiate the political, economic, and cultural upheavals of the postapartheid era.

The level of architectural diversity to be found in the northern suburbs is a reflection of what Chipkin (1993:149 and 175) interprets as a typical Johannesburg lack of contextual sensibility. This fundamental disregard for either the past or the immediate built environment has encouraged Johannesburgers to produce a kaleidoscope of structures regardless of wider aesthetic sensibilities. The result is that the houses of the northern suburbs lack a sense of cohesiveness from street to street—just as their decorative elements represent a mixture of stylistic influences.[23] Although there is no single Johannesburg model of architecture, one can recognize a Johannesburg style through the combination of a number of spatial and architectural elements. This historic vocabulary includes both design and decorative features of the built environment.

Johannesburg houses have traditionally been built around a regular, four square structural design employing single storey construction with monocourse brick walls and corrugated iron roofs. The four square design no longer holds true for many homes in the northern suburbs, although the simple brick walls (with their low insulation coefficients) remain a surprisingly durable feature given the combination of local disposable income and chilly winter nights. The use of corrugated iron has its roots in the early days of Johannesburg when it had to be brought in from Durban by ox-wagon (Barry and Law, 1985:15). At the time, this relatively durable material was used for general—not just roofing—construction, because of the high cost of wood and a general shortage of building bricks (Kallaway and Pearson, 1986:32). Even *Hohenheim*, the rambling Tudoresque mansion that dominated the Parktown ridge until its demolishment for the sake of the new Johannesburg Hospital complex (constructed in 1972–1978), possessed a corrugated iron roof (Benjamin, 1979:8). In spite of the wide range of roofing materials available today, this relatively inexpensive material remains in popular use throughout the city as a whole, although in the northern suburbs a variety of clay (especially terracotta) tile has increasingly found favor in its stead.

Many of the early Johannesburg houses incorporated *stoeps* (porch verandahs), cast iron exterior fittings,[24] and pressed tin ceilings. The older houses of the northern suburbs are no exception to this, with original tin ceilings (and even their replicas) being considered an important, city-specific selling point in the real estate market today. Like many of their predecessors, newer homes often have separate accommodation on the premises for "live-in" domestic servants—a normative feature of middle- and upper-class life in South Africa. The servants traditionally occupy a shack at the back end of a property. In the face of economic contraction, many of these shacks have recently been enlarged and/or renovated, turning them into so-called "cottages" that are rented out as a secondary revenue stream for the family or used to accommodate older children or elderly family members wishing a measure of residential independence.

In some of the newest houses of the northern suburbs, the domestic servants—invariably "black" South Africans (and often southern Africans[25])— are frequently housed in an accommodation that is architecturally linked to the main house by way of an enclosed corridor or courtyard. The courtyard often serves as a drying area for washing, a domestic function that points toward the roles that such servants play in this residential context. Domestic servants fulfill a variety of roles: women are generally employed as combination cook, cleaner, and child-minder, whereas men are often given responsibility for house repair and gardening—the last of which is a important role given the size and splendor of residential gardens.[26] The place of the domestic worker in the household is both integral and contractual. They are an indispensable component of the well-to-do South African life-style while being objectified in a manner all too redolent of history's misfortunes: if the property is sold the servant(s) of long standing may be understood (informally) to literally "come with the house."[27] As one young couple living in the upmarket area of Morningside pointed out to me in reference to their domestic servant: "We didn't really have much of a choice [morally]; she can't get a job elsewhere. . . . We more or less *inherited* [my emphasis] her when we took over the lease."[28]

Modern additions to the northern suburbs' architectural vocabulary increasingly include a number of recreational facilities: swimming pool, tennis court, children's play gym, and patio *braai* pit.[29] Such facilities go far beyond the mere sporting and culinary functions for which they are nominally designed. Along with the increasingly sophisticated "home entertainment centers" (comprising television set, video-cassette recorder, etc.) one finds in northern suburb homes, they signal a shift of interaction that draws the family into the house and its grounds to create an independent and self-contained environment for leisure and recreation. In these terms, the new features of the northern suburb home insulate and separate the family from direct experience of the wider social and cultural life of the city. They replicate recreational facilities (e.g., sports grounds, playgrounds, the local cinema/theater) that might otherwise act as a focus of community involvement. At the same time, they intensify existing relationships of acquaintance (narrowed down to a smaller circle of long-term friends) as well as focus established trajectories of social interaction.

Ironically, given current physical and social mobility, the idea of familial independence echoes the isolationism of early settler societies. In the South African context, it finds an odd parallel in a local interest in and appreciation for the American West of the nineteenth century. Neighborhood auctions in the northern suburbs often feature copies of Frederick Remington bronze sculptures and perhaps the most popular "home library" series of books is Time-Life's *The Old West*.[30] One local book-dealer explained this interest to me as representing the independent, "pioneer spirit" to be found in South African history!

In effect, the northern suburbs' family has increasingly cut itself off from social interaction on a haphazard basis. In this respect, neighborhood children are rarely seen playing in the streets, although this is not the case in more disadvantaged neighborhoods. Instead, it is the custom for suburban parents to take their children by automobile to a friend's house— often until such time as the children are themselves of driving age. On a psychological level, such a practice reinforces egocentric tendencies while limiting the range of social interaction open to children. As one former child of the northern suburbs related to me: "My mother was always ready to take me whereever I wanted to go. I thought [that] this was *normal*."[31]

When children do play in the local park, an adult—often the same domestic servant who is the major childcare provider in the family—invariably supervises them. When it comes to the rare case of a family excursion, a visit to the "local park" is frequently not local at all. There are very few parks in the northern suburbs that remain the focus of family recreation: Zoo Lake and Emmarentia Dam (see Figure 1) seem to be the major exceptions—and then primarily during the weekend. Such essentially supervised activities speak of the contemporary parent's concerns regarding personal safety. These concerns are part of a wider response to the postapartheid growth in urban crime that is architecturally manifest in what Lindsey Bremner (1999:B2) has titled the trend toward "fortification" in South Africa.

In architectural terms, fortification is exemplified in the seemingly mandatory erection of high perimeter walls around properties and the invariable installation of an automatic driveway gate (to reduce the risk of attack when entering or leaving the premises). The high perimeter walls are often topped by metal spikes, razor wire, and, more recently, electrified wiring connected to emergency alarms. In conjunction with portable "panic button" devices, the house alarms are electronically connected to "armed response" security companies. The surreal nature of such implicit violence was highlighted in my mind one day when walking with a colleague in Westdene, one of the more middle-class neighborhoods of the northern suburbs. On the street was parked a minivan from a local security company that boasted in large letters on the vehicle's side panel that they respond with "firearms and explosives."[32] Explosives?

More disturbing than the proliferation of security companies in contemporary South Africa is the way in which safety and security concerns have impacted social considerations in the context of the built environment. By such considerations, I refer not only to the restricted physical contact enforced upon people by 2- to 3-meter-high walls, but also the underlying mentality that presupposes the use of certain security devices in specific contexts. Consider the physical level at which specialty wires can be found in suburban Johannesburg. Electrified wire is commonly installed well above the reach of a child. In the suburb of Parkwood, however, I have found razor wire as low as ground level, a height fully accessible to a passing, curious child or even a toddler.

What does such evidence tell us about contemporary paranoia in suburban Johannesburg? It informs us of the very basic consequences of certain underlying processes in the built environment, specifically, of the way in which personal concerns (formerly relegated to a secondary private sphere) seem to have—as in the case above—infringed on community concerns at the wider, public level of interest and activity. This issue lies at the very crux of the dichotomy between urban and suburban environments. Not surprisingly, it focuses attention on questions of locality, community, and identity with which this discussion concludes.

CONCLUSION: IDENTITY AND BELONGING

Within the last fifteen years, suburban development in the northern suburbs has become the major reference point for wider shifts in land usage in the greater metropolitan area. These shifts have acted as a virtual weathervane of social change on the national level, concentrating a diversity of established and nascent social and political forces particular to the postapartheid era. In addition, they have increasingly impacted the city and its residents by overturning established trajectories of municipal planning and growth. In the process, inequities of urban development have been reinforced in spite of official efforts to realign the unevenness of investment in the built environment. What might such confrontation mean for Johannesburg and the other major cities of the new South Africa?

A curious relationship exists today among residents of the northern suburbs: they commonly identify themselves as living in "X, a suburb of Johannesburg" rather than referring to the city's name in the *first* instance. In parallel terms of association, the city center is commonly referred to as "Town" or simply as "Johannesburg," while suburban areas possess a popular identity *without* direct reference to the greater metropolitan area. Quite clearly, the widening gap between the urban poor and the suburban wealthy has not been able to encompass suburban expansion—an associative tension that increasingly exacerbates the public identity of the city and encourages autonomy from authority at several levels. This last point emphasizes the way in which spatial geography in Johannesburg has been historically predicated on differences of economic class (Frescura, 1995:72) and race (Cell, 1982; Robinson, 1996), and but also the salience of ideas of locality and belonging for (sub)urban areas such as the northern suburbs. Although this salience interrogates all South Africans, it is most pertinent with respect to the predominantly upper class, white residents living in the relatively isolated suburban environments of Johannesburg, Cape Town, Durban, and other cities.

In questioning residents of the northern suburbs about how they confirm a sense of local community identity, too often the answer has been that "unlike other neighbourhoods, we have a local street party" on an annual basis. In spite of such statements (or rather because of them), the sense of local identity may often be a superficial one—not withstanding the range of commendable "events" and publications that some local groups have sponsored in order to *organize* a sense of community. To some extent, this difficulty exists because community sensibilities on an entrenched and permanent basis are rarely generated out of formal efforts but evolve through that complex web of repeated social interaction found in everyday public life. Yet, with commercial domination by depersonalized shopping malls and other large-scale complexes and with the erection of perimeter walls around residential spaces, all with a bias against pedestrian locomotion, less opportunity exists in the suburbs to meet those people who constitute the proverbial "neighborhood" of community identity.

Although social interaction with one's neighbors is an important issue in the context of the northern suburbs, another is surely the experience of locality, of physical place itself. Perspectives on locality are a question of rootedness, the sense of personal, historical attachment that connects an individual to his or her immediate surroundings. In this respect, Johannesburgers have always had an awkward and impermanent association to their city, its suburban areas included. In its first decades, this was largely because of the town's fraught relationship with the Transvaal government: when Paul Kruger visited Johannesburg in February 1887, "he probably began to realize that this rapidly growing settlement could threaten his government in Pretoria" (Barry and Law, 1985:22). Even before then, it was evident that

> [The Transvaal Republic] remained opposed to the concept of the town of Johannesburg with its Uitlanders, its womanizing, its drinking and gamboling population. The government did not wish to recognise its permanency as a community and, therefore, an entitlement to the protection of the state and to essentials such as food, water, housing, lighting, waste disposal and a good communications system, which was so necessary for business. (Barry and Law, 1985:22)

The issue of municipal status is now part of the city's past, but notions of local identity in terms of belonging remain an important challenge for contemporary Johannesburg residents. Belonging "evokes the notion of loyalty to a place, a loyalty that may be expressed through oral or written histories, narratives of origin as belonging, the focality of certain objects, myths, religious and ritual performances, or the setting up of shrines such as museums and exhibitions" (Lovell, 1998:1). In terms of civic loyalty, Johannesburgers often mitigate against such a sensibility whenever they emphasize the city's commercially oriented mining heritage and then quickly add the local dictum that "no one actually comes to Johannesburg to live, only to make their money and leave."

Yet Lovell reminds us that belonging is not just a question of loyalty, but also a form of association that "is fundamentally defined through a sense of experience, a phenomenology of locality which serves to create, mould and reflect perceived ideals" (Lovell, 1998:1). The challenge for Johannesburg is exactly that common experience of the city that first colonial, and then apartheid, policies of segregation once denied the majority of citizens. The challenge focuses on equality of access *to* the built environment as much as on equality of access to resources *for* the built environment. In Johannesburg, and in South Africa as a whole, this access entails addressing the fact that the postapartheid city is becoming newly liberated for some of its residents just as for others it may seem newly limited. This is a twinned process involving political and spatial democratization for the majority but simultaneously resulting in the effective decline of spatial privileges for a minority of others. These "others" may not form a decisive electoral block within the new nation, but their continued support for the country is not least contingent upon a shared sense of physical as well as electoral freedom, an interchange of formative experience that knits together the fabric of society in informal and thereby doubly effective ways.

Such a recognition necessitates an emphasis on accepting (and if necessary constructing) commonalities of lived experience in much the same way that the more formalized "nation-building" projects of the postapartheid state seek to provide basic and equal rights, access to resources, and joint symbolic anchors for every citizen. From this perspective, the social and cultural dynamics of the built environment have as much to say about South Africa's future trajectories as each of its citizens with their new opportunity to vote. These dynamics enunciate the segregation of the past while projecting new environments of potential separation that disengage suburb from city center, resident from larger locality, and individual from the state. Whether they will continue to do so remains one of the primary questions for South Africa's future.

NOTES

1. The author wishes to thank Lindsey Bremner, David Coplan, Rehana Ebr-Vally, and Hayley Kodesh for their comments on ideas expressed in this chapter. For reasons of confidentiality, pseudonyms are used for all ethnographic informants in the text.

2. As a direct consequence, an important part of the research methodology involved walking around the northern suburbs of the city on foot, an activity that many Johannesburg residents view with considerable apprehension, if not alarm. Interviews conducted with local residents and textual research have given temporal depth to visual impressions, while the visiting of homes for sale during so-called "show" days or "exhibition" days has allowed for more detailed examination of the internal premises. In the course of these and other visits there have been additional opportunities to speak to property-owners (as well as the real estate agents representing them) about home, neighborhood, and the city. For reasons of comparative analysis, other major South African cities such as Cape Town and Durban have been studied within the wider perspective of the research project.

3. Population figure according to Greater Johannesburg Metropolitan Council statistics: http://www.igoli.gov.za. This population figure represents some 48.8% of the overall population for the province of Gauteng, of which Johannesburg is the capital. The name Gauteng derives from the Sotho term "place of gold."

4. Until the Transvaal Republic's *volksraad* officially conferred municipal status on Johannesburg in September 1897, Marshall's Town, Ferreira's Town, and other mining camps remained independent settlements (Zeederburg, 1972a:12).

5. Source: Greater Johannesburg Metropolitan Council publications at http://www.joburg.org.za.

6. The population percentile is according to the most recent (1995) national household census. The Statistical Office points out, however, that "Although half of the South African population lives in rural areas, the distribution of people in urban and non-urban areas varies according to race. Almost two thirds (63%) of Africans live in non-urban areas as against a far smaller proportion of coloureds (16%), Indians (5%) and whites (9%)." Source: http://www.statssa.gov.za.

7. The latter is such a staple of South African living that the country's most popular comic strip (*Madame & Eve*) is a satirical commentary on the relationship between domestic servants and their employers.

8. See Bohlin (1998) and McEachern (1998) for a discussion of the District Six Museum in Cape Town in relation to issues of community and memory.

9. In this case, Mrs. Blake is referring to her youth in the 1940s. Similar to other residents of the northern suburbs, she discontinued regular trips to Johannesburg's CBD by the mid-1980s (Research Fieldnotes: November 1998).

10. Research Fieldnotes: July 2000.

11. The residential and commercial migration northward to suburban Johannesburg climaxed in the early 1990s, but was stimulated by changing government policies between 1984 and 1987 (Cloete, 1991:93).

12. For an analysis of this impact in Hillbrow, see Morris (1999).

13. Comprising Midrand, Sandton, and the other northern suburbs of Johannesburg (with the exception of Parktown), Randburg's average per capita income stands at 86,600 South African rands. The second wealthiest provincial district is Pretoria, with an average per capita income of 45,021 rands. The poorest Gauteng district is Alexandra (12,500), a township directly adjacent to Sandton. Figures from the *Sunday Times*, courtesy of Bureau of Market Research, University of South Africa.

14. Likewise, Lippert's Plantation, comprising a good deal of the Parktown ridge, originally served a similar purpose (Rosenthal, 1946:153).

15. Frescura notes that "Initially this trend to the north was relatively modest, the better homes being located in Doornfontein and Belgravia, below the Braamfontein ridge, and in Berea above it" (1995:74).

16. This feature of early Johannesburg (Rosenthal, 1946:106, 1974:4; Robertson, 1991:117) persists to the current day, albeit in diluted form due to the intense planting of trees—particularly in the northern suburbs.

17. For a volume on Baker's contributions to South African architecture, see Keath (ND).

18. These include the *iGoli 2002* and *iGoli 2010* plans, the former of which addresses the financial and management dimensions of municipal development while the latter focuses on developing Johannesburg into "Africa's world class city."

19. A reminder of the city's mining history is the Johannesburg term "stand," signifying a plot of land.

20. This opinion is contra that of Chipkin (1993:vii) who nevertheless does not seem to clarify a distinctive style himself but prefers to enumerate a procession of locally utilized design forms.

21. Baker's interest in the use of local *kopje* stone (which he had quarried on site) is a distinctive feature of many northern suburbs houses and their grounds. It is a rare example of a distinctly local element within the city's architectural design canon. For a discussion of Spanish style architecture in South Africa, see Wicht (1964).

22. For the American equivalent, see Blakely and Snyder (1999).

23. In the case of the first luxury residences of Doornfontein, Zeederburg (1972b:34) notes that "The builders were Europeans who used their native building materials and methods whenever possible. The great variety of materials, usually of inferior quality, obtained locally or imported, made for even more confusion of style and often downright shoddiness."

24. Popularly known as "iron lace," the early cast iron fittings were chiefly imported from Britain, particularly by way of *Macfarlane's Catalogue*, which sold the prefabricated metal products from Walter Macfarlane's Saracen Foundry in Glasgow throughout the British Empire (Benjamin, 1979:54).

25. There is a distinct preference for non-South African domestic servants among those property-owners to whom I have spoken. The reasoning for this preference generally revolves around an argument that South Africans are "lazy" and "do not possess the same work ethic" as other African nationalities. An unspoken dimension to the preference may be that the weaker economic and legal position of foreigners invariably leads to greater acquiescence in employment relationships—which can be financially (and otherwise) exploited by the employer.

26. Domestic workers play an important role in the residential life of northern suburbs families. For a sociological treatment of domestic service in South Africa, see Cock (1980).

27. This proposition—although decreasingly part of the commercial landscape according to sources in the real estate industry—is often justified under the aegis of paternalist welfare by the former owner.

28. Research Fieldnotes: March 1999.

29. *Braai* is the Afrikaans term for a "barbecue" food grill utilizing an open, charcoal flame. According to Eliovson's tourist-oriented account: "The campfire of the old trekking days, when Boers

gathered together to grill meat and enjoy each other's company, has been transformed into the modern braaivlais. This is an informal and traditional way of entertaining in South Africa" (ND:75).
30. Frederick Remington is generally considered the premier artist of the American West by virtue of his acclaimed paintings and sculptures of cowboys and Native Indians.
31. Research Fieldnotes: July 2000.
32. Research Fieldnotes: March 1998.

REFERENCES

Barry, M., and N. Law. 1985. *Magnates and Mansions: Johannesburg 1886–1914.* Johannesburg: Lowry.
Beauregard, R. 1995. "Edge Cities: Peripheralizing the Centre." *Urban Geography* 16(8):708–721.
Benjamin, A. 1979. *Lost Johannesburg.* Johannesburg: Macmillan.
Blakely, E. J., and M. G. Synder. 1999. *Fortress America: Gated Communities in the United States.* Washington D.C.: Brookings Institution.
Bohlin, A. 1998. "The Politics of Locality: Memories of District Six in Cape Town," in N. Lovell, editor, *Locality and Belonging.* London: Routledge, pp. 168–188.
Bourdieu, P. 1990. "The Kabyle House or the World Reversed," in P. Bourdieu, editor, *The Logic of Practice.* Cambridge: Polity, pp. 271–319.
Bremner, L. 1999. "Crime and the Emerging Landscape of Post-Apartheid Johannesburg," in H. Judin and I. Vladislavic, editors, *blank__Architecture, apartheid and after.* Amsterdam: NAI Publishers, p. B2.
Cell, J. W. 1982. *The Highest Stage of White Supremacy: The Origin of Segregation in South Africa and the American South.* Cambridge: Cambridge University Press.
Chipkin, C. 1993. *Johannesburg Style: Architecture and Society, 1880s–1960s.* Cape Town: David Philip.
———. 1999. "The Great Apartheid Building Boom: The Transformation of Johannesburg in the 1960s," in H. Judin and I. Vladislavic, editors, *blank__Architecture, apartheid and after.* Amsterdam: NAI Publishers, p. F10.
Christopher, A. J. 1994. *The Atlas of Apartheid.* Johannesburg: Witwatersrand University Press.
Cloete, F. 1991. "Greying and Free Settlement," in M. Swilling, H. Humphries, and K. Shubane, editors, *Apartheid City in Transition.* Oxford: Oxford University Press, pp. 91–107.
Cock, J. 1980. *Maids and Madams: A Study in the Politics of Exploitation.* Johannesburg: Raven Press.
Czeglédy, A. P. 1999. "Villas of Wealth: A Historical Perspective on New Residence in Post-Socialist Hungary." *City & Society* (Annual Review 1998):245–268.
———. 2000. "Villa szigetcsoport: hálózati kapitalizmus és a posztszocializmus új lakóházai." *Café Bábel* 36(Summer):86–97.
Davies, R. J. 1981. "The Spatial Formation of the South African City." *GeoJournal* (Suppl. Issue), 2:59–72.
Dewar, D. 1992. "Urbanization and the South African City: A Manifesto for Change," in D. Smith, editor, *The Apartheid City and Beyond: Urbanization and Social Change in South Africa.* London: Routledge, pp. 243–254.
Drakakis-Smith, D. 1995. *The Third World City.* London and New York: Routledge.
Du Toit, B. M. 1968. "Cultural Continuity and African Urbanization," in E. M. Eddy, editor, *Urban Anthropology: Research Perspectives and Strategies.* Athens, GA: University of Georgia Press, pp. 58–74.
Eliovson, E. ND. *Johannesburg: The Fabulous City.* Cape Town: Howard Timmins.
Epstein, A. L. 1967. "Urbanization and Social Change in Africa." *Current Anthropology* 8(4):275–84.
Fiske, J., B. Hodge, and G. Turner. 1987. *Myths of Oz.* Sydney: Allen & Unwin.
Forde, D. 1963. "Background and Approaches," in *Urbanization in African Social Change.* Edinburgh: Centre of African Studies, pp. 1–6.
Foster, G. M., and R. Kemper. 1988. "Anthropological Fieldwork in Cities," in G. Gmelch and W. P. Zenner, editors, *Urban Anthropology: Readings in Urban Anthropology,* 2nd ed. Prospect Heights, IL: Waveland Press, pp. 89–101.
Frescura, F. 1995. "The Spatial Geography of Urban Apartheid." *Between the Chains: Journal of the Johannesburg Historical Foundation* 16:72–89.
Garreau, J. 1991. *Edge City: Life on the New Frontier.* New York: Doubleday.
Kallaway, P., and P. Pearson. 1986. *Johannesburg: Images and Continuities: A History of Working Class Life Through Pictures 1885–1935.* Braamfontein: Raven Press.
Keath, M. ND. *Herbert Baker: Architectural Idealism 1892–1913, The South African Years.* Gibraltar: Ashanti.

King, A. 1976. *Colonial Urban Development.* London: Routledge & Kegan Paul.

Lovell, N., editor. 1998. *Locality and Belonging.* London: Routledge.

Maud, J. 1937. *Johannesburg and the Art of Self-Government: An Essay.* Johannesburg: R. L. Esson & C

Mayer, P. 1971. *Townsmen or Tribesmen,* 2nd ed. Cape Town: Oxford University Press. (Originally published in 1961 as *Xhosa in Town: Studies of the Bantu-speaking Population of East London Cape Province Vol. 2.*)

McEachern, C. 1998. "Mapping the Memories: Politics, Place and Identity in the District Six Museum Cape Town." *Social Identities* 4(3):499–520.

Mitchell, J. C. 1966. "Theoretical Orientations in African Urban Studies," in M. Banton, editor, *The Social Anthropology of Complex Societies.* London: Tavistock, pp. 37–68.

———. 1987. *Cities, Society and Social Perception: A Central African Perspective.* Oxford: Clarendon Press.

Morris, A. 1999. *Bleakness and Light; Inner City Transition in Hillbrow, Johannesburg.* Johannesburg: Witwatersrand University Press.

Palestrant, E. ND. *Johannesburg—One Hundred: A Pictoral History.* Johannesburg: A. D. Donker.

Robertson, M. 1991. "Investing Talent in the Witwatersrand: Jewish Traders, Craftsmen and Small Entrepreneurs," in K. Mandel and M. Robertson, editors, *Founders and Followers: Johannesburg Jewry 1887–1915.* Cape Town: Vlaeberg, pp. 115–132.

Robinson, J. 1996. *The Power of Apartheid: State, Power and Space in South African Cities.* Oxford: Butterworth-Heinemann.

Rosenthal, E. 1946. *Gold Bricks and Mortar: 60 Years of Johannesburg.* Johannesburg: Housel.

———. 1974. *The Rand Rush 1886–1911, Johannesburg's First 25 Years in Pictures.* Johannesburg: A. D. Donker.

———. 1979. *Memories and Sketches: The Autobiography of Eric Rosenthal.* Johannesburg: A. D. Donker.

Sunday Times. 2001. "Where the Money Lives" (Metro section), 18 March 2001, p. 8.

Thomas, J. 1997. "Covetable Clusters." *Style Magazine* (April Issue):16.

Wicht, H. 1964. *Spanish Houses of South Africa.* Cape Town: Howard Timmins.

Zeederburg, H. 1972a. *Golden Days.* Johannesburg: Van Riebeck.

———. 1972b. *Down Memory Lane.* Johannesburg: Archivist.

3

The Race, Class, and Space of Shopping

RICHARD TOMLINSON AND PAULINE LARSEN

Foreign visitors flying into Johannesburg International Airport would find much that is familiar. Driving along the freeway toward the central business district (CBD), with skyscrapers picturesquely framed by the hills against the afternoon light, they would carefully circumnavigate the center, as advised by their travel agent, and connect with another freeway heading north. Arriving at their hotel, an extension of an up-market mall, they would unpack, download their e-mails, and, seeking local color, go for a walk in the mall. Amidst all the stores with the familiar brands, they would find shops selling artifacts and khaki clothes for the bush and perhaps also encounter some street trading zones included in the mall's shopping experience. Settling down in a coffee bar they would find the coffee perhaps a bit strong, but they would relax, feeling at home.

With time to observe the shoppers, our visitors would notice a large number of black shoppers in the mall. Knowing that the suburbs close to their hotel are legendary for their wealth, they would be surprised by the intermingling of races. Were they to find out that the malls in the north have begun to draw clients from Soweto, they might also find the intermingling of classes somewhat surprising.

The visitors would be most unlikely to go the CBD. If they were to venture downtown, they would encounter the street traders crowding the sidewalks at transport termini and along busy shopping streets or have to walk in the street for lack of space on the sidewalk. They would note the absence of white shoppers and the dirt and clutter. They would be unlikely to discern that a good number of the traders and shoppers are from Mozambique, Zimbabwe, and other countries north of the border. What the visitors would

experience is but one-half of the parallel Americanization and Africaniza-
tion of shopping in Johannesburg.[1]

And so to notions of race, class, and space. The coincidence of race and
space, evident since the 1960s in the decentralization of shopping out of
the CBD, is giving way to integration along class lines. Significant changes
have occurred in the racial and class character of CBD shopping as well as
in the nationality of the shoppers and the traders. Whether in large stores,
wholesalers, or street traders, shopping is now focused on the low-income
predominantly black market. Retail in the townships, which is small in
volume and viewed among township dwellers as second-rate, receives only
passing mention. Migrant traders also warrant mention, if only because
they may offer the opportunity of reintegrating shopping in the CBD
around crafts, entertainment, and food.

DECENTRALIZATION: RACE AND SPACE

The primary cause of the racial character of shopping is the decentralization
of retail activities in search of the white consumer. Figure 1, derived from
Beavon (2000), demonstrates the process of decentralization. The dots identify all shopping development with gross lettable area (GLA) over 30,000 m².
There has been a steady increase in the number of malls, mostly located in
the north, such that in 1999 the amount of decentralized retail space greatly
exceeded the retail space in the CBD.

In Table 1, Sandton City is shown as having 110,000 m² of gross
lettable area, an amount added to the initial investment of 30,000 m².
The first really large mall, opened in 1979, was Eastgate, which provided
90,000 m². The only large development in the south is Southgate. It was
established in 1990 and is located at a freeway junction. It is, notably, not
in Soweto.

Decentralization started in the early 1960s, dates that cast doubt on the
current tendency to blame "crime and grime" for pushing retail out of the
CBD (Tomlinson, 1999). Together with the construction of the freeways,
pull factors rather than push factors propelled decentralization.

The most significant pull factor was the relative wealth of whites. The
residential expansion north and later east of the CBD began in the early
1890s, soon after the discovery of gold in 1886, and never stopped. The
continued northern suburbanization of the white population is evident
from the 1996 census. In 1996, the suburbs north and east of the CBD contained 70% of the city's white population and whites, Alexandra excluded,
constituted 60% of the suburban population.[2]

Figure 1

Large retail centers.

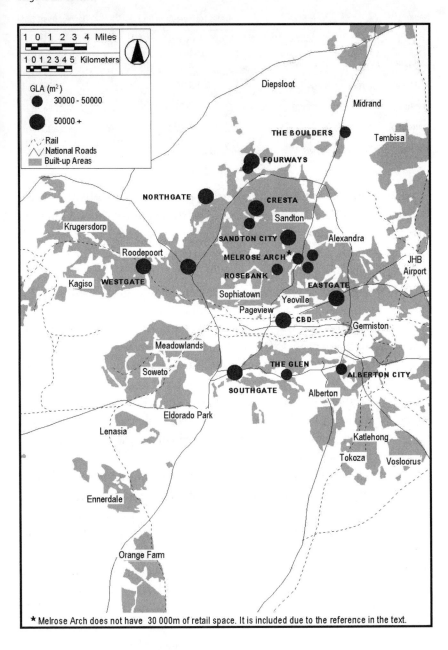

* Melrose Arch does not have 30 000m of retail space. It is included due to the reference in the text.

Table 1

Five Largest Shopping Centers in Johannesburg

Name	Node	Gross Lettable Area (m²)	Development Date
Sandton City	Sandton	110,000	1973
Eastgate	Bedfordview	105,000	1979
Westgate	Roodepoort	100,000	1985
Cresta	Randburg	73,321	1986
Southgate	Mondeor	65,000	1990

Source: Shopping Centre Directory 1999, South African Council of Shopping Centres.

The building of malls in the north was further abetted by the decentralization of commercial development that started shortly after retail facilities began to relocate. This office decentralization went north; very little went east and almost nothing went west or south. As was the case with retail, Beavon (2000) found that in 1999 the amount of decentralized office space exceeded CBD office space by a factor of 1.6, paralleling the retail ratio of 1.6.

This decentralization is also associated with the racial restructuring of the CBD labor force. In 1960, it was 7:1 white to black. Nowadays, as a result of the relocation of most business and professional services to the north, the large majority of office workers in the CBD are black. The decentralization of offices led not only to the decentralization of linked retail and wholesale services and the consolidation of white shoppers in the north, it also introduced black office workers to the malls in the north.[3] An "outdoor" meal at lunchtime bustle makes the point.

A further consideration when it comes to explaining decentralization is the National Party's creation in 1969 of separate local governments in Randburg and Sandton. In the case of Sandton, the National Party's motives included the fear that it would lose Johannesburg to the political opposition. Thereafter, the Sandton Council was controlled by business people. They set out, with some animus, to compete against the Johannesburg CBD through offering considerably lower property tax rates and the ready zoning of additional land for commercial and retail development (Tomlinson, 1999). [In June 1995, the assessment rate (cents per Rand of land value) was 2.65 in Sandton and 6.43 in Johannesburg.] They succeeded all too well.

Should we expect that this racial exclusivity will persist? The answer, in the case of Southgate, is that it will. The 1990 date of Southgate's development reflects the commitment of resources to the increasing wealth of blacks prior to the demise of apartheid. More recently, despite formal sector unemployment in excess of 40% in Soweto, household incomes in the area have grown at 10% per annum since 1993 (Prinsloo, 1999). As will be evident when we turn to the informal sector, an increasing divide exists within the black population between those with and those without skills. For the former, employment opportunities are increasing somewhat and wages are increasing sharply. For the latter, employment opportunities are declining precipitously (Bhorat and Hodge, 1999).

Less likely is that racial exclusivity will persist in the northern suburbs, but not because integration is proceeding in proportion to the increase in black purchasing power within the catchment area of the malls. Beavon (2000) reports that only between 1 and 2% of houses in the former whites-only northern suburbs had been sold to blacks by the end of the 1990s. On the other hand, Bremner (2002) has found a much higher percentage of blacks moving into the newer, edge city suburbs, in particular, new walled estates.

The increase in the number of blacks shopping in malls cannot be explained solely by residential proximity. Instead one must, first, refer to the parallel decentralization of shopping and offices that has introduced blacks into malls in the north as their share of employment in the business and professional services sector increases. Second, the catchment areas of malls that traditionally have been calculated on a 5-km to 10-km radius from a mall are often much wider. An obvious example is that of the Eastgate. It serves a regional market and even draws shoppers from north of the border.[4] Similarly, many of the weekend shoppers in, say, Sandton City, come from Soweto. (By 1999 a third of Sowetan households had cars.) Rather than the residential location of the shopper being the determining feature of where he or she shops, the employment status of the shopper may be more important.

CBD: FROM RACE TO CLASS

The establishment of suburban malls did not lead to the demise of retail in the CBD. On the contrary, the restructuring of the CBD retail market to serve the black consumer was, for a time, viewed as its savior. Although certain major retailers, and nearly all speciality stores and stores serving tourists as well as upmarket hotels, relocated to the suburbs, other major retailers reoriented their operations to serve the black market. The focus was on food and furniture.

On the negative side, racial restructuring was designed to accommodate the exit of whites from CBD residential areas and their replacement by blacks (Bremner, 2000) and to facilitate commercial decentralization and the exit of most white office employees. On the positive side, the racial restructuring attracted increasingly well-to-do weekend consumers from Soweto and further afield and served the market arising from the convergence of transport routes. [Approximately 790,000 commuters enter the CBD every working day.[5]]

With white shopping largely occurring in malls and most black shopping occurring in the CBD, the correlation between race and space was especially stark in the early 1990s. Then, push factors came to accentuate decentralization. Contributing to the changed retail mix was crime. During the course of a survey concerning the inner city economy, one property manager observed that "every single retailer has been robbed, held up or had an attempted break in" (Tomlinson *et al.*, 1995). The white client base for shopping and for office visits was scared and preferred the managed and more secure mall environment. The perspective of the CBD from the northern suburbs is captured in a cartoon taken from the *Mail & Guardian* (July 4, 1997) that shows a skyline festooned with "For sale" and "To let" signs, and the streets being walked by dinosaurs that, themselves, are being mugged.

Unfortunately the success of the CBD was short-lived. When surveyed in 1999 to ascertain why they were shopping there, 74% of consumers indicated that work was the main purpose of their visit (JHI Real Estate, 1999). Shopping (12%) and entertainment (5%) figured much lower on their list of reasons for coming to the inner city. Indicative of the small downtown population relative to the population in the townships, only 4% of consumers indicated that they were shopping there because this was where they lived. In effect, the future of retailing is closely tied to the state of the CBD's nonretailing economy.

This takes us back to the shopping experience. When black shoppers were asked what they disliked about CBD shopping, they mentioned crime, overcrowded streets, dirt and disorder, and, unfortunately, considerable dissatisfaction with the presence of foreigners on the streets and in the shops.[6] Those with sufficient income have taken their shopping elsewhere, mostly to Southgate, Highgate, and Eastgate, and increasingly also to Sandton City.

It was always a possibility that shopping opportunities for blacks would diversify out of the CBD. Observers anticipated that this would happen with the creation of malls in the townships. The diversification process is taking a different form, though, with the better-off households in the townships taking their business to malls located elsewhere. Although there is increasing integration in many of the malls, lower-income families still depend on the CBD.

Thus, established chain stores in the CBD are experiencing falling turnover, demand for retail space is contracting, and retail rentals are declining. The retailers that are now taking up space are geared to the low-income market and operate downmarket cash businesses.

TOWNSHIPS: CLASS AND RACIAL EXCLUSION

The third, rather obvious, overlay of race and space has to do with retail in the townships. Sizable retail developments were prohibited under apartheid and still are inhibited by the apartheid heritage. The convoluted layout of the townships was intended to inhibit public gatherings and enhance "security" by making internal circulation difficult. The result was that it is easier to conduct daily shopping in the CBD, what came to be known as a "two bag center" (that which you can comfortably carry on the way home). The consequence is that retail development in the townships has remained meager.

The largest shopping center located in a former township is the Dobsonville Shopping Centre in Soweto, with a gross lettable area of 17,320 m². This is similar in size to a neighborhood shopping center, despite serving one of the most densely populated communities in metropolitan Johannesburg. Property managers at retail centers in the townships argue that local centers are perceived by consumers to be "second-rate" or of an inferior quality, with the result that township shoppers often prefer to pay for a taxi trip to one of the regional malls. For this reason, the major competitors for Dobsonville Shopping Centre shoppers are said to include Westgate and Southgate regional malls.[7] CBD and township shopping becomes the preserve of the poor and the measure of a township inhabitant's wealth is how far they travel to purchase goods and services.

THE INFORMAL SECTOR: RACE, SPACE, AND CLASS

The context for the rapid increase in street trading and in the informal sector more generally is the sharp drop in formal employment opportunities. At the end of the 1970s, when informal enterprises were banished from the streets of Johannesburg, there were thought to be about 200 to 250 hawkers on the sidewalks of the CBD. Two decades later, there were about 15,000 street traders there (Rogerson, 2000). The growth of street trading does not represent traders formerly "bottled up" in the townships now "surfacing" in the inner city. Instead we are witnessing the struggle to survive amidst far-reaching structural changes in the country's economy arising from its exposure to the global market (Bhorat and Hodge, 1999). The informal sector is becoming increasingly characteristic of countries to the north where an ever-larger proportion of the labor force will never find formal employment (Halfani, 1996).

Rogerson (2000:679) reports that in Gauteng province 46% of informal entrepreneurs are engaged in retail activity, with 31% being involved in services and 23% in manufacturing. Street traders are by far the largest group, but an unknown proportion are not traders in their own right so much as employees of wholesalers. The main retail sectors are food vending, clothing, and soft goods textiles.

The bulk of informal trading is concentrated around bus and minivan taxi termini and along formal retailing streets. It serves the black market. Traders follow the market to transport termini throughout the city and to other areas (e.g., building sites) where a high number of black consumers is concentrated. Most visible for the visitor are the many persons selling goods at stoplights (called "robots" in South Africa). In this case the market comprises all persons passing by, regardless of race.

Considerable overtrading exists due to low barriers to entry. There is also a tremendous sameness of products, with tray after tray of tomatoes or whatever is in season. The significance of the street traders lies not only in the fact that they mostly serve the low-income black market, and so again point to the race, class, and space of shopping, but also that they actively, albeit inadvertently, contribute to the racialization of shopping through pushing both businesses and consumers out of the CBD. Where formal jobs are few and informal opportunities require ingenuity, competition frequently involves circumventing the rules. Where law enforcement is weak, the competition also entails theft and violence.

Informal retail activities all too often intrude on nonretail activities. Informal retailers often inconvenience office workers and diminish the sales of formal retailers. Taxi groups take over a particular street, spur informal retailing, and drive out the activities that formerly were there. Contestation over "turf" easily turns violent. But competition also takes place in more subtle ways. For example, reportedly a major problem,[8] the formal retailers and wholesalers face competition from other "wholesalers," invariably identified as Asians, who import goods without paying import duties and who commission the hawkers who position themselves on the streets outside the formal retailers.

MIGRANT TRADERS: SPACE

A less well-understood feature of the retail market is the migrant traders from southern Africa whose numbers increased sharply after 1990. These traders reportedly contributed R1.4 billion to Johannesburg's retail turnover for export purposes in 1998.[9] This turnover is generated throughout metropolitan Johannesburg, but most of it originated in the CBD. For stores targeting this market, the sales to migrant traders from Mozambique and Zimbabwe generally constitute around 25–30% of their turnover. The number of people coming to Johannesburg in 1998 is shown in Table 2. A great many more people are said to come by air from the rest of Africa, although data are unavailable.

The relevance of transport is evident in the items purchased by people using different transport modes. Air traders concentrate on "big ticket items," including cars, industrial items, computers, and chemicals such as plastics and colorants. Rail users purchase groceries, plastics, and clothing. Road users tend to purchase appliances and the typical groceries, plastics, and clothing. Many of these passengers constitute repeat business. Particularly in the case of train users, they bring goods (crafts and ethnic clothing) south to generate the revenue to purchase items to take home to sell.[10] There are increasing reports of their following the tourist market around the country (Peberdy, 2000). Migrant traders are also increasingly becoming a part of the formal sector, as illustrated by the construction of an African crafts center at the Rosebank mall.

Recent anecdotal reports suggest that the relative position of the CBD in serving the migrant market is declining. A major wholesaler and retailer complained that his migrant trade has disappeared due to violence directed at migrants. He claims that the migrants have shifted to Fordsburg.[11] Indeed, reports of community and police harassment of African foreigners are widespread. In a sense, it was to be expected that the migrant traders would initially come to the CBD and that as their familiarity with Johannesburg

Table 2

Passenger Flows from the Rest of Africa, 1998

Mode	Total Passengers	Estimated Traders
Train	350,000	250,000
Road	150,000	100,000

Source: Spoornet for total passengers and Sheny Medani for estimated traders.

increased, they would disperse. It appears that this process is being accelerated by violence arising from xenophobia.

The tourist shopper business, however, may well be a temporary phenomenon. It arises from the wars to the north, poor infrastructure in these countries, and barriers to trade. South African retailers show every sign of wishing to extend their activities northward. Once this happens, that is, once peace returns to the subcontinent, there will be less profit to be made by coming south to buy items such as baked beans, coffee, underwear, and jeans for resale back home.

Two opportunities will likely remain from the migrant trader phenomenon. The first is the market for "big ticket items," that is, if the intended "Abu Dhabi" environment of low-cost shopping can be created at the Johannesburg International Airport. The second arises from the fact that persons making a living selling crafts to tourists will continue to come south. An opportunity exists to create a craft market representing South Africa and other African nations and cultures that could serve as a market, a form of entertainment, and a tourist attraction.

Neither activity suggests the racialization of space. On the contrary, these are activities that might reintegrate retailing in the city with craft markets, eating establishments from diverse countries, and entertainment enticing whites back into the CBD (Tomlinson, 1998).

WHERE THINGS ARE HEADED: CLASS AND SPACE

Two underlying trends ensure that class and not race becomes the distinguishing feature of shopping in Johannesburg. The first trend concerns the continued decentralization of shopping and the second the attempt to reverse this trend as part of efforts to revitalize the CBD.

Melrose Arch is a spectacular example of the persistence of decentralization. Located alongside the M1 freeway, it is immediately south of Sandton and north of the CBD. The development integrates housing, shops, workplaces, and recreational areas. On completion in 2008, the almost 300,000 m^2 development will be a home, workplace, or leisure precinct for 20,000 knowledge workers. The development is planned to unfold in seven phases, the first of which opened in 2001.[12] Over 9,200 m^2 of retail space will be provided in phase one. Although Melrose Arch has given rise to many misgivings, for example, that it will swamp the commercial market when demand is weakening (*Financial Mail,* 12 January 2001), and is adding space to an already overtraded retail environment,[13] it reinforces the trend of upmarket decentralized retail to all that can afford it, regardless of race.

The second trend involves the attempt to better manage the CBD. This includes relocating taxis to parking garages and moving street traders into managed markets. Metromall is a good example of the former. It is one of three developments, the other being Jack Mincer and Park City, that have formal and informal retailing as a component of investments in CBD transport termini. Metromall will serve 2,800 taxis and upward of 100,000 commuters through providing 1,700 taxi bays and 30 bus bays. The investments are intended to reverse urban decay through removing taxis from the streets. Providing a range of retail and some entertainment facilities for commuters and residents is seen as revitalizing the CBD. Metromall will serve black commuters who rely on public transport and, while helping to "clean up" the CBD, will perpetuate the role of the CBD in serving low-income black shoppers.

In the case of street trading, the aim is to shepherd traders to designated spaces and to ensure that the trading is accompanied by greater cleanliness and hygiene. An Informal Trade Management Company has been established to rent spaces to traders by recycling well-positioned, but boarded-up, public and private buildings and by filling up spaces with markets. Most markets will be integral to taxi ranks so that the traders retain their links to their commuter customers. The intention is to provide 3,000 spaces by the end of 2002. In the light of considerable opposition by traders to the first scheme (Rockey Street Market) in Yeoville and their boycotting of the facilities, it is unclear whether that this target will be achieved. The success of this scheme is uncertain.

The city's attempt to manage informal trading will not affect the characteristics of the market being served. Instead, its broader significance will depend on whether it reduces crime and grime. The same is true of Metromall. If the negative effects of both can be minimized, then the push factors so detrimental in the past will diminish, but not so as to reverse the trend toward decentralization. Instead, this effect will be determined by whether a more favorable environment entices additional companies to locate offices in the CBD. An increasing number of office workers might well reverse retailing going downmarket.

Finally, it is interesting to reflect on the role of apartheid and its demise in these changes. Apartheid had a limited *direct* role and the decentralization process was largely market driven. Instead, the link to apartheid derives from racial "group areas" and education and labor legislation that predetermined who became rich and where they lived, with the wealth of those living in the northern suburbs drawing shopping northward. For example, in 1978, white shoppers contributed to 70% of Johannesburg's retail turnover and one-third of all white shoppers were shopping only in the decentralized malls (Beavon, 2000). In addition, the retail changes underway since the 1980s arise from residential desegregation in the CBD and the parallel enrichment of skilled and professional blacks and the impoverishment of unskilled blacks.

In the CBD itself, probably the most visible change has been the growth of the informal sector. This was made possible by government's ceasing to restrict such trading, coupled with rapidly increasing formal sector unemployment among the unskilled that left the informal sector as the only alternative means of survival. Of course, the structural changes that caused the job loss were accentuated by the postapartheid government's removing protections against imports and seeking to create an export-based economy.

The converse of this economic restructuring has been the slow growth of the service sector, though its growth in the black community has been significantly more rapid due to affirmative action policy. The latter, in turn, contributes to increasing integration in the malls. Mixed shopping has less to do with integration than with the increasing wealth of professional and skilled blacks and their greater mobility. Apartheid and its demise put in place a new class and racial context and the market has responded accordingly.

NOTES

1. The idea for this type of introductory paragraph follows a similar format adopted by Beavon (1997).
2. The percentage would be even greater were it not for domestic help living in the houses of the rich.
3. Dirk Prinsloo, CEO of Urban Studies, provided the latter information.
4. Information provided by Sheny Medani, CEO of Market Decisions, who has undertaken a number of retail market surveys in the CBD.
5. Transport data were obtained from the then Greater Johannesburg Metropolitan Council for inclusion in the CBD economic strategy. See Tomlinson and Rogerson (1999).
6. Sheny Medani provided this and the subsequent information on retail trends, as well as considerable insights.
7. Information provided by Viruly Consulting.
8. This was the view mentioned to Tomlinson by traders during the preparation of the CBD economic strategy (Tomlinson et al., 1995).
9. The data for this section were supplied by Sheny Medani and the consultant's insight into this question largely arises from conversations with her. Her views are consistent with other migrant surveys.
10. Sheny Medani reports that very often these traders come south in groups of three: one to sell the artifacts, one to buy, and one to guard the goods and to package the goods for the trip home.
11. Various interviews with Rees Mann during the course of preparing the inner city economic strategy.
12. Information obtained from www.melrosearch.co.za.
13. Personal communication, Francois Viruly of Viruly Consulting, http://www.viruly.co.za.

REFERENCES

Beavon, K. 2000. "Northern Johannesburg: Part of the 'Rainbow' Nation or Neo-apartheid City in the Making?" Mots Pluriels, no. 13 (http://www.arts.uwa.edu.au/MotsPluriels/MP1300kb.html).
Beavon, K. S. O. 1997. "Johannesburg: A City and Metropolitan Area in Transformation," in C. Rakodi, editor, The Urban Challenge in Africa: Growth and Management of Its Large Cities. Tokyo: United Nations University, pp. 150–191.

Bhorat, H., and J. Hodge. 1999. "Decomposing Shifts in Labour Demand in South Africa." *The South African Journal of Economics* 67(3):348–380.

Bremner, L. 2000. "Post-apartheid Urban Geography: A Case Study of Greater Johannesburg's Rapid Land Development Programme." *Development Southern Africa* 17(1):87–104.

———. 2002. "Sunday Times Bessie Head Award Research." *Sunday Times,* 10 February 2002.

Halfani, M. 1996. "Marginality and Dynamism: Prospect for the Sub-Saharan African City," in M. A. Cohen, A. Ruble, J. S. Tulchin, and A. M. Garland, editors, *Preparing for the Urban Future: Global Pressures and Local Forces.* Washington D.C.: Woodrow Wilson Press, pp. 83–107

JHI Real Estate. 1999. "Johannesburg City Centre Retail Survey," Rosebank.

Peberdy, S. 2000. "Mobile Entrepreneurship: Informal Sector Cross Border Trade and Street Trade in South Africa." *Development Southern Africa* 18(2): 201–219.

Prinsloo, D. 1999. "Know Your Market—Understanding the Changes." *Urban Studies Newsletter* No. 11, November.

Rogerson, C. 2000. "Emerging from Apartheid's Shadow: South Africa's Informal Economy." *Journal of International Affairs:* 53(2):673–695.

Tomlinson, R. 1998. "Jo'burg's Dynamo Will Hum in 2010." *Mail & Guardian* June 26 to July 2.

———. 1999. "Ten Years in the Making: A History of Metropolitan Government in Johannesburg." *Urban Forum* 10(1):1–40.

Tomlinson, R., and C. Rogerson. 1999. "An Economic Development Strategy for the Johannesburg CBD." Strategy prepared as part of the UNDP City Consultation Process on behalf of the CBD Section 59 Committee.

Tomlinson, R. (Project Manager), R. Hunter, M. Jonker, C. Rogerson, and J. Rogerson. 1995. "Johannesburg Inner-City Strategic Development Framework: Economic Analysis." Consultant report prepared for the Greater Johannesburg Metropolitan Council.

4

New Forms of Class and Racial Segregation

GHETTOS OR ETHNIC ENCLAVES?

ULRICH JÜRGENS, MARTIN GNAD, AND JÜRGEN BÄHR

The South African city is a product of apartheid policies. Under apartheid, *sociospatial engineering* created a spatial separation of residential areas on the basis of ethnic or racial criteria applied in accordance with a dogmatic model of the apartheid city (cf. Bähr and Schröder-Patelay, 1982; Jürgens and Bähr, 1998:4 ff.). Demographic and ethnic segregation, invasion, and succession on the basis of free decisions made by individuals, as posited by the Chicago School, were observable only in the part of the city that was declared "white." Population groups with other skin colors, in contrast, were restricted to constrained housing markets by state-imposed negative discrimination.

In the aftermath of the political changes in South Africa, the question arises whether suppressed socioecological processes have been released. The increasing differentiation and spatial mobility of nonwhites (e.g., because of preferential hiring of blacks[1] in the public and private sector), liberalized and often underutilized housing markets in the previously "white" city, and the rising expectations of the nonwhite population as to housing conditions, residential location, and security are factors that suggest that European and African patterns of development are likely to change.

What residential structures will arise from these developments? Will the transformation process lead to stable mixed racial neighborhoods or will desegregation processes be only temporary? In this context the important question will be whether the white population will be willing to live with nonwhites at the level of their immediate residential neighborhood. Are new ghetto-like residential areas emerging at the edge of the cities in the form of gated communities? Will Johannesburg continue to exist as a fragmented city, characterized by residential islands of varying ethnic homogeneity and differing standards of security?

The purpose of this chapter is to identify the analogies between the South African situation, and, with its many particular circumstances, international experience. We will look at developments in the inner city and at the northern edge of the formerly "white" part of Johannesburg. The two localities have a very different social status reflected in property prices, housing conditions, and residential density. It is therefore to be expected that the two types of localities will be affected differently by the processes of social transformation.

International examples frequently show that developments at the edge of the city are interconnected with and merely a way to flee from the problems of the inner city. Areas that were originally public are being privatized. Social problems that could adversely affect the value of the property are externalized. Since the 1950s, when such developments began in the United States, shopping centers and office complexes have been affected (Kowinski, 1985) and, since the 1980s, increasingly also residential areas (Blakely and Snyder, 1998). Municipal or government functions are being taken over by private agencies and incorporated into building and street complexes (Sack, 1990). In South Africa, the spread of shopping centers in the 1970s (along with residential suburbanization) and the emergence of so-called *security villages* and *enclosed neighborhoods* since the end of the 1980s have promoted analogous developments.

GHETTOS AND ETHNIC ENCLAVES

In the following discussion we want to determine whether in South Africa—as in the United States—residential areas develop as ghettos or ethnic enclaves. We begin with previous research and how it fits into the South African context.

Apart from so-called *township ghettos* (like Soweto), it is the inner city areas that, ten years after the end of the apartheid era, are in danger of developing into new black ghettos (Hart, 1996; Bähr *et al.*, 1998; Morris, 1999). On the basis of the invasion–succession theory, Hart (1989), Schlemmer (1989), and Schlemmer and Stack (1989, 1990) hypothesized at a very early stage of the transformation process that new forms of ethnic homogeneity would supersede a temporary phase of mixed racial living. The formal segregation of South Africa would then be replaced by informal ghettos as in the United States. This view has been disputed. Critics such as Saff (1995) point out that the historical experiences of South Africa and the United States differ so greatly that they cannot be compared. Therefore, terms such · as ghetto should not be used in the South African context. Some authors have resorted to semantic devices, such as contrasting "voluntary" and "involuntary" segregation, to avoid the pejorative connotations of the term "ghetto" (Marcuse, 1997a).

According to the classic socioecological model of invasion–succession, a ghetto develops in three phases (Duncan and Duncan, 1957:108 ff., cf. also Deskins, 1981). In the first stage, individual black households move into a white area resulting in a scattered pattern of black residence. In the second stage, the separate black residences coalesce to form a compact cluster (ghetto). In the third phase, the ghetto expands and spills over into adjacent areas, though the diffusion process does not occur evenly in all directions (Morrill, 1965).

The concept of the ghetto, the segregation of population groups within cities, emerged primarily from the scientific analysis of the black ghettos in the United States. In many studies, the ghetto is understood to be an Afro-American residential area in which more than a certain proportion of the population is Black; other social characteristics are not considered significant (Rose, 1971; Ford and Griffin, 1979; Cutler et al., 1999). Massey and Denton (1993) expressly emphasized the dominant criterion of race and resulting external factors, such as prejudice and discrimination, that have continued to lead to extreme segregation structures in North American cities up to the 1970s independent of other socioeconomic variables (cf. also Marshall and Jiobu, 1975). According to Ford and Griffin (1979), the critical proportion of Blacks in a residential area is 50% (others suggest only 15–30%). Then, whites begin to avoid living in that area (cf. also Rose, 1971).

Other authors expand the concept of ghetto. They replace the spatial distribution of ethnic groups (*race and space*) by the relationship between minority status and socioeconomic discrimination (*race–class dichotomy*) (Wilson, 1987; Marcuse, 1997b). Exploitation, unemployment, and marginalization lead to poverty, which results in a disproportionate concentration of impoverished persons, the *ghetto poor* (Wilson, 1987, 1991). Jargowsky (1996:14) refers to a residential area as a ghetto if it is inhabited primarily by Blacks and at least 40% of the inhabitants are living below the poverty line. Poor white areas, in contrast, are referred to as *white slums.*

According to this approach the Black ghetto population consists of structurally disadvantaged persons who cannot keep pace with the restructuring of the economy and consequently are cut off economically, socially, and spatially from mainstream society. As a result of the horizontal (i.e., spatial) segregation that is occurring in addition to vertical (i.e., social) segmentation, the ghetto inhabitants represent an *underclass* even within the Black population, because the Black middle class is moving out of these areas (Wilson, 1987:58). In the resulting *ethclass ghettos,* we find specific milieus emerging that are only loosely connected to the formal labor market, dependent on state welfare, and marked by broken homes and drug and alcohol problems (Wilson, 1987). Marcuse (1997a) and Marcuse and van Kempen (2000) characterize these residential areas as "ghettos of exclusion." Mainstream society's perception of these areas, with their hopelessness, social tensions, and separate economic and political life, is negative. These authors contrast this type of ghetto with so-called ethnic enclaves, for which voluntary, "positive" segregation is characteristic. Ethnic enclaves are distinguished by a strong feeling of solidarity among the population and by informal social networks that give rise to a feeling of community (Marcuse and van Kempen, 2000:18).

THREE RESIDENTIAL AREAS

To document the social and ethnic transformations occurring within Johannesburg and determine to what extent terms derived from the U.S. experience are applicable, one inner city residential area (Yeoville) and two new housing developments (Forestdale and Santa Cruz) at the edge of the city were analyzed in detail. (For study methods, see the Appendix.) Yeoville will be contrasted with the situation at the northeastern edge of the city where new housing developments are increasingly being planned as so-called gated communities. These are residential areas completely surrounded by walls or closed off from the public space in some other manner, most of which would be categorized as security communities (Blakely and Snyder, 1997).

Yeoville is a good example of the atmospheric changes in the inner city during the transition from an originally European to an African residential city. Located at the northeastern edge of the inner city, it is one of Johannesburg's oldest residential areas. Over the years it became a densely built-up area characterized by multistoreyed buildings (up to 5 storeys) merging with single, unattached houses. Traditionally Yeoville was considered a residential area for the English-speaking middle class. Originally it had a strong concentration of orthodox Jewish residents in the northern part, many one-person households (retired persons and students), and a high proportion of academics and European immigrants (Portuguese and Greeks).

Forestdale and Santa Cruz were planned and constructed by the Sage Schachat Group. In contrast to the types of gated communities distinguished in the literature as life-style or prestige communities, which represent only a small segment of the total housing market, they were designed for the middle class. They reflect the social and spatial spread of gated communities in South African society. Our choice of study areas was limited from the outset by the fact that scientific investigations cannot be carried out in gated communities—their professed goal is to withdraw from "public life"—without prior arrangement with the management. However, another reason played an equally important role in the choice of Forestdale and Santa Cruz. Contrary to earlier discussions in South Africa we wanted to look for signs indicating whether gated communities are merely a "white" phenomenon, meaning that residential apartheid could develop anew, or whether the affluent black population is sharing in this new residential seclusion.

Forestdale lies in a belt of single-family houses at the northwestern periphery of the city in the district of Douglasdale. It borders directly on a freeway. Within a radius of a few kilometers there are high schools and shopping centers. The area was designed and established as a compact settlement in 1994 after the land had remained unsalable since 1990, even though the necessary infrastructure was available. The area encompasses 75 lots (14 of which were vacant in June 1999) with a maximum size of 500–600 m². The entire settlement is surrounded by a wall and the entrance is watched by guards. Forestdale includes various communal facilities, such as a club house, swimming pool, and tennis courts. Its price level (approx. R200,000) is such that it attracts the middle level of the middle class.

In March 1997 the first inhabitants moved into their freehold flats in the densely designed settlement of Santa Cruz (a total of 65 units; in June 1999 six were vacant). Santa Cruz is part of the residential area of Country View, which belongs to the booming community of Midrand and lies halfway between Pretoria and Johannesburg. Country View was originally a unique area in the apartheid system. It was founded in 1989 as a so-called *free settlement area,* i.e., an area in which a population of mixed racial composition was allowed to live. From the outset the area attracted primarily Indians and Blacks. The average lot size is 800 m² and individually designed houses are permitted. After the Group Areas Act was repealed in 1991, when Indian households moved out, Country View began to develop into a *gilded ghetto* for the Black middle class (Bähr *et al.,* 1998). The question arises whether Santa Cruz, as a walled island within Country View, will end up being inhabited by the same Black residential population. The housing units are smaller (50–80 m^2) than in Forestdale (87–143 m^2) and are all built almost exactly alike. Prices of approximately R130,000 per unit lie at the lower limit for *security villages.*

YEOVILLE—AN INNER CITY GHETTO?

At the beginning of the 1980s, one of the important features of the South African city, aside from the apartheid zoning, was an extremely puritanical way of life. One exception to this was Yeoville, whose white inhabitants were distinguished by a highly liberal mentality. As a result, the influx of nonwhites or mixed race couples, though illegal until the repeal of the Group Areas Act, was initially considered an enrichment of the local culture and was explicitly welcomed. Because the white population was decreasing from natural causes or moving into the suburbs, nonwhites profited from the growing number of centrally located vacant housing units.

The "graying" began primarily in the western and southwestern part of Yeoville. The nonwhite households were initially concentrated in a few buildings (so-called pockets) (Jürgens, 1991). An evaluation of real estate transfers reveals that nonwhite buyers spread from the southwest to the northeast, though not because of differences in price levels, which vary only slightly within Yeoville. The most important reason was a spillover effect from the very densely settled *gray areas* of Hillbrow and Berea where the cores of the expansion existed.

As in other *gray areas* of Johannesburg, the pioneers consisted of well-educated coloreds, Indians, and Blacks on their way up the social ladder who were moving in from all over the country and whose social status was as high as that of the white inhabitants or even higher (cf. Fick *et al.,* 1987; Rule, 1989). Nevertheless, the proportion of nonwhite and mixed race households was only 3.4% in 1989. The multiculturalism in Yeoville at the end of the 1980s was still characterized by white dominance and was based primarily on the various European immigrant groups and the strong Jewish population. This was the first phase of invasion and desegregation.

The second phase, of succession and resegregation, began after apartheid was repealed in 1991. Many nonwhites now moved legally into the area and the population of Yeoville rose considerably. The population has almost doubled since 1991 (Table 1), whereas the number of housing units has remained approximately constant. After the elections in 1994 the influx of nonwhite households increased, causing the proportion of this population group to rise to 84.1% by 1998.

Table 1

Population Groups in Yeoville 1970–1998

	1970	1980	1985	1991	1996	1998
Whites	9,722	9,012	7,290	6,517	3,243	2,507
Coloreds	18	17	28	279	582	883
Indians/Asians	2	31	46	177	259	378
Blacks	1,489	983	878	1,237	6,809	11,999
Total	11,231	10,043	8,242	8,210	10,893	15,767

Source: Central Statistical Service and Statistics South Africa (1970–96); 1998: estimate: Department of Geography, University of Kiel (1998).

Because fewer young whites are moving in, the white population is aging in place. On one hand, there is a residual white population that is immobile because of age and, on the other hand, a growing young black population. Further evidence of *white flight* and strong migration dynamics among non-whites is the average duration of stay. Among whites it has more than doubled, from seven (1989) to seventeen years (1998), whereas the nonwhite population records only a slight increase from two to three years. Altogether almost 70% of the inhabitants have lived in Yeoville for five years or less. Of these, nonwhites represent the overwhelming majority (94%).

The influx of nonwhites is primarily an intraurban migration flow. In total, 74.2% of the persons who moved to Yeoville in the past three years came from other parts of Johannesburg. The majority of the in-migrants (55.5%) came from the neighboring residential areas, which points to a spillover effect. Reasons given for moving to Yeoville are its central location and lower crime rates than the place of origin.

The socioeconomic contrasts among the inhabitants have intensified, although the nonwhites are by no means a typical *underclass* or *outcast ghetto* population (cf. Marcuse, 1997a). The income and educational levels of non-whites are that of the middle class, if we take the South African nonwhite average as our standard (Tables 2 and 3). A comparison of the socioeconomic structure of the nonwhite with that of the white population reveals considerable disparities, however (Table 3). According to Saff (1995:795) this is the reason why whites increasingly avoid the area: "many whites are not opposed to living in the same neighborhood as blacks as long as they perceive them to come from a similar class." This assessment is losing its validity, though, and the population is shifting. About half of the nonwhite households have

Table 2

Educational Level of Inhabitants over 20 Years of Age in Yeoville 1998 (in %)

| | Population Group | | | |
| | White | | Black | |
	Yeoville	Republic of South Africa	Yeoville	Republic of South Africa
No schooling	0.8	1.2	1.0	24.3
Some/complete primary (up to Standard 5)	0.8	1.2	5.2	27.2
Some secondary (Standard 6–9)	17.5	32.8	40.8	32.8
Matric	55.8	40.7	40.6	12.1
Higher degree	25.0	24.1	12.3	3.0

Source: Survey by Department of Geography, University of Kiel (1998); Statistics South Africa (1998).

an available income of less than R3,000 per month. The average net income of white households is 2.5 times as high as that of nonwhite households.

Among the population a *climate of fear* has spread. Violent crimes, such as murder, armed robbery, and rape, more than doubled in the district covered by the Yeoville police station between 1994 and 1998. The number of persons directly affected by crime is very high. In every second white household, at least one person has already been the victim of an attack. Among the nonwhite households, the figure is one in three. Physiognomically the area is decaying due to overcrowding of housing units and slumlording. Almost one-fifth of the units housing nonwhite households is occupied by five or more persons.

Ever since real estate agents and financial institutions began to anticipate in 1995 that Yeoville would be the "next Hillbrow" (Minogue, 1995), the flow of capital has been considerably reduced, though it has not entirely

Table 3

Average Net Household Income in Rand per Month

	White	Nonwhite
Yeoville	7,751 (5,490)[1]	3,254 (2,217)
Republic of South Africa[2]	8,583	1,917 (Blacks) 5,917 (Indians)

Source: Survey by Department of Geography, University of Kiel (1998).
[1]Standard deviation in parentheses.
[2]South African Institute of Race Relations (1998): status as of 1995.

dried out. Because they fear a progressive loss of value of the real estate, more and more often the social and physical decline induces finance companies to deny mortgage loans (Tomlinson, 1998; Interview, 1999). So-called *redlining* contributes considerably to Yeoville's physical decay. Although there is a demand for residential property (almost exclusively for blacks), buyers are lacking because they cannot secure financing. On the other hand, because of falling real estate prices, many white property-owners would like to sell their property. Property can only be sold at a great loss, if at all (*The New Republic*, 2 December 1996). In this situation, many owners divide their houses or flats into several units, and hope to amortize the property in a short period of time through usurious rents. This strategy means that the owners no longer invest in maintenance and tacitly accept that the buildings will decay (*North Eastern Tribune*, 29 March 1996). Beyond the physical decay of the buildings the negative structural change is apparent in a noticeable increase in street prostitution, illegal taxi stands, car repair shops in backyards, and *shebeens* (houses or flats in which alcohol is sold without a licence). The economic structure is adapting to the transformation process and many restaurants and shops are closing. These business people are moving to safer parts of town farther to the north (cf. also Beavon, 1998).

A third wave of invasion and succession began in the mid-1990s. Since the mid-1990s, Blacks from South Africa's neighboring countries and from western and central Africa have been moving into the inner city of Johannesburg (Morris, 1998). Many are illegal migrants who are in danger of being deported. Like the nonwhite South Africans at the beginning of the 1980s, they are also willing to pay higher rents. In Yeoville, the street scene and especially the informal sector are increasingly being dominated by French speaking persons with conspicuously darker skin. Of the residential population, 8% come from southern Africa and 6% from other parts of Africa. Unlike black South Africans, the pioneers of this new invasion are not persons who are rising up the social ladder and who have a socioeconomic status comparable to that of the old-established residential population (Table 4). The unemployment rate of the migrants from Black Africa and the southern African states is above average and the majority of those who are gainfully employed work in the informal sector.

GATED COMMUNITIES AT THE URBAN PERIPHERY?

The paranoia arising from insecurity and political uncertainty has led to a great number of security measures in the white-dominated cities since the middle of the 1980s (Jürgens and Gnad, 2000). The market for *walled com-*

Table 4

Selected Socioeconomic Indicators for Residents of Yeoville

	SADC States (N = 89)	Black Africans[1] (N = 66)	South Africans[2] (N = 880)
Unemployment rate[3] (%)	33.3	21.1	17.4
Persons employed in the informal sector[3] (%)	12.3	36.8	3.6
Net household income per month	R2,319	R2,969	R4,012

Source: Survey by Department of Geography, University of Kiel (1998).
[1]Migrants from Central and West Africa.
[2]South African citizens.
[3]Proportion of the employed population >15 years.

munities includes not only the upper class, but also members of the white middle class and the emerging black middle class. An especially booming market is the so-called *gash market* (*gash* = good address, small home), which appeals to young couples, single persons, and single parents. In addition, so-called empty-nesters (i.e., couples whose children have left their childhood home and who cultivate a so-called "lock-and-leave" life-style) are mentioned as potential customers. Gated communities accommodate people's desires not only for residential security but also for convenience. A communal association can organize responsibility better and lessen individual burdens (e.g., cleaning of the communal swimming pool).

Both Forestdale and Santa Cruz are mixed racial residential areas, though with very different ethnic predominance. Whereas in Forestdale 85.6% (of 132 persons) are whites and 14.4% nonwhites (5.5% Blacks), in Santa Cruz 80.2% are nonwhites (of these 67.9% are Blacks) and 19.2% whites (of a total of 156 persons). Both areas are changing and adapting to the ethnic structure and image of their surroundings.

From the beginning of its development as a *free settlement area*, Country View, where Santa Cruz is located, suffered because it was perceived as a special type of Black township. Consequently white buyers tended to avoid the area and still prefer to move into traditionally white suburbs in which the real estate prices are prohibitively high for interested Blacks. Even Blacks who can afford them are not necessarily looking for new densely built areas such as the *matchbox* houses so familiar from Soweto or Katlehong. Instead they prefer larger lots. This may explain the initial disinterest in Forestdale among Blacks.

A conspicuous feature of both study areas is the very young age structure. The average age in Santa Cruz is approximately 25 years. In Forestdale, it is around 31 years. In 22 of 56 households in Santa Cruz there were children six years of age or younger, in Forestdale in 14 of 55 households. Only 8.4% of all persons included in the survey (N = 286) in the two areas were older than 50 years of age. Because the houses tend to be rather cramped, most are inhabited by small nuclear families. The average household size ranges from 2.3 (for white households in Forestdale) to 2.9 (for Black households in Santa Cruz). Additionally, in Forestdale 22% of the households (11% in Santa Cruz) are one-person households, consisting mostly of young, employed single persons. Only one of these persons was older than 60. Both study areas are not so much refuges for retired persons as for persons actively involved in work and family life.

The main reason for moving into the community for more than half of the interviewed households was the security aspect. "For the first time I feel secure about my children's safety" was one mother's reply to the question as to what had improved for her after moving into a walled settlement (Safety in clusters, 1997:53). The children can play on the private roads of the settlement where they are safe from crime.

The cost of buying or renting a house or lot and the monthly basic fees are such that the population of *security villages* tends to be a socially select group. In Forestdale, the average net monthly household income was approximately R13,200 and in Santa Cruz R8,000. The income of Blacks in Santa Cruz was thus two and a half times as high as in Yeoville. The incomes are also related to the higher level of education: only 7% in Forestdale and 15% in Santa Cruz of these surveyed had an educational level lower than matric. As a consequence, no one in Forestdale and only six of 102 persons in Santa Cruz of employable age professed to be unemployed. In comparison, the official rate in 1996 for the province of Gauteng was 28.2%.

Nevertheless the satisfaction with *walled communities* is not total. The interviewees especially criticized a "lack of privacy." The houses were not well soundproofed and were too close together. Particularly in Santa Cruz cultural differences have led to complaints by whites about their Black neighbors. For such reasons at least one white household wants to move out and rent their house. This practice is already widespread; in 20 of 54 cases the housing units are not inhabited by their owners. In fact, 60% of the households in Forestdale and approximately 45% in Santa Cruz have to make do with a smaller average number of rooms than in their previous housing. As a rule they formerly lived in single, unattached houses in Greater Johannesburg. In fact, the majority of nonwhite buyers and renters moved in from other "white" areas and not from black townships.

CONCLUSIONS AND OUTLOOK

The white population's perception of Yeoville is negative and they avoid it as a residential area. Only the nonwhite population considers it attractive. Whereas the pioneers of the nonwhite influx were comparable in social status to the white population, in the further course of the succession process a socioeconomic gap appeared. A process of filtering is occurring, which will end in the concentration of a nonwhite *urban underclass.* A mixed racial composition is merely the transition from one ethnic homogeneity to the next.

Parallel to this, the commercial environment has changed dramatically. Because of the altered consumer behavior of black residents compared with the original white population and their generally lower spending power, not only various shops but also branch banks and public services have been forced to close in the past years. Informal activities, be they legal or illegal, dominate life on the sidewalks. Nevertheless, the traditionally strong self-image of Yeovillites stemming from its progressive apartheid era cosmopolitanism still persists. Citizen groups composed of blacks and whites attempt various activities (e.g., establishment of a so-called Business Improvement District, cleanliness campaigns, setting up a market for street traders) and lobby together with the city hall to improve the social and economic qualities of the area.

In contrast to the strategy of seclusion practiced by gated communities, Yeoville does not function as an island. Social problems such as prostitution and drug trafficking have spread to Yeoville from neighboring, more anonymous areas that lack Yeoville's self-healing forces. Both the pace at which the exchange of population is proceeding and the conspicuous changes in the residential environment allow us to speak of ghetto elements (cf. definition by Morrill, 1965), though in some cases only on the basis of a single block or house. Even if we apply more complex definitions (Wacquant, 1998) that emphasize how the inhabitants identify with their residential area, the criteria are partially fulfilled. If, in contrast, we stress the high degree of satisfaction of persons who are moving in, Yeoville is closer to an "ethnic enclave" (Marcuse and van Kempen, 2000:19) than a "ghetto of exclusion" with its pejorative connotations.

Perceived insecurity encourages the construction of gated communities at the edge of the city and our initial research shows that these are not mixed racial communities. In the process of social assimilation that has been occurring since the end of apartheid, members of nonwhite ethnic groups have begun to make the same high financial expenditures for security made originally by whites. In Santa Cruz, we can already see a tendency toward succession emerging, with Black households succeeding the original white inhabitants. In contrast to the experience in the inner city, this is not leading to social tensions or deteriorated buildings. Indeed these residential areas are so artificial in design, without connections to public transport and, except for a service station in Country View/Santa Cruz, without places to purchase supplies, that no feeling of belonging develops. A completely new type of structure is occurring with (at least according to our case studies) a population structure characterized by a pronounced race–class dichotomy. In a descriptive sense, this may be considered an ethnic enclave but without the functional linkages and deeply rooted social base of Yeoville.

Contrary to the vision of a *rainbow nation*, the white population has been reacting to the in-migration of the black population with *white flight*. Johannesburg is disintegrating into residential islands with different degrees of security, different images, and dominated by individual ethnic groups. The changes in Forestdale and Santa Cruz that can already be observed do not exclude the possibility that the different ethnic groups in South Africa will create their own *gated communities*. Social segregation is supplemented by a new type of racial segregation [Beavon's (1998) "neo-apartheid city"] that is emerging on the basis of varying price levels and images of individual residential areas. Gilded ghettos (middle-class enclaves) or fortified citadels, as they are referred to by Marcuse (1997b), and decaying ghettos (sometimes only on a street-block basis as in Yeoville) exist side by side.

APPENDIX: STUDY METHODS

Our empirical findings for Yeoville are based primarily on two statistical surveys carried out in 1989 and 1998 by the Department of Geography of the University of Kiel. The first involved a single random sample of 420 standardized interviews and the second a stratified sample of 350 standardized interviews. Additionally the transfers of real estate registered in the Johannesburg Deeds Office were documented and those for Yeoville in the period from 1989 to July 1999 were analyzed. The family name was used to identify the ethnic group of the persons involved in the transfers. In the case of Indians and Blacks, the results are highly accurate. Because members of the colored population have Afrikaans or English names, these were disregarded in our analysis of the transfers. In 1999, representatives of Sage Schachat Ltd. were interviewed and all inhabitants of Forestdale and Santa Cruz were surveyed. By means of a standardized questionnaire, 55 of the 60 units household in Forestdale and 56 (of a total of 59) households in Santa Cruz were interviewed. Information on the ethnic structure of the missing households was obtained from neighbors.

NOTE

1. In this chapter, Blacks refers to Africans in the South African context and Afro-Americans in the U.S. context while blacks (lower case) refers to Africans, coloreds, and Indians in South Africa as a whole.

REFERENCES

Bähr, J., and A. Schröder-Patelay. 1982. "Die südafrikanische Großstadt. Ihre funktionale und sozialräumliche Struktur am Beispiel der 'Metropolitan Area Johannesburg.' " *Geographische Rundschau* 34:489–497.

Bähr, J., U. Jürgens, and S. Bock, 1998. "Auflösung der Segregation in der Post-Apartheid-Stadt?—Diskutiert anhand kleinräumiger Wohnungsmarktanalysen im Großraum Johannesburg." *Petermanns Geographische Mitteilungen* 142:3–18.

Beavon, K. 1998. *Nearer My Mall to Thee: The Decline of the Johannesburg Central Business District and the Emergence of the Neo-Apartheid City.* Seminar Paper from 05/10/1998, Johannesburg (unpublished).

Blakely, E., and M. Snyder. 1997. *Fortress America: Gated Communities in the United States.* Washington, D.C.: Brookings Institution Press.

———. 1998. "Forting Up: Gated Communities in the United States." *Journal of Architectural and Planning Research* 15:61–72.

Cutler, M., E. L. Glaeser, and J. L. Vigdor. 1999. "The Rise and Decline of the American Ghetto." *Journal of Political Economy* 107:455–506.

Deskins, D. R. 1981. "Morphogenesis of a Black Ghetto." *Urban Geography* 2:95–114.

Duncan, O. D., and B. Duncan. 1957. *The Negro Population of Chicago. A Study of Residential Succession.* Chicago: University of Chicago Press.

Fick, J., C. de Coning, and N. Olivier. 1987. "Residential Settlement Patterns: A Pilot Study of Socio-political Perceptions in Grey Areas of Johannesburg." *South Africa International* 17:121–137.

Ford, L., and E. Griffin. 1979. "The Ghettoization of Paradise." *Geographical Review* 69:140–158.

Hart, G. H. T. 1989. "On Grey Areas." *South African Geographical Journal* 71:81–87.

———. 1996. "Resegregation within a Process of Desegregation: Social Polarization in Johannesburg," in J. O'Loughlin and J. Friedrichs, editors, *Social Polarization in Post-industrial Metropolises: Berlin–New York.* Berlin: de Gruyter, pp. 195–206.

Interview. 1999. Maurice Smithers, Chairman of the Yeoville Community Development Forum, June.

Jargowsky, P. A. 1996. *Poverty and Place. Ghettos, Barrios and the American City.* New York: Russell Sage Foundation.

Jürgens, U. 1991. *Gemischtrassige Wohngebiete in südafrikanischen Städten.* (Kieler Geographische Schriften 82). Kiel: Kiel University Press.

Jürgens, U., and J. Bähr. 1998. *Johannesburg: Stadtgeographische Transformationsprozesse nach dem Ende der Apartheid.* (Kieler Arbeitspapiere zur Landeskunde und Raumordnung 38). Kiel: Kiel University Press.

Jürgens, U., and M. Gnad. 2000. "Gated Communities in Südafrika—Untersuchungen im Großraum Johannesburg." *Erdkunde* 54:198–207.

Kowinski, W. S. 1985. *The Malling of America.* New York: Morrow.

Marcuse, P. 1997a. "The Ghetto of Exclusion and the Fortified Enclave. New Patterns in the United States." *American Behavioral Scientist* 41:311–326.

———. 1997b. "The Enclave, the Citadel, and the Ghetto. What Has Changed in the Post-Fordist US-City?" *Urban Affairs Review* 33:228–264.

Marcuse, P., and R. van Kempen. 2000. "Introduction," in P. Marcuse and R. van Kempen, editors, *Globalizing Cities–A New Spatial Order?* Oxford: Blackwell, pp. 1–21.

Marshall, H., and R. Jiobu. 1975. "Residential Segregation in United States Cities: A Casual Analysis." *Social Forces* 53:449–459.

Massey, D. S., and N. A. Denton. 1993. *American Apartheid: Segregation and the Making of the Underclass.* Harvard University Press: Cambridge.

Minogue, R. 1995. "The Spirit Fights the Stork." *Sidelines* 3:62–69.

Morrill, R. L. 1965. "The Negro Ghetto: Problems and Alternatives." *Geographical Review* 55: 339–361.

Morris, A. 1998. "Our Fellow Africans Make our Lives Hell: the Lives of Congolese and Nigerians Living in Johannesburg." *Ethnic and Racial Studies* 12:1116–1136.

———. 1999. *Bleakness and Light. Inner-city Transition in Hillbrow, Johannesburg.* University of Witwatersrand Press: Johannesburg.

Rose, H. M. 1971. *The Black Ghetto. A Spatial Behavioral Perspective.* New York: McGraw-Hill.

Rule, S. P. 1989. "The Emergence of a Racially Mixed Residential Suburb in Johannesburg: Demise of the Apartheid City?" *Geographical Journal* 155:196–203.

Sack, M. 1990. "Stadt als 'Intérieur'?" *Werk, Bauen + Wohnen* 44(6):46–53.

Safety in Clusters. 1997. *Housing in SA* (November/December):53–55.

Saff, G. 1995. "Residential Segregation in Postapartheid South Africa—What Can Be Learned from the United States Experience." *Urban Affairs Review* 30:782–808.

Schlemmer, L. 1989. *Racial Zoning and Problems of Policy Change.* Johannesburg: Centre for Policy Studies.

Schlemmer, L., and S. L. Stack. 1989. *Black, White and Shades of Grey: A Study of Responses to Residential Segregation in the Pretoria-Witwatersrand Region.* Johannesburg: Centre for Policy Studies (Research Report #90).

———. 1990. "The Elusive Deal: International Experience of Desegregation," in A. Bernstein and J. McCarthy, editors, *Opening the Cities: Comparative Studies on Desegregation.* Durban: Centre for Development Studies, pp. 43–53.

South African Institute of Race Relations (SAIRR), editor. 1998. *South Africa Survey 1997/98.* Johannesburg: SAIRR.

Statistics South Africa, editor. 1998. *Census in Brief.* Pretoria: Government Printer.

Tomlinson, M. 1998. *A Position Paper on the Perception of "Redlining" by the Financial Institutions When Granting Mortgage Bonds.* Johannesburg: The Banking Council (unpublished).

Wacquant, L. 1998. "Drei irreführende Prämissen bei der Untersuchung der amerikanischen Ghettos," in W. Heitmeyer *et al.*, editors, *Die Krise der Städte.* Frankfurt/Main: Suhrkamp, pp. 195–210.

Wilson, W. J. 1987. *The Truly Disadvantaged.* Chicago: University of Chicago Press.

———. 1991. "Studying Inner-City Social Dislocations: The Challenge of Public Agenda Research." *American Sociological Review* 56:1–14.

5

Property Investors
and Decentralization

A CASE OF FALSE COMPETITION?

SORAYA GOGA

B etween 1975 and 1992, owners of office real estate in the central business district (CBD) of Johannesburg began investing in decentralized
locations. Their motives did not stem from user demand arising from economic growth, the most common motivation (Wheaton, 1987; Barras and
Ferguson, 1985, 1987). The Johannesburg economy, like the South African
economy, was undergoing an economic crisis. An average 10% gross geographic product (GGP) growth in Johannesburg in the 1970s dropped to
0.6% in the 1980s before increasing to 1.7% between 1991 and 1994
(Gelb, 1999; Marais, 1998). Office construction, however, increased 42%
between 1975 and 1995, thus casting doubt on the alleged relationship
between economic growth, office demand, and concomitant supply. Johannesburg's office supply was not reacting to demand generated from economic growth. The city had an oversupply of space and rental income was
declining across the region.

Instead, real estate investors and owners in Johannesburg reacted to user
demand for relocation. Thus, oversupply had a spatial manifestation as owners invested in offices in decentralized locations to which users then relocated.
By 1993, the vacancy rates of A Grade space, the rental indicators, and the
number of building plans approved in the CBD began to diverge from those
in Sandton and Midrand where decentralization was occurring.

Increased car usage within the region, parking problems in the central
business district, residential suburbanization, and the concomitant "boss factor" where managers suburbanized offices in an effort to close the home–work

divide, were crucial for relocated demand in South Africa.[1] Furthermore, real estate investors pointed to a demand for a different type of space due to changing work type, concomitant changing technological needs, and changes such as low-rise buildings and park-like settings. Finally, certain real estate investors noted the effect of political changes, subsequent deracialization of the CBD, and declining standards of public health and safety in the area as push factors (Goga, forthcoming).

In Johannesburg, the existing CBD owners market also provided the capital for decentralized investment. Even though returns and rentals in decentralized areas were better than in the CBD, general yields and rentals across the metropolitan area were declining. This meant that CBD owners, through investing in real estate in new locations in a period of economic decline, undermined their investments in the CBD and also their metropolitan real estate portfolios. Second, and more importantly, both the owners and the investment markets were not competitive, but oligopolistic. About twenty institutions (Tomlinson *et al.,* 1995) controlled the existing owners market in the CBD, while approximately six institutions controlled the investment markets. This oligopolistic market structure should have made it possible for these institutions to manage relocated demand. Response to normal relocated demand could have continued, but the demand from perceived technological change could have been controlled through refurbishment of CBD buildings. Stylistic changes, as supply-driven issues, could also have been controlled. As to political changes, a commitment to the CBD could have created an impetus for local government to improve services.

Thus, the issue in Johannesburg is not so much that relocation occurred but the extent of relocation. Given that CBD owners could have dampened decentralized demand, relocation seems to have been excessive. Why did CBD owners undermine their own investments? What explains their investments in decentralized locations rather than in the traditional CBD?

This chapter argues for the importance of supply-side factors affecting investor decisions. It will describe how an excess of capital in search of investment and held by insurance houses and pension funds (that is, long-term financial institutions) acted as a necessary condition in driving investment to decentralized areas. It will further describe how the oligopolistic industry structure of these institutions drove a "false" competition within the market to exacerbate conditions of oversupply.[2] Finally, it will note how poor internal organization and management within the investment institutions contributed to oversupply across the metropolitan area.

Decentralized development was a "spatial fix" for institutions possessing overaccumulated capital in Johannesburg.[3] This, in turn, interacted with a CBD that, because of its age, was quickly losing its value, even while the suburbs were gaining value. The valorization of suburbs and devalorization of the CBD interacted with investment imperatives to boost market values and appreciation. These aspects of the property market when coupled with the land ownership structure impelled the decentralized location of investment.

CAPITAL AVAILABILITY IN SOUTH AFRICA

In the late 1980s, the South African economy underwent an overaccumulation crisis with an excess of capital relative to investment opportunities (Bond, 1990). This did not arise out of a changing configuration of financial markets as had occurred internationally (Renaud, 1995; Fainstein, 1994; Coakley, 1994). Instead, it was a result of the growing assets of long-term financial institutions in the country culminating in a pension fund economy.

The pension fund economy arose from international macroeconomic and political conditions. In the 1980s, high inflation rates (about 13% for most of the 1970s and 1980s) led to increased investment by households in insurance and pension funds as a form of savings (Jones and Muller, 1992:336). The assets of these institutions increased more than 74-fold between 1961 and 1988 (Jones and Muller, 1992:337).

This overaccumulated capital then sought investment. Vos (1993:19) describes the choices available to European institutions as "choice between several asset categories; choice between countries and currencies; within a certain category . . . choice of product market (location and sector)." In South Africa, choices between countries and currencies for the long-term financial institutions were limited due to sanctions and exchange controls.[4] Furthermore, choices between asset categories were curtailed by a crisis in commodity production. In manufacturing, average capital output fell by nearly 40% between 1970 and 1986 (McCarthy, 1987:7 in Black, 1991:169). Subsequently, "during 1986, real GDFI in manufacturing was only 40% of the 1980' figure, and at its lowest level for 16 years" (Black, 1991:169). The long-term financial institutions thus switched investment into real estate.[5] Real estate transactions doubled in value from around R10 billion a year in the mid-1980s to R20 billion in the late 1980s.

The property assets of institutions grew thirty times between 1960 and 1996. More importantly, property assets as a percentage of total assets grew 12% between 1960, reaching a high point in 1987. This occurred even within a sluggish economy and despite the poor performance of property.

Real property only marginally outperformed the consumer price index, and underperformed vis-á-vis the Johannesburg Stock Exchange.

Property investors confirmed the overaccumulation of capital by long-term financial institutions and the subsequent use of the real estate market as a sponge for overaccumulated capital. "The South African market was like a pressure cooker by the end of apartheid. There was more and more money chasing less and less investments, so we pushed it into property" (Interview with Gil Da Silva, Research Manager: Old Mutual Properties). The availability of capital set conditions for oversupply.

The conditions were amplified in Johannesburg. A large proportion of this overaccumulated capital went into the Johannesburg market. The Rand value of approved building plans reveals a relatively high percentage in Sandton, Midrand, and Johannesburg (see Table 1). Furthermore, an examination of the portfolios of the major institutions reveals that they

Table 1

Johannesburg, Midrand, and Sandton: Value of Building Plans Passed

Year	Percentage of National Total
1974	22.2
1976	33.4
1978	38.1
1980	15.8
1982	27.9
1984	18.1
1985	24.1
1988	28.6
1989	33.5
1990	38.7
1991	37.3
1992	39.8
1993	33.5
1994	21.6
1995	42.8
1996	35.1

Source: Central Statistical Services, *Building Plans Passed*, various years.

hold most of their property assets in the Johannesburg region. For example, Rand Merchant Bank has over one-half its property holdings there, and the Ampros portfolio is similarly concentrated with 72% of its holdings in the region. The necessary condition for oversupply (i.e., the availability of capital) interacted with the particular investment imperatives of the capital suppliers themselves and the institutional structure to produce a final oversupply.

The long-term financial institutions, and thus the real estate investment and ownership industry, in South Africa had a peculiar institutional composition. The industry is oligopolistic, but in their real estate investment decisions acts not as an oligopoly but as a false competitive investment market. Competition was not based entirely on response to demand, as should be the case in a perfectly competitive market, but rather on the actions of other investors.

The industry is extremely concentrated. A few large players control most of the pension funds and thus investment. The largest two, Old Mutual and Sanlam, receive two-thirds of the total premium income of all long-term insurers and the top seven about 90%. These few provided a large percentage of the money for real estate development, specifically in the Johannesburg region. The oligopolistic structure of the industry coupled with the small user market created an extremely competitive investment and owners market. These investors/owners consistently monitored and mimicked each other's decisions. As Gil Da Silva of Old Mutual argued:

> This was the problem facing institutions. Say if we had said that morally we could not invest in property in the suburbs, as there really was enough space in Johannesburg, the problem was that someone else would have invested. They would then have produced better returns and we would just be sitting with vacancies so we would lose business. Once it started, we had to compete. We had to look at the weight of our portfolios compared to our competitors. Thus, everyone in Johannesburg was looking over their shoulders to see where everyone else was locating and what they were investing in and all invested in decentralized office development in Johannesburg. Decentralized office investment in Johannesburg thus became the "flavor of the month." (Gil da Silva—Old Mutual)

An investment analyst describes the situation at the time:

> It was a prosperous time for developers. They would put together new properties that had a weight of funds flowing in by institutions and they would

sell back to the institutions on the strength of one head lease. Very often, they were not fully let. Thus, developers would put together a development, fax it to five of them (institutions), three would call back and say, "yes—we want it." (Interview with Kelly Clinton, Rand Merchant Bank Properties)

The constant competition effectively created a "herd mentality" akin to what Fainstein (1994) describes within the London developments in the late 1980s and early 1990s. Decisions within the investment market were not based on sound investment logic.

Competition interacted with a second institutional factor. This was the way in which the imperatives of the long-term financial institutions began to structure the owner's market and the investment market, and the concomitant way in which long-term financial institutions realized exchange. In South Africa, the long-term financial institutions took over the role of short-term financiers (usually banks) and thus structured the short-term investment market. However, they did not immediately "sell on" real estate as short-term financiers usually do. Their investment was of a long-term nature. Thus, one saw integration of the short-term investment market that financed both development and the owner's market. This had implications for oversupply.

First, the fact that long-term financial institutions replaced short-term financiers weakened a measure of control between point of construction and point of final sale. Thus, the market was less responsive to demand.

Second, the long-term financial institutions as owners had very different investment criteria than private owners who had previously owned real estate in Johannesburg. For most private investors (who often buy property with loan financing), a constant rate of return in the form of annualized rent is an important determination of value, as it is required to repay loans (Haila, 1991; Harvey, 1983, 1989).

The value of property over a long term (that is, the capital gain at the end) and thus current valuation, rather than value derived from a steady income stream (rent), was the most important reflection of the value of the property.[6] As Cloete argues: "since they were investing their own (or their savers') money rather than relying on borrowed funds, capital appreciation was as important to them as income. They were not troubled by the income deficit problems of property owners who rely on borrowed money" (Janowitz, 1997:12). The emphasis on capital gain makes the owners less responsive to the immediate dynamics of supply and demand within the property market, thus implying an inherently speculative market.

Ultimately, this market structure led to less concern for actual present value based in the immediate sale price. Second, long-term financial institutions as owners also led to a lack of concern on medium-term exchange, i.e., rent. The lack of controls on short- and medium-term exchange exacerbated conditions for oversupply within the region.

A third institutional problem exacerbated this decreased concern, a deep lack of systematic knowledge of the profitability of the real estate industry. This stemmed from poor management. At an institutional level, the South African property industry had none of the four operating management principles—market research, portfolio management, property management, and property funding—attributed by Vos (1993) to good portfolio management.

Although property firms in South Africa do engage in market research, the state of data within the market is poor. Until 1998, South Africa did not have a uniform property index to measure property performance.[7] Thus, the portfolio managers failed in the first operational task of a property fund, that is, in distinguishing market trends with the aim of carrying out in-time mutations in the investment portfolio (Vos, 1993:17). This severely affected investment policy (defined as making decisions on the markets in which the company will operate and the investment and selection of specific buildings).

Furthermore, in the late 1980s property companies did not utilize portfolio management. In fact, they have done so only since the opening of South Africa to international investments in the early to mid-1990s. As a property economist (David Green of JHI) pointed out:

> Investments at the time were not viewed as part of the broader investment environment. Boards would just decide on what they wanted to invest in based on what they thought. There was very little holistic management.

Property management was also ad-hoc. In many property divisions, the same division that bought and sold the property would also provide management and refurbishment of the property. Because the focus of the division was often on acquisition, refurbishment and tenant satisfaction received little attention (Interview with Alan Kerdachi—McCreedy & Friedlander). Only recently (since about 1993) has property management, as distinct from sales and purchases, emerged as a distinct field. Lack of property management also meant less of a focus on what users needed in their individual buildings; the goal was acquisition. Subsequently, the needs of the existing clients were subsumed by those of future clients.

The fourth principle is property funding. This refers to attracting equity or debt money in a certain combination. However, because of the over-abundance of capital for development from the institutions, this principle, which ostensibly could act as a block on investment, failed to function as a check in the process. Investors, developers, and owners did not take invest-ment decisions in property generally and investment in decentralized loca-tions in Johannesburg specifically in a way that considered sustainable demand versus sustainable supply.

Finally, flawed mechanisms of book valuation also contributed to poor investment decisions.[8] The flawed valuation mechanism is particularly sig-nificant in light of the fact that the owner's market regarded market value in terms of capital gain and capital valuation rather than as a steady rental income stream.

Most South African institutions use the income approach to valuation rather than the comparable method or the replacement value.[9] The basis of the income approach is the capitalization rate. However, the mechanism for determining capital rates within South Africa is extremely subjective. Tenant mix, lease term, and location determine capital rate. Furthermore, South Africa has no legal requirement or mechanisms for independent assessment of property for accounting valuation within the company. Institutions thus value their own buildings rather than use an independent assessor. The implications for the oversupply of space in decentralized areas in Johannesburg arising out of the changing conception of property are profound.

For the most part, then, capital values reported on the books of insti-tutions are largely fictitious. This makes owners less responsive to user mar-ket demand and more responsive to their own needs of proving their profitability. As one property analyst (asking for anonymity) stated:

> a big problem is that the big boys, and in fact everyone values their proper-ties themselves. So they can create what values they want. Everyone says "oh well the auditors will pick it up," but the truth is that most of the auditors don't like to challenge the books of the big boys—after all they need work next year.

What it meant for Johannesburg was that the perception of property as a "good return" continued even in the face of poor returns. This was propped up by fictitious values reflected on the books of institutions. A lack of connection between the valuation and a steady income stream further exacerbated conditions for oversupply and thus fueled oversupply and decentralized investment.

EXPLAINING DECENTRALIZATION

Finally, what of decentralization? The final investment location for overaccumulated capital links directly to the way in which the built environment is valorized and devalorized by capital over time. Fixed capital in the CBD, in the form of the office buildings, was devalorizing due to its age (Smith, 1996).

The Johannesburg CBD had long been the location of the majority of investment within the office sector in South Africa. Its role as the primary and most valuable location for office investment began with the discovery of gold in the late 1890s. Past investments thus "imprisoned and inhibited" further accumulation for the institutions (Harvey, 1989; Smith, 1996). As such, it had reached the apogee of its profitability and investment capital was slowly devalorizing (Smith, 1996:84). Most of the buildings in the CBD were over 20 years old. Individual institutions wrote such properties off their books. As one developer from Stocks and Stocks noted:

> The institutions shouldn't complain. If they had been doing their job right, they should have been devaluing their buildings every year. Most of them were built a long time ago. Everything devalues—buildings are the same. The CBD buildings are only bringing in profit now—they don't have to pay any expenses except for operating expenses—but they should be written off the books. It's all cyclical you know. If the CBD comes right, we will develop there again.

The new suburban developments, on the other hand, represented new capital investments and new profitability within an area that was fast becoming valorized rather than devalorized. A flawed property management system did not see declining rent across the metropolitan area, but only increasing returns from decentralized areas. A spatial fix, through relocating investment to valorizing decentralized areas, was an obvious solution.

The gap between a devalorizing CBD and valorizing decentralized locations was widened by the ownership structure within the property market. Most institutions were "exposed" within the CBD. Thus, for them, investing in new accumulation opportunities while spreading risk meant investing in decentralized areas. However, they were investing in decentralized areas in periods of little or no economic growth. As areas of new accumulation and, subsequently, areas of increasing valorization, the decentralized developments were under pressure to perform. The manner of investment increased this pressure; the institutions concentrated their investments in particular decentralized areas. Thus, Rand Merchant Bank concentrated ownership in Sandton and Sunnighill, Ampros in Bruma Lake, Liberty Life in Sandton, and AFC Holdings in Parktown (Rogerson, 1997:3). The institutions were thus under pressure to ensure that these particular locations, and not just particular investments, performed. Thus, vacancies within these areas could not be high, as confidence in the area would drop. In response to this threat, these institutions were not averse to offer rentals in the decentralized area to their clients in the CBD. Poaching of other CBD customers was also a major strategy.

> We negotiated with tenants sometimes in our CBD buildings to move out of the CBD and into Sandton. We have mixed feelings about this—but we would do it anyhow. (Interview with George de Bie, Liberty Life Property)

Finally, the concentrated ownership structure in the CBD also meant that a single institution attempting to sell its CBD portfolio has an enormous effect on the confidence of other investors. Once one institution had applied the decentralization strategy, the confidence of the remaining owners plummeted, stimulating further capital flight.

CONCLUSION

The initial question posed at the outset was why CBD owners invested in decentralized development in Johannesburg in the context of economic crisis, thus undermining their own developments in the CBD, particularly when they could effectively control the market?

The availability of capital often determines what is built where and when. In South Africa and in Johannesburg in the late 1980s, this was the case. South Africa saw an overaccumulation of capital in the economy. A large proportion of this capital was switched into investment in real property in the presence of a crisis in manufacturing.

Capital availability for property development provided a necessary condition for decentralized investment in Johannesburg. In and of itself, it did not lead to oversupply and subsequent decentralized investment. Instead, institutional factors acted as conditions to oversupply office space. The locus of the capital availability (i.e., in the oligopolistic market of the long-term financial institutions) drove the creation of a "false" competition where the property divisions of long-term investors reacted to their competitors' actions rather than to user demand. Furthermore, the integration of the long-term and short-term financial markets removed a control in the supply process, thus also acting as a contributing factor to oversupply.

The fact that long-term financial institutions as owners measured value as long-term capital meant that they were thus less in touch with demand across the metropolitan area as measured by rental income, even though as owners in the CBD and in decentralized areas one would have expected them to be so. This stimulated oversupply.

Poor property management, with measures of profitability almost nonexistent, also contributed to the oversupply. Portfolio management was not undertaken. This led to a conception of property returns as returns from each individual geographic area and created a false perception of the actual returns on property. This exacerbated oversupply in the first instance and decentralized investment in the second.

Concerning decentralization, the interaction between the measurement of value as capital gain and the way in which property is valorized and devalorized was a primary factor. The fact that the CBD was devalorizing meant that in terms of capital gain (and as reflected in profit and loss statements or annual reports) these properties were of decreasing value while suburban developments were increasing in value. Thus, investors located their development in decentralized areas. This condition was exacerbated by the ownership structure of large properties. The concentration of one investor in one suburban location ensured that the investor was committed to making that area "work." Poaching from the CBD was a mechanism to ensure full occupancy.

Finally, the concentrated ownership structure of the CBD meant that when one investor disinvested it led to a massive devaluation of CBD property thus reinforcing the concept of a devalorizing CBD. Consequently, the decline of the Johannesburg CBD and the growth of its northern suburbs cannot be solely, or even accurately, explained by demand factors. Investors in office space did not flee the CBD simply because demand had fallen there and risen to the north. Institutional investors and large property owners made these decisions on other grounds.

NOTES

1. Barras and Ferguson (1985, 1987) argue for such externalities as a secondary factor in supply, whereas neoclassical economists cast them as distortions.
2. True competition involves many investment actors in one market reacting to user demand. False competition occurs when these self-same investment actors react also to each others' actions.
3. A spatial fix occurs when investment escapes to new locations in times of a profitability crisis. Harvey (1983, 1989) describes three conditions under which it can occur: to escape a falling rate of profit, during underconsumption crisis, and to find new markets.
4. South Africa had strict exchange controls from the early 1960s. These were relaxed in the 1980s. However, conditions were still largely restrictive until the 1990s (Kahn et al., 1992).
5. In the South African economy, overaccumulation of capital also occurred. Here, the excess soon found two investment homes, one in the Johannesburg Stock Exchange and the other in South African real estate (Bond, 1990:48). Within this period, the Johannesburg Stock Exchange saw a phenomenal growth with value of shares rising from around R50 billion in 1982, following the gold slump, to nearly R400 billion in 1990 (Bond, 1990:47–48).

6. This is not to say that rents were not important. They were, and the CBD buildings had lower rentals than the decentralized buildings.
7. Certain individual firms such as Richard Ellis properties did measure returns.
8. This valuation relates not to the valuation for property tax purposes, but rather to the value used for assessing the overall worth of the business.
9. The comparable method bases value on comparable sales. The replacement value approach is more commonly used for insurance purposes.

REFERENCES

Barras, M. 1983. "A Simple Theoretical Model of the Office Development Cycle." *Environment and Planning A* 17:1382–1391.
Barras, M., and D. Ferguson. 1985. "A Spectral Analysis of Building Cycles." *Environment and Planning A* 19:439–520.
———. "Dynamic Modeling of the Building Cycle: 1. Theoretical Framework." *Environment and Planning A* 19:493–530.
Black, Anthony. 1991. "Manufacturing Decline in South Africa," in Stephen Gelb, editor, *South Africa's Economic Crisis.* Cape Town: David Philip.
Bond, Patrick. 1990. *Commanding Heights & Community Control: New Economics for a New South Africa.* Johannesburg: Ravan Press.
Coakley, J. 1994. "The Integration of Property and Financial Markets." *Environment and Planning A* 26:697–713.
Fainstein, Susan. 1994. *The City Builders: Property, Politics and Planning in London and New York.* Cambridge: Blackwell.
Gelb, Stephen. 1999. *South Africa's Economic Crisis.* Cape Town: David Philip.
Goga, Soraya. Forthcoming. *Property Power vs Peoples Power.* Ph.D. Thesis, Rutgers, the State University of New Jersey.
Haila, Anne. 1991. "Four Types of Investment in Land and Property." *International Journal of Urban and Regional Research* 15:343–365.
Harvey, David. 1983. "The Urban Processes under Capitalism: A Framework," in Robert Lake, editor, *Readings in Urban Analysis.* W. New Brunswick: Center for Urban Policy Research.
———. 1989. *The Urban Experience.* Baltimore and London: Johns Hopkins University Press.
Jankowitz, Carl Alexis. 1997. *Institutional Investment in Fixed Property.* Johannesburg: University of Witwatersrand Press.
Jones, Stuart, and Andre Muller. 1992. *The South African Economy 1910–1990.* London: Macmillan.
Kahn, Brian, Abdel Senhadji, and Micheal Walton. 1992. *South Africa: Macro-economic Issues for Transition.* Washington D.C: World Bank. Informal Discussion Paper of the Economy of South Africa, 2.
Marais, Hein. 1998. *South Africa Limits to Change: The Political Economy of Transformation.* London: Zed Books.
Renaud, Bertrand. 1995. "The 1985–94 Global Real Estate Cycle: Its Causes and Consequences." Washington, D.C.: The World Bank.
Rogerson, Jayne. 1997. "The Central Witwatersrand: Post-Election Investment Outlook for the Built Environment." *Urban Forum* 8:97–108.
Smith, Neil. 1996. *The New Urban Frontier.* London: Routledge.
Tomlinson, Richard, Roland Hunter, Marzia Jonker, Jayne Rogerson, and Chris Rogerson. 1995. *Johannesburg Inner-City Strategic Development Framework: Economic Analysis.* Johannesburg: Greater Johannesburg Transitional Metropolitan Council.
Vos, G. 1993. "International Real Estate Portfolios," in Jim Berry, Stanley McGreal, and Bill Deddis, editors, *Urban Regeneration: Property Investment and Development.* London: E&FN Spon.
Wheaton, W. C. 1987. "The Cyclical Behaviour of the National Office Market." *American Real Estate and Urban Economics Association Journal* 15:281–299.

Section II

EXPERIENCING CHANGE

The postapartheid era has yet to witness a burgeoning economy generating well-paying jobs for blacks and whites alike. Rather, employment has declined sharply and GDP has grown only slowly. Although life in Johannesburg might offer more opportunities than life in the townships, making a living is quite onerous for large numbers of people.

Given the weaknesses in the formal economy, many turn to street trading or attempt to develop small businesses. Kesper describes in great detail the fate of those who decide to become business owners. Focusing specifically on clothing manufacturing, though also providing a glimpse of furniture making, she shows us the precarious nature of these small enterprises. Small businesses are not the savior of the Johannesburg inner city, as others claim, though they play an important role for those who operate and work within them.

With opportunities meager and the city in flux, a rise in crime is predictable. Violent crime is a particular problem in Johannesburg. The social and spatial transformation of the city has severed social relations and the city's spaces have become more problematic, Palmary, Rauch, and Simpson argue. Women and migrants are particularly at risk, and youth violence is significant. Marginalization and social exclusion are the root causes.

In effect, the rights to space—what Gotz and Simone call "belonging"— are more highly contested. If Johannesburg, though, is to prosper and provide a place where people can live decently, it must allow people to belong as well as to "become," that is, enable people to make connections and associations across space and social groups. In tension, these two forces perplex local government as it attempts to gain control over businesses—specifically street trading and taxi services—it views as unwieldy yet ripe with possibilities.

It is through their daily lives—going to work, running a small business, shopping, walking the streets—that people experience the city. In post-apartheid Johannesburg, that experience is more rather than less difficult.

6

Making a Living in the City

THE CASE OF CLOTHING MANUFACTURERS

ANNA KESPER

Johannesburg has been South Africa's manufacturing heartland for more than a century. In the early 1980s, however, large and formal production exited from the inner city and revealed a multitude of small and informal manufacturing activities. The experience of prospering localities in industrialized and developing countries suggests that having a predominence of smaller firms is advantageous (van Dijk, 1993). Small, medium-sized enterprises, and microenterprises (SMMEs) have the potential to resolve the persistent problem of insufficient employment growth and build a competitive advantage in an increasingly globalized economy (Sengenberger and Pyke, 1992). The end result is local prosperity.

For Johannesburg's inner city, SMMEs may or may not function as a driving force in urban regeneration.[1] South Africa's small firms have long been isolated and protected from the global economy and are now exposed to a business environment shaped by new industrial policies and labor regulations, rapid shifts in market demand, and increased competition both in the domestic and export markets.

Based on case studies of small business growth trajectories in the Johannesburg inner city, this chapter concludes that SMMEs play an important role as urban survival strategies, but are no panacea against inner city decline. This argument is developed in four parts, beginning with a discussion of the potential role of South Africa's SMMEs in the social and economic transition after apartheid. The chapter then turns to the revival strategy for the Johannesburg inner city of which SMME promotion in general and the creation of a clothing garment district in particular form an important part. The third section then introduces the reader to clothing production in the Johannesburg inner city. This serves as background for the presentation and interpretation of case study material and survey findings from the inner city SMME clothing industry.

SOCIOECONOMIC DEVELOPMENT THROUGH SMMEs?

Since the first democratic elections of April 1994, the issues of black economic empowerment and a more equal income distribution have been high on the political agenda of the "New South Africa" (Rogerson and Rogerson, 1995b). However, the need to take the South African economy onto "a higher road"—a diversified economy in which productivity and international competitiveness are enhanced, wage levels are high, investment is robust, and entrepreneurship flourishes—is recognized as a condition to address these issues successfully (Kesper, 2000d). The South African government suggests that the SMME sector, with government support, is capable of fulfilling these objectives. The White Paper on National Strategy for Development and Promotion of Small Business in South Africa assigned the SMME sector at least three socioeconomic roles: job creation, fostering international competitiveness, and income redistribution. In addition, it promulgated supply-side measures to promote this formerly neglected sector (Republic of South Africa, 1995). For example, Khula Finance Ltd. administers public funds to address the overall SMME constraint of access to finance, while the Ntsika Enterprise Promotion Agency designs and implements nonfinancial SMME support programs (Ntsika Enterprise Promotion Agency, 1999; Kesper, 2000d).

More critical voices argue that the role of SMMEs in South Africa's socioeconomic development will be rather limited. Not only is the SMME sector highly diverse, but the majority of firms are survivalist enterprises engaged in retail activities with little prospect of growth (Ntsika, 1999).[2] A lack of dynamism among microenterprises and the adoption of a "jobless growth strategy" by formal SMMEs means that the actual labor-absorptive capacity of the SMME economy will be rather low (Bloch and Kesper, 2000a, b; Kesper, 2000a–d). Likewise, and in contrast to international evidence, South Africa's manufacturing SMMEs are seemingly noncompetitive in both domestic and foreign markets (Kaplinsky and Morris, 1999). As a consequence, trade liberalization in the mid-1990s has caused a wave of SMME closures (Ntsika Enterprise Promotion Agency, 1999).

Research evidence suggests that only more established and larger, viz. white-owed, SMMEs have the potential to contribute to employment and economic growth and hence long-lasting elimination of poverty (Levy, 1996; Rogerson, 1999a; Kesper, 2002). If this potential is to be realized, a clear differentiation has to be made between promoting these dynamic firms and simply providing for their survival, the latter being the focus of welfare policies with the aim of alleviation of poverty (Kesper, 2000d).

SMME PROMOTION AS A REVIVAL STRATEGY FOR THE JOHANNESBURG INNER CITY

The traditional role of the Johannesburg inner city as the hub of the manufacturing, banking, and public sectors has declined continuously since the late 1980s (Tomlinson and Rogerson, 1999). A lack of investment by developers in new forms of office and industrial space on the one hand and the development of new, outlying industrial townships on the other fueled suburbanization (Rogerson, 2000a). As property investment flowed to the suburbs, the benefits of agglomeration were lost. This caused additional firms to leave. The exodus of the corporate sector was accelerated by failures in public service delivery and rising crime levels. Inner city decay as well as economic and social change brought about by trade liberalization and "black economic empowerment" provided the impetus for a common revival strategy for the thirteen once racially separate local government structures of the Johannesburg inner city. The resultant economic development strategy suggested how the Greater Johannesburg Metropolitan Council (GJMC) and other stakeholders represented on the Inner City Section 59 Committee might promote economic development there (Tomlinson and Rogerson, 1999). As both retail and manufacturing SMMEs numerically dominate the inner city economy, their promotion is considered crucial. The growth of more formal manufacturing SMMEs is expected to absorb those retrenched by large firms, while the informal sector provides employment for other job seekers (Rogerson, 2000a).

With regards to more formal SMMEs, the inner city revival strategy aims to slow down or even reverse the process of disinvestment in property and fixed assets by limiting the supply of commercial development rights outside the inner city, improving service delivery, and dealing with crime and physical decay. Identifying and opening up new market opportunities lie at the heart of the promotion of the inner city's informal manufacturing firms.

The leading industry—clothing—in both the formal and informal manufacturing economy of the inner city receives particular attention for its potential for growth and benefits of agglomeration. Indeed, the Johannesburg Inner City Office initiated the Inner City Fashion Project to investigate the potential and suggest strategies for the creation of an inner city garment district. Whether Johannesburg's small-scale clothing activities provide opportunities for growth or mere survival is highly debated (for example, Kesper, 1999; Rogerson, 2000b).

CLOTHING PRODUCTION

For more than a century, the Johannesburg inner city has been the site of manufacturing in predominantly light industries such as clothing, food processing, printing, jewelery, and furniture (Rogerson and Rogerson, 1995a; 1997). With regards to clothing, Johannesburg is the third most important area following the Western Cape and Durban for formal manufacturing in the country. More than half (3,225) of the currently 6,232 workers in the formal clothing industry of Gauteng province produce garments in the eastern inner city of Johannesburg and an estimated 80% of formal production takes place in the inner city. Small and medium-sized enterprises predominate in numbers and provide the majority of employment. Indeed, Industrial Council data (see Table 1) suggest that only four out of 167 formal establishments employ more than 200 workers, the size limit for a medium-sized manufacturing firm (Republic of South Africa, 1996).

Historically, high wage levels in Johannesburg and decentralization incentives have been the root cause of the demise of large-scale garment production since the early 1960s (Altman, 1996; Kesper, 1999). Throughout

Table 1

Clothing Establishments and Employment by Size Class[1] for Gauteng Province

Employment Size Category	Number of Establishments	Employment Total
Micro (1–5)	19	60
Very small (6–15)	52	558
Small (16–50)	60	1,612
Smaller medium (51–100)	24	1,633
Larger medium (101–150)	7	824
Smaller large (201–500)	4	862
Large (>500)	1	683
Total	167	6,232

Source: Industrial Council Register, March 2001.
[1]The terminology adapted does not correspond to the definition of size classes by the National Small Business Act (Republic of South Africa, 1996).

the 1980–2001 period, the entire formal clothing economy in Johannesburg showed a consistent pattern of decline in terms of the numbers of establishments and total employment. Indeed, formal employment in Gauteng's clothing industry shrank from its peak of 25,000 in the early 1980s to just over 6,000 in 2001 (interview with Theresa Daniels, Clothing Bargaining Council). Nevertheless, clothing production still dominates the formal manufacturing economy, much of it done by SMMEs that prefer to stay close to their markets (Rogerson and Rogerson, 1995a, b; Kesper, 1999). It ranges from basic cut-make-and-trim (CMT) work to high-quality custom-made product.

The downturn in formal clothing production needs to be understood in the context of South Africa's clothing industry. Market conditions are shifting rapidly. On the one hand, the end of apartheid has given rise to an aspiring and fashion-oriented black middle class. On the other, trade liberalization has opened the lower end of the clothing market to massive import penetration from China, Pakistan, and India via Southern Africa Customs Union countries. Hence, closures of clothing SMMEs in the inner city are often a consequence of increasing competitive pressures in consumer markets, while rising levels of crime and violence there push small clothing designer workshops closer to their customers in suburban locations (Rogerson, 2000a). The inner city has lost much of its locational advantage for SMMEs that target the higher end of the clothing market.

The decline of formal clothing employment and establishments has paralleled a surge of black-run microenterprises that are estimated to employ at least half as many workers as the formal industry (Clothing Federation of South Africa, 1998; Rogerson, 2000b). The relocation of the central business district (CBD) to the northern suburbs has freed office space in the center and this space has been converted into affordable, central, and safe premises for informal manufacturers (Kesper, 1999). In addition, inexpensive rentals and the increased "informalization" of clothing as a labor-intensive production have encouraged informal clothing SMMEs to move from township locations into the inner city (Rogerson and Rogerson, 1997). Unlike formal clothing manufacturers, many informal garment producers are attracted to the inner city *because* of its relative safety and lack of violence as compared to township areas (Rogerson, 2000b). In contrast to their formal counterparts, which departed from the inner city, the informal clothing SMMEs are those that have "made it" in the townships or the low-income flatlands on the edge of the inner city. Moreover, the inner city hosts a wide range of suppliers such as fabric and accessory wholesalers and market opportunities for budget garments. The latter includes hawkers, factory shops, and commuters to neighboring countries.

The inner city has become a potential "incubator and nursery" for informal enterprises in general and a clothing cluster with intense networking in particular (Rogerson and Rogerson, 1996a, b; Upstart Business Strategies, 1999). At the same time, the majority of informal small entrepreneurs operate their businesses out of necessity rather than choice and the high replication of activities with low entry barriers leads to market saturation and few returns (Kesper, 1999; Rogerson, 2000b). Moreover, inter-enterprise cooperation, what little there is, is defined by ethnicity rather than common interests. Indeed, xenophobia precludes black South African clothing entrepreneurs from collaborating with their generally more skilled and capital-endowed immigrant counterparts from sub-Saharan Africa (Rogerson, 1999b). Subcontracting to informal clothing SMMEs does occur, but does not hold any other advantage than "keeping the workshop busy under sweatshop conditions" (Interview with Bibi Dlamini, Cindy Africa Fashions). Likewise, lack of trust prevents cooperation between black, white, and Asian-owned clothing SMMEs. As a consequence, many informal clothing SMMEs in the Johannesburg inner city are "stuck" at the low end of the market. Hence, the likelihood of emerging entrepreneurs "sprouting" from the microenterprise seedbed *without any assistance* appears to be low.

TRACING TRAJECTORIES OF CLOTHING SMMES

The SMME clothing economy in the Johannesburg inner city is far from homogeneous. It is hence not surprising that evidence of SMME success and failure coexists. This section presents comparative findings from a survey of clothing and furniture SMMEs in the inner city and, using two case studies, notes the elusive potential for new growth opportunities within the garment sector.

NATURE OF THE STUDY

The survey and case studies of furniture and clothing SMMEs in Johannesburg's inner city were undertaken in 1998 and repeated in 2000 to shed light on the potential contribution of SMMEs to socioeconomic and industrial regeneration. The personally administered questionnaire was designed to generate information about the growth trajectory of the firm and its characteristics and those of the entrepreneur. The survey focused on forty SMMEs involved in the clothing and forty SMME's in the furniture industry. It paid particular attention to whether horizontal and vertical networking was part of firm growth strategies and whether networking was facilitated by geographic agglomeration. The sample was nonrandom and targeted successful SMMEs, i.e., firms that had experienced growth in numbers of employees, net assets, and/or profits between 1994 and 1998. Official company registers and referrals were used to identify potential interview participants and generate the sample of formal and informal, as well as white- and black-run SMMEs.

SURVEY FINDINGS

Although the large majority of the SMMEs had experienced turnover and profit increases, about one-third of these firms grew "jobless." This is quite surprising since clothing is considered a labor-intensive industry. Growth instead involved casualization of highly seasonal fashion production, subcontracting, and introduction of numerically controlled machinery in knitting and embroidery.

The assumption that clothing SMMEs that have experienced growth in turnover and/or employment in the recent past will continue to do so in the future is unwarranted. Table 2 indicates that 25% of the clothing SMMEs had to close between 1998 and 2000. This contributed to the loss of 621 jobs. Although clothing SMME entrepreneurs complained about "cut-throat competition" at the lower end and shrinking demand at the higher end of the clothing market, the furniture SMMEs interviewed were seemingly less hard hit or better prepared against these threats.

The survey data also suggest that clothing firms are generally younger and smaller than furniture firms (Table 3). Likely, entry barriers are lower in the clothing industry. This makes it easier for PDI[3] entrepreneurs, female in particular, to operate clothing rather than furniture SMMEs. Indeed, women are scarcely found as owner-managers of furniture SMMEs. With male immigrants from sub-Saharan Africa, they dominate the group of black SMME clothing entrepreneurs in inner city Johannesburg.

Table 2

Employment Losses and Firm Closures (Clothing and Furniture SMMEs), 1998–2000 ($N = 80$)

	Clothing ($N = 40$)	Furniture ($N = 40$)	Total ($N = 80$)
Employees in 1998	1,620	2,277	3,897
Employees in 2000	999	2,068	3,067
Net employment loss	621 (38%)	209 (9%)	830 (21%)
Firm closures	10 (25%)	2 (5%)	12 (15%)

Table 3

Comparison of Selected Characteristics of Clothing and Furniture SMMEs (*N* = 80)

	Clothing (*N* = 40)	Furniture (*N* = 40)	Total Sample (*N* = 80)
Average firm age (years)	14.3	23.4	18.95
Average size (employees)	41	58	49
Male entrepreneurs	70%	97.5%	84%
PDI entrepreneurs	73%	13%	43%

There was little difference in the determinants of growth between the furniture and clothing SMMEs.[4] Neither the age of the firm nor the numbers of employees working there could be related to success or failure. Moreover, the initial motivation of the owner-manager (e.g., unemployment or perceived opportunity) had no explanatory value for growth. Nevertheless, SMMEs whose owners had worked before or that had been taken over from the family had generally a higher asset base and grew faster. Moreover, the human capital incorporated in the entrepreneur seemingly contributes to business performance. Whether higher education or business experience is more crucial for success is difficult to determine. The educational level of the entrepreneur was related to growth in profits and assets, while SMMEs run by older and more experienced owner-managers showed slower growth rates.

The skill levels of the labor employed ranged from "no skills" to "skilled craftsmen." Workers' skills per se do not explain SMME growth, but do so if labor is motivated to apply their skills to the benefit of the firm. Most firms reported that they adjusted workers' skills to their needs, either training them in-house or externally. Further training and good labor relations were positively related to profit growth, while labor mobility had no explanatory value.

In terms of equipment, most clothing SMMEs were operating with basic machinery, viz. industrial sewing machines. Neither the degree of automation nor the age of the equipment could explain growth or lack of it. Lead times for production delivery ranged from one to 180 days, but long lead times had seemingly no negative impact on the expansion of the sample SMMEs. Missing delivery dates and the lack of standardized quality control procedures did seem, by contrast, related to SMME stagnation. Moreover, production targets for each worker, departmental and personal supervision, or team work contribute positively to growth.

The SMMEs that focused on a certain market segment performed better than those that took on "anything which comes their way." Higher profits were associated with serving the higher end of markets and direct sales to the customer (not necessarily the end-user) with employment growth. Advertising or a high customization of production, by contrast, had little explanatory value. Most business was acquired through referrals while other traditional advertising techniques like "mail shots" (brochure or information sheets sent to existing or potential customers) and newspaper advertisements were not considered cost effective.

Networking of sample SMMEs took various forms, but seemed to be based on ethnicity rather than proximity. Both subcontracting and being a subcontractor were positively related to profit growth and provided the necessary flexibility to respond to the highly volatile product and rigid labor markets. Other forms of horizontal networking, such as information exchange with peers or being a member of the industry association, were insignificant for SMME growth. Close relationships to customers could not explain higher profits either, while good relationships with suppliers became seemingly more and more important in the attempt to minimize inventories but still meet delivery dates. Overall, clothing SMME entrepreneurs agreed that "networking heals the symptoms, but does not tackle the root cause of problems."

The Johannesburg inner city is still favored as a production location by clothing SMMEs because of the availability of cheap premises and because it is perceived to be the "centre of Gauteng's clothing industry" (Kesper, 1999; Rogerson, 2000b). For furniture manufacturers, by contrast, the inner city has lost its locational advantage. Premises on street level are ideal for furniture production, but highly affected by crime. For all SMMEs serving the higher end of the market, the inner city has lost its attractiveness for their major customers. Many have located their retail outlets farther north.

CASE STUDIES

The two emblematic SMME cases from the clothing industry indicate that, apart from the more technical competencies of the firm discussed above, the "drive" or "aspirations" of the owner-manager have a great stake in shaping the SMME trajectory.

Firm A

Firm A is located on the third floor in one of the inner city's office blocks. It was founded in 1995 in a flat in Yeoville by a Nigerian dentist who wanted to realize his mother's dream of having her own clothing business. In the beginning, the firm specialized in traditional African wear using the mother's designs. The dentist then got married and his wife was interested in enlarging the microbusiness. Two Nigerians and the dentist's sister were hired to sew. Customers were identified by word of mouth. The company registered as a Closed Corporation with an additional nonfamily owner so that it would always be accountable to an outsider. Although the couple discovered a niche in the black South African boutique market, it lacked the financial resources to establish and supply a retail outlet in the northern suburbs. Their loan applications at commercial banks and the then Small Business Development Corporation (SBDC) were turned down, while private money lenders asked exorbitant interest rates.

To generate more money, the firm continued to manufacture high-quality ethnic and Western wear and contracted with black schools to provide uniforms. In return, the firm would donate 10 cents for each uniform purchased to the school. The owners encountered distrust and insecurity on the part of the schools, presumably because they were foreigners. A sales person was hence employed to market their concept. When schools placed their first orders, the dentist used an overdraft option in his medical aid package to purchase the necessary machines and fabric and to employ additional skilled workers.

The firm currently employs 15 male and 10 female machinists. Recruitment of workers has been easy; current workers know about skilled job seekers and promise to ensure that they work well. Men are preferred to woman because they are found to work at least equally well and women frequently request time off because of their children.

The dentist is part of a well-organized network of clothing producers in Johannesburg bound by ethnicity rather than proximity. Due to a bad experience in the past, the firm prefers not to share orders with other firms. Rather, it commissions a whole order to another "well-known" firm. The responsibility of completing the order in time and with the desired quality rests then with one firm only.

Despite the amicable relationship to other clothing manufacturers, all transactions are based on a written contract. The owners of the firm perceive that customers are relatively strict with them because they are foreigners. Several customers have not been willing to pay the customary deposit on placement of an order, but the firm has waived the deposit in these instances to prove itself in a highly competitive market. The resulting cash flow problems are balanced with the dentist's personal savings. Deliveries are made by the firm. This is perceived to be beneficial both to the customers, who do not need to travel to the inner city, and to the firm in terms of customer satisfaction and feedback.

The inner city of Johannesburg seems the most strategic location for the firm. Workers largely live in the area, suppliers are close, and the rent is reasonable for the space the firm occupies. Nevertheless, a planned retail outlet will be located in the northern suburbs close to where the firm's owners live. Although the particular building hosts a number of other clothing manufacturers, there is no collaboration among them. Instead, the dentist had tried to take over another clothing company in order to increase his production capacity and obtain a retail outlet. The owners refused, though, because the dentist proposed to pay in instalments. Overall, the dentist is optimistic about the future and is still focused on his initial goal of opening an "African boutique": "We will start in Johannesburg with a retail outlet, but aim to distribute nation-wide. We are foreigners, but we have our role to play in the New South Africa." However, in mid-2000, the company was closed. The dentist had used money out of the business to open his own practice in the northern suburbs. This pushed the business into cash flow problems that it could not bear.

Firm B

Firm B was founded in 1985 as an informal business. Irene, the owner, had been working for 16 years as a machinist in a clothing factory that went into liquidation in 1985. Irene had started there after she had failed her Matric. Not knowing anything about clothing production, over the years she progressed from a table hand to a department manager. She used her lunch breaks to teach herself how to operate and repair sewing machines. Despite her relatively high wages, Irene had not been able to accumulate savings; she was the only breadwinner of her family. Nevertheless, Irene managed to buy several second-hand machines from her previous employer when she decided to set up her own business.

Irene started with two former colleagues in her backyard in Kathlehong. As there was no disposable income to purchase fabric, she accepted any order under the condition that the customer supplied the fabric. Working from home was forbidden in those days, and apart from police raids, power failures were disturbing her production. Hence, when the Kathlehong Industrial Park was built and subsidized by a large South African corporation, Irene was among the first to move in. The park was secure, had regular electricity supply, and hosted an NGO consulting SMMEs. With its help, Irene managed to win an order to manufacture work wear for Colgate. This gave her "a real push." Apart from the supply of fabric, she received a monthly payment for the pieces as they were completed instead of after the delivery of the full order. She could thus pay her workers monthly and employ even additional ones. Nevertheless, the concessionary order was not repeated, and Irene had grown bigger and had to look for larger orders than before to be able to keep all her workers.

She moved to the Johannesburg inner city in the early 1990s. It was more central and had become known for its clothing production. Irene started to work for other clothing manufacturers and established her own company contacts at the same time. She found, though, that her workshop never managed to finish orders in time. Subcontractors then paid her less for the garments. As the business was becoming less and less profitable, Irene designed work sheets and worked out daily targets for each of her workers. She spent more time with the workers on the floor than visiting buyers of large corporations. Since the mid-1990s, Irene has never run late with her orders. Her workers have a 40-hour week, but might have to come on Saturdays and on public holidays if an order has not been completed.

In 1996, Irene obtained a loan from Standard Bank after she had taken part in one of its management training workshops. At first, she felt threatened by the additional payment obligation each month. She then learned to appreciate the loan: "When I have got a commitment to make, I work even harder because I am more focused." Nevertheless, Irene has never again applied for a loan for her business (only for a bond on her house) as she expects the business to finance its growth. By 1999, Irene employed 28 male and female machinists. She also had moved out of her previous location in Market Street to a neighboring block where she occupies more space. Irene's business is well known to agents as a reliable manufacturer of all types of work wear. She no longer needs to go out to look for work, but receives samples by courier for price quotations. Irene can agree to manufacture any quantity as she knows a number of other informal clothing manufacturers to whom she could subcontract part of the order. Nevertheless, she prefers to complete her orders in her own workshop where she can supervise and monitor production. At times, Irene is asked to quote on corporate wear, but she never does as her workers find it difficult to work as efficiently with the more complicated patterns and softer material.

Irene is confident that her business will grow in the future. However, her firm is not yet registered for value-added tax (VAT) and hence is excluded from directly bidding for government tenders. Scheduled for the year 2000 were the firm's VAT registration and the purchase of a company van for deliveries. In July 2000, Irene purchased a company bakkie (pick-up truck) for deliveries, but has not registered it yet.

"Making a living" as an SMME clothing entrepreneur in the Johannesburg inner city is an ongoing process of overcoming problems and taking advantage of opportunities. Along their trajectories, both firms were able to identify so-called "turning points," i.e., incidents that had a deep impact on their firm. For firm B it was a large contract won by luck that facilitated its expansion, while the owner-manager of firm A chose another way of "making his living" when the opportunity arose.

The reasons for success or failure of an SMME can be many. The case studies suggest that general trends such as macroeconomic decline, shrinking market demand, and import competition do impact upon SMMEs in general, but incidents termed "luck" or "bad luck" and the "human capital" of the owner-manager decide which trajectory the SMME takes. These factors make it difficult to unequivocally pin down determinants of SMME success.

SMMEs are expected if not to drive at least contribute to the revival of the Johannesburg inner city by creating employment opportunities and wealth. Being the largest dominant industry in both formal and informal manufacturing, clothing production has been identified as a key target.

Still, it is questionable whether clothing SMMEs can fulfill the roles of employment and income generators. SMMEs permit the individuals employed in them to make a living. Nevertheless, these livelihoods form a continuum from mere survival to wealth. As regards the potential to contribute to employment growth, the survey findings indicate that more profitable operations do not necessarily generate their proportional share of employment, while most microenterprises remain small throughout their existence. This lack of dynamism of microenterprises and jobless growth of more formal and larger SMMEs is not a problem inherent to small enterprises in Johannesburg but a common feature of SMMEs throughout South Africa (Kesper, 2000 a, b, d). Critical voices argue therefore that only a small segment of the SMME economy will ever fulfill the roles of employment and wealth creators and be the impetus for urban regeneration. By contrast, what is typical and problematic for the Johannesburg inner city is the fact that more successful firms—those targeting the higher end of the clothing market—tend to move out, mainly because of distance to high-income consumers and the prevalence of crime. Although the viability of creating and promoting an

SMME garment district in the inner city of Johannesburg is less clear, there is no doubt that reducing crime levels and attracting long-term investments will have positive impact on the regeneration of the inner city.

CONCLUSION

In the beginning of the third millennium, South Africa is operating in a new, more open, trading climate that translates into the need to build competitive advantage and prosperity at the local level and in a global context. International experience suggests that SMMEs have a high labor-absorptive capacity and are more able than large firms to react to increasing market segmentation. A shift in size distribution toward smaller units is therefore considered desirable. Nevertheless, it is not the enterprise size per se that decides success or failure. Indeed, SMME activities undertaken as last-ditch attempts to provide a livelihood to the founder are reported to have abnormally high failure rates and employment conditions in these SMMEs are insecure and often very poor (Sengenberger and Pyke, 1992).

Light manufacturing activities have long been a feature of Johannesburg's economic landscape. Most recently, though, there has been a structural change from large-scale formal production to small-scale and partly informal production. Successful and failing SMMEs coexist, and the formers' growth trajectories have been all but smooth. The comparative findings from the survey suggest that enterprise age and size are not decisive regarding success or failure, and neither is the type of equipment in use. Rather, it is the owner-manager, and his or her human capital, who guides the SMME toward success or failure with the aspiration to stay in business or even expand it further. Nevertheless, external factors such as incidents termed "luck" or "bad luck" also impact upon the trajectory of an SMME and take the form of winning a large order or being a victim of crime.

Making a living as an SMME clothing entrepreneur in the Johannesburg inner city is problematic. Given the multiplicity of potential determinants of growth, it is difficult to determine *ex ante* which form this will take. This adds a question mark to the viability of creating and promoting an SMME garment district. Less in doubt is the fact that living and working conditions in the Johannesburg inner city are deteriorating.

NOTES

1. In the South African context the term SMMEs refers to small, medium, and microenterprise. The Small Business Act defines four size categories within this sector using turnover, fixed assets, and employment figures as indicators. In terms of numbers of employees, survivalist and micromanufacturing enterprises are defined as employing 0–4, very small 10–20, small 20–50, and medium-sized firms between 50 and 200 workers (Republic of South Africa, 1996).

2. Survivalist enterprises are one-man operations that allow the owner-manager some kind of survival, but have no prospect of growth.
3. PDI stands for "previously disadvantaged individual" and includes blacks, Indians, and coloreds as classified by South Africa's historical race classifications.
4. The following paragraphs are based on the author's interviews with forty successful clothing and furniture SMMEs in late 1998. These firms were reinterviewed in August 2000.

REFERENCES

Altman, M. 1996. "Labour Regulation and Enterprise Strategies in the South African Clothing Industry." *Regional Studies* 30:387–399.

Bloch, R., and A. Kesper. 2000a. *Supporting the Small and Medium Manufacturing Sector in the Western Cape.* Unpublished report for the Council for Scientific and Industrial Research, Pretoria.

———. 2000b. *Supporting the Small and Medium Manufacturing Sector in the Highveld Production Region.* Unpublished report for the Council for Scientific and Industrial Research, Pretoria.

Clothing Federation of South Africa. 1998. *Overview of the South African Clothing Industry.* Unpublished mimeographed report, Johannesburg.

Kaplinsky, R., and M. Morris. 1999. "Trade Policy Reform and the Competitive Response in Kwazulu Natal Province, South Africa." *World Development* 27(4):717–737.

Kesper, A. 1999. "Small Clothing Manufacturers in the Johannesburg Inner City: Facing the Global Challenge." *Urban Forum* 10(2):137–164.

———. 2000a. *Manufacturing SMMEs in the Vaal Triangle. Survey Results.* Unpublished report prepared for the Office of Industrial and Policy Research, Johannesburg.

———. 2000b. "Making a Living in the City: Success and Failure of Manufacturing SMMEs in the Johannesburg Inner City." *Africa Insight Special Issue.*

———. 2000c. *Failing or Not Aiming to Grow? Manufacturing SMMEs and Their Contribution to Employment Growth in South Africa. TIPS Working Paper* 15-2000, Johannesburg.

———. 2000d. *Promoting SMMEs in South Africa: Review of Experience and Agenda.* Report prepared for TIPS and the Department of Trade and Industry, Johannesburg.

———. 2002. *Tracing Trajectories of Successful SMMEs in Gauteng.* Unpublished Ph.D. dissertation, University of the Witwatersrand, Johannesburg.

Levy, B. 1996. *South Africa—The Business Environment for South Africa's Industrial SMMEs, World Bank Discussion Paper 11.* Washington D.C.: The World Bank.

Ntsika Enterprise Promotion Agency. 1999. *The State of Small Business in South Africa: Annual Review 1998.* Pretoria: Centre for Small Business Promotion.

Republic of South Africa. 1995. *National Strategy for the Development and Promotion of Small Business in South Africa.* Cape Town, White Paper of the Department of Trade and Industry.

———. 1996. National Small Business Act, Pretoria, Government Printer.

Rogerson, C. M. 1999a. "Small Enterprise Development in Post-apartheid South Africa," in K. King and S. McGrath, editors, *Enterprise in Africa: Between Poverty and Growth.* London: IT Publications, pp. 83–94.

———. 1999b. *Johannesburg's Clothing Industry: The Role of the African Immigrant Entrepreneurs.* Report prepared for BEES Consulting Group, Johannesburg.

———. 2000a. "Manufacturing Change in the Witwatersrand." *Urban Forum* 11(2):311–340.

———. 2000b. "Successful SMMEs in South Africa: The Case of Clothing Producers in the Witwatersrand." *Development Southern Africa.* 17(5):687–716.

Rogerson, C. M., and J. M. Rogerson. 1995a. "South Africa's Economic Heartland: Crisis, Decline or Restructuring." *Africa Insight.* 25(4):241–247.

———. 1995b. "The Decline of Manufacturing in Inner-City Johannesburg, 1980–1994." *Urban Forum* 6(1):17–42.

———. 1996a. "The Changing Post-apartheid City: Emergent Black-Owned Small Enterprises in Johannesburg." *Urban Studies* 34(1):85–103.

———. 1996b. "The Metropolis as Incubator: Small-Scale Enterprise Development in Johannesburg." *Geojournal* 39(5):33–40.

———. 1997. "Intra-metropolitan Industrial Change in the Witwatersrand, 1980–1994." *Urban Forum* 8(2):195–223.

Sengenberger, W., and F. Pyke. 1992. "Local Economic Regeneration: Research and Policy Issues," in W. Sengenberger and F. Pyke, editors, *Industrial Districts and Local Economic Regeneration,* Geneva: International Institute for Labour Studies, pp. 3–29.

Tomlinson, R., and C. M. Rogerson. 1999. "An Economic Development Strategy for the Johannesburg Inner City." Strategy prepared as part of the UNDP City Consultation Process on behalf of the Inner City Section 59 Committee, Johannesburg.

Upstart Business Strategies. 1999. *Report on the Support Needs of SMME Manufacturers in Gauteng.* Report prepared for the Office of Industrial and Metropolitan Research, Johannesburg.

van Dijk, M. P. 1993. "Industrial Districts and Urban Economic Development." *Third World Planning Review* 15(2):175–186.

7

Violent Crime in Johannesburg

INGRID PALMARY, JANINE RAUCH, AND GRAEME SIMPSON

Johannesburg is popularly believed to be the "crime capital" of South Africa. Like all apartheid cities, it was characterized by the intense management of public space focused primarily on regulating the access of black people. This management of the urban space of Johannesburg resulted in the *ghettoization* of urban townships, dramatic disparities between areas of extreme wealth and intense poverty, and palpable divergences in the quality of services offered (including policing and security services) in black townships as compared to white suburbs. Apartheid urban management and social engineering have had severe consequences throughout Johannesburg's history, and much of this has played itself out in a dramatic fashion immediately before and during the transition to democracy in the 1990s (Van Onselen, 1982).

This chapter addresses forces evident in the past two decades in Johannesburg in terms of their relationship to the types of violent crime characterizing the city today. In particular, we consider the consequences of transition on the social relationships that people have to urban space and the conditions that these factors create for violent crime.

We look at the often neglected dimensions of violence against women, violence against foreigners, and youth violence in the city in an effort to understand the factors that facilitate the city's high rates of mortality and injury. We touch briefly on alcohol abuse, firearms, and the cycle of repeat victimization.

THE ORIGINS OF PRESENT DAY VIOLENT CRIME

Despite the formal enfranchisement of the previously disenfranchised majority, patterns of social and economic exclusion as well as the distribution of social services are very slow to change (Simpson, 2001). In Johannesburg this is particularly pertinent if one considers the extent to which previously marginalized groups such as women, young people, and migrants are still excluded from access to economic power. Goldberg (1993) has noted the extent to which complex hierarchies exist in respect of access to the benefits of urban life and the extent to which these hierarchies uphold the status quo. Informal mechanisms have replaced formal ones, ensuring that access to the benefits of urban life remains the domain of a privileged few.

The struggle against social and economic marginalization can be seen in the patterns of conflict that manifest themselves within marginalized communities, for example, the illegal occupation of land and houses, taxi wars, and disputes among street hawkers and squatters. Much of the experience of sustained marginalization in Johannesburg erupts through violent struggles over rights to access the basic benefits and amenities of urban life. For example, violent attacks on migrants have been justified because migrants are perceived to be responsible for taking South Africans' jobs. In this way, a language of legitimacy regulates access to urban spaces while systems of hierarchy remain intact.

Families and community networks (alongside educational structures and other institutions such as churches) play a central role in the socialization process, particularly of young people. One of the primary consequences of the disruption of these socializing agents is a high crime rate, particularly violent crime (World Bank, Online).[1] In addition to the economic exclusion facing many people in Johannesburg, systems of migrant labor and experiences of intense urban squalor have harmed family systems and community networks. An example of this can be seen in the historical establishment of single-sex hostels that created and skewed economic, social, cultural, and community relationships. The experience of migration to the city is but one factor in the process of marginalization, a process experienced differently by different groups. For example, the system of pass laws was implemented differently for men and women resulting in gendered patterns of migration. These experiences play themselves out through violent conflicts within and between marginalized communities, rather than just in the explicit conflicts between those in power and those on the margins. For this reason, violence is far more prevalent within poorer communities of the city and does not simply manifest in the victimization of Johannesburg's middle classes (Louw et al., 1998).

The regulation of social space and the resulting hierarchies of access to urban comforts need to be understood in terms of their impact on the safety of the city's residents. The International Centre for the Prevention of Crime

(ICPC, 1997:20–21) has identified a number of factors that lead to increased levels of crime. They include

- poverty and unemployment deriving from social exclusion, particularly among young people;
- dysfunctional families with poor parenting, domestic violence, and parental conflicts;
- social valuation of a culture of violence—where violence has become sanctioned as a normative vehicle for the assertion of power or to attain change in a society or community[2];
- easy access to facilitating factors in violent crime, such as firearms, alcohol, and drugs;
- discrimination and exclusion deriving from sexist, racist, or other forms of oppression; and
- degradation of urban environments and social bonds.

Johannesburg is a city sharing many of these characteristics, sometimes referred to as a breakdown of social capital. Defined by the World Bank as "the institutions, relationships, and norms that shape the quality and quantity of a society's social interactions," *social capital* has become a useful way with which to analyze the local impact of the large-scale erosion of the social fabric. Social capital is particular kinds of social networks that enable a group of people to access goods or services that they could not otherwise access. It is considered a major factor in protecting a society against crime, and a breakdown of social capital is often associated with an increase in crime (World Bank, Online).

Through its repressive manipulation and crude social engineering, the system of apartheid and the city planning associated with it did enormous damage to the fabric of social relations within urban communities. This was wrought not only by the system of institutionalized labor migrancy, but by the devastation of family life more generally, the destruction of schools as places of cohesion and creative learning, the denial of basic facilities and amenities, the nondelivery of recreational opportunities, and the pervasive experiences of unemployment or underemployment. Ever-increasing and more untenable disparities between the very rich minority and the very poor majority resulted. The legacy of structural violence embedded in both the destruction of the urban social fabric and the politics of exclusion was further complicated by processes associated with the transition to democracy.

One of the liabilities of the negotiated settlement of the South African transition was the inheritance, largely intact and with an embedded institutional culture, of the key institutions of the state. Of particular relevance are the institutions of the criminal justice system (the police, judiciary, and prisons) and welfare institutions. Furthermore, the amnesty provisions contained within the Truth and Reconciliation Commission legislation (Office of the President, 1995) and the "sunset clause," which protected the jobs of incumbent government officials who staffed these inherited institutions, also ensured that little change took place in the immediate relationships between ordinary South Africans and the institutions of the new democratic state. Yet, as important as this may have been, popular expectations of change associated with democratization did not stand or fall only on the basis of relationships of trust between civil society and state institutions. Equally important was the test of whether a new government could effectively translate its visionary policymaking into effective delivery of social services (Simpson, 1999).

In this context, the snail's pace of change in the post-1994 period meant that few expectations of urban renewal, housing development, schooling improvement, job creation, or service delivery were realized. Frustrated expectations of change entrenched social conflict in the cities, frequently resulting in increasing, rather than decreasing, levels of violence and crime. The transition to democracy therefore appears to have compounded the problems associated with apartheid's historical decimation of the social fabric of the cities.

Nowhere were these trends more apparent than in the difficulties experienced in attempts to transform policing and criminal justice practice. Not only was a relationship-building endeavor associated with the shift to a more locally accountable "community policing" paradigm limited in its impact, but government failed to translate visionary crime prevention and crime-fighting policy into effective local level service delivery. In the mid-1990s, sustained levels of violence and the prevailing fear of urban crime dominated the urban public psyche. This occurred at precisely the time when the government shifted the locus of public safety service delivery to the city level; the 1998 White Paper on Safety and Security required city administrations to take responsibility for city-level crime prevention and bylaw enforcement.

Consequently, patterns of violence within the city reveal much continuity amidst all the change that characterized the transition to democracy. Although the violence that emanated from experiences of marginalization during the apartheid era was largely understood and framed as political (and therefore as *socially functional*), continuity in the experiences of sustained marginalization during the early phase of democracy resulted in high levels of violence. These *new* patterns of urban violence were selectively relabeled as "criminal" rather than "political" and as "antisocial" rather than "social."

Crime in South Africa is often claimed to have reached epidemic proportions following the 1994 elections, a reputation that is perhaps most starkly manifested in Johannesburg. However, when considering the factors underlying violence, this is less of a postapartheid phenomenon than it appears. It cannot be clearly differentiated from the political violence of the past twenty years.

As a point of illustration, one can compare the forces underlying political violence in the 1980s and 1990s and the current criminal violence. Much political violence took place at flashpoints; that is, points of displacement and social and familial disruption such as the single-sex hostels. The victims and perpetrators were (and still are) predominantly young black men, and one's life chances were (and still are) largely determined by one's race, gender, and nationality. In other words, many of the underlying factors that fuelled earlier political violence continue today. They underpin much of the current violence in most South African urban centers. Rather than a decrease in political violence being replaced with an increase in criminal violence following the politically negotiated settlement, there has occurred qualitative shifts in a continuing pattern of violence personified in struggles over access to urban benefits.

Clearly, the transformation of Johannesburg has impacted different social groups in different ways. On a purely visual level, the prevalence of road closures and gated communities in middle-class suburbs, the new boom in high-security fencing around township homes, and the emergence of "cardboard cities" of homeless people in the central business district are evidence of the varied responses to new trends in safety and security. The socioeconomic rifts are indeed some of the factors that predict high levels of crime.

VICTIMS OF VIOLENT CRIME

A variety of methodological problems are associated with analyzing existing data on violence in Johannesburg. First, the South African Police Service (SAPS), compilers of official crime statistics, use a geographic definition of Johannesburg that excludes more than half the area of the metropole. In an attempt to rectify this situation, we use statistics from Johannesburg and Soweto (a total of 38 police station areas), although even the amalgamation of these places excludes approximately 10 police stations to the north, east, and west of the city. Second, the SAPS crime statistics reflect only those crimes that were reported to the police. Recent research on violent victimization (SALC, Online) shows that many victims of violence never report the incidents, instead merely reporting to hospitals for medical treatment.

Third, the internal processes for compiling crime statistics in the SAPS are, by the government's own admission, unreliable. Although new methodologies are now being used, no statistics on crime in Johannesburg have been made available since these were implemented. The most recent crime statistics available to the public show recorded crime figures from 1999. Fourth, although victimization surveys are generally considered a more reliable indicator of crime trends, no official victim survey has been conducted in Johannesburg. The Institute for Security Studies (ISS), a nongovernmental organization, conducted a victim survey in Johannesburg in 1997. This was the first city-wide victim survey conducted in South Africa and it suffers from a variety of methodological limitations, not least of which is the fact that victim surveys are most useful when they are conducted repeatedly, enabling comparisons over time. This has not been the case in Johannesburg. Finally, much of the research that we cite was conducted only in particular parts of Johannesburg. Given the significant differentials that characterize the city, they cannot be generalized to describe violence in all of the city.[3]

In 1997, 46% of reported crimes were directed against people and 44% against property (Louw *et al.,* 1998). The highest levels of violent crime tend to be concentrated in black townships and poorer parts of the city. In this respect, these trends in Johannesburg are congruent with patterns of interpersonal violence in other cities. The high levels of reported property crime are inconsistent with trends in other developing countries. This finding is partially due to the reporting requirements associated with property crime— a requirement for insurance claims.

In the victimization survey conducted by the ISS in 1997, the most common violent crimes reported by residents of Johannesburg were mugging and robbery. Of residents, 16.5% reported that they had been victims of mugging or some other form of robbery in the preceding five years and 15.5% reported that they had been victims of assault (Louw *et al.,* 1998).

In Johannesburg (as in many other parts of South Africa) race is still strongly associated with income level and with place of residence, which renders particular racial groups statistically more vulnerable to violent crime than others. The white and Asian communities are the predominant victims of property crimes due to their relative wealth and resulting reporting patterns. However, according to the ISS survey, violence was more likely to be used during the course of property crimes if the victim was black. One of four Africans who had experienced property crimes in the Johannesburg survey reported that the incident involved violence, while only 9% of white victims reported the same. In contrast to frequently sensationalized popular media constructions of *carjackings* in Johannesburg, the ISS survey found that 73% of the victims of carjackings were black. In addition, more than 75% of the total assault, rape, and murder reported in the ISS survey was committed against black people, particularly those living in townships. Those marginalized and disadvantaged by apartheid continue to suffer the burden of violent crime in spite of media discourse concerns regarding the middle classes.

CASE STUDY 1: WOMEN'S SAFETY IN THE CITY

In addition to the racial variation, strong gender variations exist in both the frequency and forms of violent crime in Johannesburg. The ISS survey suggests that men are at greater risk of being victims of crime, but levels of violence against women are notoriously difficult to establish reliably. Violent crimes against women are often poorly reported in surveys and to police, particularly when the crimes are of a sexual nature or when they have been committed by perpetrators known to the victim.

In the ISS victim survey, only 42% of the victims of violent crimes knew their attackers. This is significantly at odds with trends in sexual assault in other parts of the world where such crimes tend to be committed by perpetrators who are known to the victims. The Johannesburg finding may relate to the high levels of seemingly motiveless violence, a feature of South Africa's "culture of violence," or it may reflect an outcome of the survey methodology in which women are less likely to disclose to a complete stranger (the interviewer) experiences of violence involving their intimate partners or family members.

Early in 1996, the Institute for Social and Behavioural Studies at the University of South Africa (UNISA) began a surveillance project on rape. Its findings were subsequently used by the Johannesburg Sexual Violence Forum to analyze rape data presented by victims at the Johannesburg Medico-legal Clinic and the Lenasia and Chris Hani Baragwanath hospitals (Vetten, 1998).[4] Distinct patterns and trends in the incidence of rape emerged from the analysis of approximately 800 cases. In particular, rape in Johannesburg seemed to vary according to time and space with the majority of rapes taking place between 6 PM and 10 PM on Saturdays, Fridays, and Sundays. Less than a third (30%) of the rapes took place in the assailant's home, 25% on open ground, and 17% in the victim's home. Far higher levels of rape in the perpetrator's home were recorded by the ISS survey, which states that 64% of rape cases involved abduction of the victim. This finding is unsurprising given that many of the women (42% in the study) knew their assailant and over half knew him by sight.

One would expect variation by race given that apartheid exposed black women to extreme poverty, rapid urbanization, and other factors thought to be linked to sexual violence. The ISS survey (Louw *et al.,* 1998) suggests that 70% of cases of rape were perpetuated against black women, a figure roughly congruent with the demographics of the city. It suggests that sexual violence is *not* specific to one or another racial group of women. Sexual violence in Johannesburg, whether or not a consequence of the disintegration of social fabric, does not respect the boundaries of race, class, or political persuasion.

Experiences of rape did vary according to other demographic factors. The women victims were most likely to be between the ages of 13 and 30 years. Also geographic location impacted on women's experiences of rape; women living in the townships were most at risk (24% according to the ISS survey) followed by women living in the inner city.

Recent work by Vetten and Dladla (2000) exploring women's experiences of sexual violence in inner city Johannesburg highlighted a continuum of coercive sexual behaviors that ranged from sexual harassment to rape. Their study indicated that women had experienced other forms of violence as well, including mugging, robbery at either gun or knifepoint, and assaults. Many women had also witnessed murders, rapes, robberies, and assaults on others. However, not all of these incidents occurred in inner city Johannesburg and not all the inner city women they interviewed shared the perception that central Johannesburg is particularly dangerous. For at least one woman, Johannesburg was safer than Soweto, from which she moved after a man attempted to rob her at knifepoint. Lesbians living in the surrounding townships also found central Johannesburg safer. Expressing or acting on an attraction to other women was seen as less risky in central Johannesburg than in the township environment.

The above studies suggest that vulnerability to rape varies among women and depends on variables such as the place where women live. This, as well as their economic insecurity, often makes homeless women or sex workers particularly high-risk groups for violence.

Dladla (2002) is currently investigating the experiences of homeless women in Johannesburg. She finds that

> The [homeless] women's perceptions of fear and safety are not simply about dangers within the public spaces, but also about danger within the private spaces, which for them are not quite private because they have to share their spaces with others. The women's *houselessness* is connected to their vulnerability to many types of danger including rape, assault, domestic violence and HIV/AIDS. In general, the women living in those places with no management and no security felt that they were unsafe and vulnerable to crime.

This vulnerability is made worse by the fact that service providers, such as the police and the ambulance services, are no longer willing to service the homeless people of certain shelters and settlements in the inner city. To access such assistance, these women have to pretend to be living elsewhere. Homeless women are perceived by service providers to be illegitimate within the city and therefore are thought to have no right to access State services (Dladla, 2002).

> The ambulance does not come to this place anymore because people get hurt all the time . . . right through. If you phone the ambulance you must go to the shop and then you have to pretend to be living elsewhere. The man next door was helping us but he does not bother anymore . . . what does it help? The people are fighting all the time.
>
> They say that they are tired of the people from this place because everyone who gets hurt is from this place. All the patients who come from here cannot even pay. It does not matter that the person is hurt badly . . . we cannot call the ambulance . . . we call the police. People fight here and then they take each other to the police station. They then pay each other compensation and then they go to the police to withdraw the charges. The police are also tired of us.

In addition to places traditionally seen as unsafe within the inner city (such as bars or taverns), homeless and lesbian women also find some places typically perceived to be safe by other women unsafe:

> Suburbs where women seek work during the day were labeled as unsafe, because they are deserted during the day. The women thought that if something should happen to them, they would not be helped by anyone, because the houses are walled off and there are very few people on the streets. Some of the women find the streets safe when they see security guards and police patrolling the place. This seems to assure them that nothing could happen to

them. In a tragic contradiction, other women, who were nearly raped by men like these (security guards and police officials), have come to associate them with being unsafe. (Dladla, 2002)

The access that wealthier Johannesburg women have to security exacerbates the insecurity that homeless women felt.

In addition, half the homeless women interviewed by Dladla reported experiences of domestic violence that include sexual, physical, economic, and emotional abuse. The extreme economic dependence of homeless women on men ensures that they usually tolerate such abuse. Ironically, abusive men were seen by the women to offer them protection.

Dladla's descriptions of women's struggles with homelessness show that many of their experiences of violence are linked to their inability to acquire safe shelter. Owning their own homes may not prevent domestic violence, but it would certainly reduce their vulnerability, which is related to their dependency on men. However, safety for these women is not only about owning a home but also about being able to access public health and protection services.

In addition, research in Soweto has shown that to a greater or lesser extent, security threats faced by women are enforced and legitimated by social norms and expectations of men. In findings from a study on sexual violence in the south of Johannesburg, one out of five men thought that he was entitled to have sex with a woman without her consent. Twenty percent of all respondents (male and female) thought that violence against women was a woman's fault and 41% said that it was both the man and the woman's fault. Forty percent felt that it was permissible for men to "punish" their wives, particularly for behaving in what they considered to be a gender-inappropriate manner (such as being unfaithful or refusing a man's sexual advances) (CIETafrica, 2000).

The notion of men's sexual entitlement is supported by recent research by the Centre for the Study of Violence and Reconciliation in Soweto schools. It found that young men explain sexual violence in terms of their entitlement to sex if they spend money on a woman (Zulu, 2001). These norms are not exclusive to violent men but are upheld by many in society thus sanctioning gender-based violence. In a study on rapists in Riverlea in Johannesburg, McKendrick and Hoffman (1990) noted that rapists often ensure a woman's cooperation by threatening to tell people about the rape, thereby ruining the woman's reputation. This indicates a double standard as well as the lack of community support for rape survivors.

If we accept (as the evidence above suggests) that the social degradation of women and the cultural valuation of sexist practices are the primary reasons for high levels of violence against women, how can this be explained? Other authors (see, for example, McKendrick and Hoffman 1990) suggest that in societies in transition, economic success and the social status that it awards men are often insufficient to achieve a strong sense of masculinity. In the absence of being able to fulfill stereotypical roles (such as economic provision) their frustration is turned on women (particularly female partners). Although women's rights have, in many instances, been secured in policy, aggressive sexual practice remains unchanged. Also, the traditional socializing agents that would act as a buffer against male violence (such as the church or elder family members) are less strongly entrenched due to the social and physical upheavals associated with this transition.

In addition, McKendrick and Hoffman (1990) and Morris (1998) have suggested that when traditional hierarchies are beginning to erode, many social groups entrench them in an even more rigid manner. A similar phenomenon has been experienced with displaced people or refugees who, in the face of threats to traditional cultures, enforce particular practices even more strongly than in their home country.

CASE STUDY 2: XENOPHOBIC VIOLENCE

Under apartheid, policies of racial exclusion strictly controlled the movement of black people into the city. This was also the case for people from other African countries, with the exception of the hundreds of thousands of foreign black people who were brought into South Africa as contract workers (Morris, 1998). With the advent of democracy, some of the formal mechanisms for excluding foreigners have been relaxed, resulting in increased immigration from other African countries. Although foreign migration is not entirely new, the level to which it rose after the democratic elections has been a source of fear for many South Africans (Sinclair, 1998), in spite of research suggesting that people perceive that there are far more migrants in the country than there actually are (Morris, 1998). Also, the levels of xenophobia currently found in South Africa do not seem to be extended to the contract workers brought to South Africa under apartheid. Xenophobia is one clear example of how informal mechanisms of exclusion replace formal ones, maintaining the city as a contested and racially exclusive site.

Although little information exists on migrants (whether documented or undocumented) in Johannesburg, some qualitative research is being undertaken. Harris (2001) focuses, *inter alia,* on the experiences of violence among migrants, refugees, and immigrants. Some early reports of this type of violence go back to the immediate postelection period, in December 1994 and January 1995, when a campaign entitled "Operation Buyelekhaya" (go back home) was run in Alexandra township by

> armed gangs of youth, claiming to be members of the local ANC, South African Communist Party and South African National Civic Organisation [who] carried out a concerted campaign of intimidation and terror to rid the township of illegal aliens. . . . They specifically targeted Shangaan-speakers and Zimbabweans and other residents with "dark complexions" by throwing them and their possessions out of their homes and flats. Some of those targeted had their homes burnt down and their possessions looted. Others were frog-marched to the local police station where it was demanded that they be removed immediately. (Minaar and Hough, 1996:188–189)

Xenophobic violence was not restricted to Alexandra or other township communities. It is as prevalent in inner city Johannesburg:

> [i]n August 1997, local hawkers in Central Johannesburg attacked their foreign counterparts for two consecutive days, scattering and looting their belongings and beating the foreign traders with sticks and knobkerries and sjamboks. . . . A few days later, on August 18, 1997, local hawkers attacked foreign hawkers at the Kerk Street Mall in Johannesburg, severely beating several Senegalese hawkers. . . . On October 23, 1997, approximately 500 hawkers marched again in Johannesburg, chanting slogans such as "chase the 'makwerekwere' out", and "down with the foreigner, up with South Africans". In November 1997, the Greater Johannesburg Hawkers Association called for a boycott of goods sold by "makwerekwere", including Pakistanis, Chinese, Indians, Senegalese, Somalis, Nigerians, Moroccans, Zimbabweans, and Mozambicans. (Minaar and Hough, 1996:129–131)

The inner city has become home to many migrant communities and the tendency for migrants to "cluster" in certain areas due to safety concerns, shared language, culture, and economic status is not unique to Johannesburg. Nonetheless, it is ironic that in the absence of formal mechanisms of race-based social engineering, informal mechanisms for ethnic clustering and social organization have appeared. Furthermore, Harris points to the paradox that where certain foreigners live in close proximity, they attract hostility and violence as a vulnerable group. Consequently, areas that are known to be occupied by foreigners may actually attract crime and violence. Quotes from Harris's (2001) informants demonstrate this:

Firstly, we're staying in the dangerous cities. Here in Johannesburg, especially Hillbrow, Yeoville, Berea, Bertrams [there are many killers]. . . . [Secondly] when those thieves, robbers, gangsters, pickpockets etc. learn that you're a foreigner, that exacerbates the situation. (Rwanda man)

Harris (2001) explains that for this respondent there is an overlap between being a foreigner and vulnerability to violence resulting in a "double burden." He suggests that foreigners (by virtue of their foreignness) are located in dangerous parts of the city and this enhances their vulnerability to "ordinary" crime and violence.

Economic crimes against foreigners are common and are often accompanied by violence, as the following extract reveals (Harris, 2001:102):

Four guys put a gun to my head and told me to get in the car. They told me that "makwerekwere" have got bucks and that I must give them money. They took my three hundred Rand and those shoes I bought. And then they were beating me. And one stabbed me here [points to scar on left side of the abdomen]. Then they told me that they would let me live on one condition: "each and every month we gonna come and fetch three hundred Rand". . . . And those men came every month for the money. They threatened me that they would kill me and I did it for three years. (Burundian man A)

Xenophobic attitudes and practices stand as a strong disincentive for foreigners to report crime and violence to the police. As is the case with homeless women, foreigners are seen to be making illegitimate claims on the resources of the city when accessing safety and related services. "I would go to the police if there was justice but my first experience was with the wrath of the police. So, where must I go if the police themselves are doing such actions on me?" (Ethiopian man, p. 71).

[The police asked for my refugee papers, which had not yet expired]. They say that "fuck you" and they just tear the paper and seize my money and cellphone. . . . So then, what they do is take me to the police station. I was shouting and crying that "Why did you tear up my paper?" One of them just removed something like a little shocker. He was shocking me, shocking me. Say that I was to shut up at once and if I wasn't shut up, he was going to shock me until I die. Then I had to keep my mouth [shut]. (Sierra Leonian man, p. 67)

For most of Harris's foreign respondents, a central feature of South African society is the prevalence of violence and crime. They comment on the insidious ways in which violence infiltrates their daily lives. For some, especially those fleeing violence and war in their home country, their encounter with violence and crime was contrary to their initial expectations about South Africa. For example, asylum seekers (Harris, 2001:128) commented that "It's not like they told us . . . in Jo'burg, when we get here, a very, very heavy life . . . even especially the crime, it's not safe. It's too much crime. They are robbing me three times here in Jo'burg" (Ethiopian man).

Ironically, migrants are often blamed for crime in spite of their vulnerability to it and crimes against foreigners (such as assault) are often justified in terms of their perceived involvement in crime. In addition, migrants have the least access to safety services. Not only are police and other service providers not of assistance, they are frequently responsible for attacks on them (Morris, 1998). For many migrants, their image of the city is not one of harmony and peace, but rather one of violence, danger, and high crime rates that they accept only because they have no alternatives. "Before I came here, people used to say that South Africa has a lot of crime, especially in Johannesburg. . . . Yes, I remember my travel agent asking me if I really wanted to go there, knowing that it was a dangerous city" (Norwegian woman, F3).

> This thing [of crime] is not really that much new for everybody else. Even the people and the families who are out of South Africa, they know that the crime rate here in South Africa, especially Johannesburg, [is high]. When we are communicating with our families, they will tell us beforehand: "So [South Africans] will thieve everyday". See, it is like that here in South Africa. (Ethiopian man, S1)

Morris (1998) suggests that South Africans may be using migrants as a scapegoat for the failings of the new South Africa and for unmet expectations, particularly the lack of employment and persistent inequality and poverty. He also suggests that the isolation that South Africans experienced during apartheid means that they have no experience in incorporating other groups and tend to be intolerant of outsiders. In addition, South African leaders have seldom challenged stereotypes that exist about foreigners and indeed many in political power have voiced such stereotypes. The government's failure to address xenophobia is in stark contrast to statements on racial integration in South Africa (Sinclair, 1998).

CASE STUDY 3: YOUTH VIOLENCE IN THE CITY
Young people are most likely to be both the perpetrators and the victims of violence in Johannesburg. Nationally, over 25,000 young men between the

ages of 16 and 25 years were serving prison sentences in 2001 (Segal *et al.*, 2001). Indeed, much of the discussion of violence in this chapter refers to youth as the primary perpetrators or victims. Moreover, the line between *antisocial* criminal youth violence and the *socially functional* violence associated with political resistance during the 1970s and 1980s is less clear-cut than is portrayed in public discourse on crime. The fluctuation between the two types of violence may well reflect the resilience of young peoples' response to their social, economic, and political circumstances. "While this resilience is a powerful indicator of the dynamism of youth sub-cultures, it is not value-specific and may even pave the way for young people who move (rather more easily than is often assumed) between political and criminal activity, as well as a range of other social, religious or cultural environments" (Simpson, 2001:118).

In his seminal piece on the rise of the Jackroller Gang in Soweto during the 1980s, Mokwena (1991) identifies the roots of gang formation in the processes of marginalization: a lack of educational opportunities and political exclusion. Mokwena points specifically to the experience of emasculation that young men felt, particularly as a consequence of pervasive racism. For the Jackrollers, the main response was a collective macho culture that espoused the attainment of male dominance by committing violent acts against young women. This aggressive assertion of manhood was clearly an antisocial response to feelings of inferiority, rooted in the gangsters' irreversible status as "boys" in apartheid South Africa. Mokwena also notes the specific role of unemployment in cementing this sense of emasculation, an experience viewed as a personal rather than a social failure.

The Jackrollers had limited interest in acquisitive crime, but became renowned for the brazen abduction (usually from public places) and gang rape of young girls. The gang became so notorious that the word *Jackroll* became part of a commonly used township vocabulary referring to rape. Furthermore, Mokwena suggests that Jackrolling became something of a male "fashion" among young boys seeking to assert their manhood or toughness and became rooted in a pervasive youth culture in parts of Johannesburg.

Far from being either active or on the fringes of political resistance, gangs often obstructed the objectives of the political movement. Conflict between the gang members and political activists was often overt and fierce. Mokwena suggests that the evolution of alternative youth structures in the form of self-defense units (particularly in Diepkloof) was frequently in response to the threats posed by such gangs. Evidence in amnesty applications and through victim testimony before the Truth and Reconciliation Commission bears out the allegation that the apartheid state actively sponsored and supported criminal gangs as a vehicle for destabilizing resistant township communities.

The youth-based political "counterrevolution" that some anticipated as a consequence of the frustration of young radicals at the slow pace of transformation in the early 1990s took the form of youth-based criminal activity and organization that, by the end of the 1990s, would present the gravest threat to the embryonic human rights dispensation.

> [T]his trend had its roots in the sometimes seamless interface between youth involvement in criminal and political violence of the preceding era—rather than being miraculously born as a twin to the new South African democracy. (Simpson, 2001:122)

Youth violence became synonymous with gangs. In a 1997 survey of police station commissioners across the Johanesburg Metropolitan area (Rauch, 1998), over two-thirds mentioned criminal organizations—of various types—as contributing to the crime problem in their area.

> I won't call it gangs. I will call it a group of people who come together to commit a crime. You see, a gang is an organised gang, with names, like the Scorpions, things like that. I don't think it's the same thing here in the black townships. They form groups to work for the syndicate. When you talk about gangs, they are controlling a whole area. And the township is not like that. It's groups operating to commit crimes.
> I still maintain there is someone else behind this. The brains behind this. And you will never see him. He just gets these little people, especially the youngsters, to commit these crimes, and give them a few bucks or whatever. (Rauch, 1998:14)

The evolution of township gangs is linked to urban migration and forced removals that have resulted in severely overcrowded urban conditions and disruption of family structures. This has meant that young people spend much of their time on the street and without family supervision. The youth of the 1990s found an alternative place of belonging and social cohesion within criminal youth gangs. Like political organizations in the previous era, the gangs offered an alternative subculture to the dominant culture that had rejected these youth. The gangs boasted their own uniform, their own language or tsotsi-taal, and their own fashions and rituals (the latter often associated with growing substance abuse). Perhaps of greatest significance was the economic opportunities offered by the gang.

One of the best windows on gang subculture is contained in research conducted by Segal *et al.* (2001). The uncensored voices of Johannesburg gangsters and imprisoned juvenile offenders reveal the subcultural mirror that gang organization holds up to the society that marginalized them. In describing guns as the tools of their trade and in referring to their criminal activities as "going on duty" or "doing business," these young gangsters mock the "economic powerhouse of Gauteng" from which they are excluded. In referring to their consumerist aspirations to acquire the most recent and fashionable goods through criminal activities, these young gangsters speak of "keeping up with the syllabus," thus mocking the world of the school from which they are also excluded. Mindful of the hierarchical organization of the gang and of criminal activity, one of Segal's informants states:

> In crime there is a hierarchy. You grow from strength to strength until you are up there doing the business where there is a lot of money. When you are there, we respect you and to us you are like someone working on the Johannesburg Stock Exchange. With that kind of business, you relax and only use your brains. (Segal *et al.*, 2001:100)

What is most powerful and perhaps most significant is the extent to which youth gang subculture in the late 1990s explicitly entrenched violence as a key means of acquiring status and thus of graduating within its hierarchy. Youth identity within the gangs has become fundamentally enmeshed with violence as a means of acquiring power and status in a context in which there are few other routes to wealth, power, and status for young people. This research also highlights the salience of family dislocation and poverty with many young gangsters relating their decision to commit crime to the injustices that they and their families suffered under apartheid.

FACILITATORS OF VIOLENCE
GUNS AND VIOLENCE

Because of inadequacies in the firearm registry system for legal firearms and because numbers of illegal firearms are hard to estimate, levels of gun ownership are difficult to determine. Gauteng province boasted the highest number of new firearm licences issued between 1994 and 1998 (37% of all licences). Johannesburg likely accounted for many of these, given the concentration of Gauteng's population in the Johannesburg Metropolitan area. Nationally, firearms were the leading cause of nonnatural death for South Africans in 1999 and the leading external cause of fatal injuries among those aged 15–64 years (Butchart, 1999). It is unlikely that this figure would be any lower for Johannesburg and, indeed, given the volume of firearms—both legal and illegal—in urban centers, it could be higher. In the study by Segal *et al.* (2001), gangsters noted the ease with which they were able to acquire firearms through housebreaking.

Not all groups are equally affected by gun violence. Nationally, 88% of victims of gun violence are male and 12% of victims are female (Butchart, 1999). However, the firearm-related violence that women face is quite specific, often being interpersonal in nature. For example, in a small study of 118 women who were murdered in 1993/94 in Johannesburg, 94 out of the 118 were shot by partners or ex-partners (Vetten, 1996).

Research into crime in Alexandra township and the suburb of Bramley showed that 78% of crime victims had their guns stolen during the crime perpetrated against them, with only 22% able to use their gun in self-defense (Altbeker, 1998). This research suggested that people in Johannesburg who carry a firearm are at greater risk of becoming a victim of violence.

Research by the CSVR has shown that the high prevalence of guns has also become common in the schools of Greater Johannesburg, with both teachers and students claiming to need firearms for protection against one another as well as from the intrusions of gangs into the schoolyards. When students were asked why they carry weapons in schools, they said that they want to defend themselves against the teachers, and the teachers likewise (CSVR, 1998). Initiatives are underway to create "gun free zones" in schools. So far, twenty one schools in Soweto are "gun free" (Gun Control Alliance, Online).

Internationally, the second most important factor predicting high levels of violent crime is gun ownership (ICPC, 1997). High levels of gun ownership not only predict high crime but make existing crime more fatal and therefore more of a strain on economic and social resources. Given the population and firearm concentrations in Johannesburg, this is likely to be a primary facilitator of violence.

REPEAT VICTIMIZATION AND VICTIM EMPOWERMENT

Repeat victimization is another increasingly important factor in the contemporary analysis of crime patterns. In cities, the risks of repeat victimization are stronger for certain types of crime (e.g., burglary, domestic assault) and a large volume of crimes is committed by a relatively small group of repeat perpetrators. One of the most striking findings of the ISS survey was that most victims of crime in Johannesburg had been a victim of the same crime more than once (Louw *et al.,* 1998). This suggests that the environment in which people live and work as well as their socioeconomic circumstances "trap" people in circumstances that compound their risk of victimization.

One of the interventions that has been shown to reduce levels of repeat victimization is the provision of adequate services to victims of violent crime (for instance by the police, health, and welfare workers), especially after their first victimization. The ISS survey found that the majority (89%) of victims of crime did not make use of specialized victim services. Rape victims in Johannesburg were most likely to make use of specialized services although only approximately 20% of them used such services due to a very low level of knowledge about them. Aside from the rape victims, there was little evidence that victims of violent crime were more likely to access victim services than victims of property-related crime. In spite of this, approximately half the sample indicated that victim services would have been useful.

In addition, victims of violent crimes expressed low levels of confidence in the ability of the SAPS to deal effectively with crime. Victims' feelings of safety and faith in the justice system are particularly important when considering localized crime prevention initiatives. A lack of faith in the justice system has often been cited as a reason for vigilante violence. Again, a lack of adequate localized information makes this a difficult issue to address. Although victim services seem (from the ISS survey) to be lacking, relatively speaking Johannesburg is one of the best resourced areas in the country. Indeed the ISS survey suggests that crime victims are simply unaware of the services available to them rather than there being a service shortage.

SUBSTANCE ABUSE

In the survey of police station commissioners across the Johannesburg Metropolitan Area mentioned previously, substance abuse, predominantly alcohol, was the second most common factor mentioned as contributing to crime. It was referred to at 23 out of the 38 stations surveyed.

> Alcohol is the number one [cause]. Alcohol, definitely number one. People are claiming [that it's] unemployment; I don't think that unemployment can contribute here. Because if you are unemployed, where will you get the money to buy alcohol?
>
> Some years ago, some people predicted that the AK47 would destroy the country. I say no, alcohol. Alcohol will. . . . Unfortunately, I don't know what the liquor companies will say, but it's a serious problem. Honestly. (Rauch, 1998:14)

The police commissioners' references to alcohol were universally made in relation to violent crime, such as assault, rape, and murder.

Illegal liquor outlets have begun to be clearly identified with various forms and sites of crime, both by police and community groups. Shebeens and unlicensed taverns flourish in all parts of Johannesburg, despite attempts by the SAPS and Council bylaw enforcement officials to crack down on the illegal liquor trade. This is a view particularly supported by the homeless women in the study by Dladla (2002).

Although the role that substance abuse plays in violent crime is not always clear, South African research has suggested that a large number of crimes are committed while a person is under the influence of alcohol, and that alcohol abuse is linked to higher rates of suicide. A person's likelihood of being involved in violet crime is increased by an early onset of alcohol use (Allan *et al.*, 2001). Similarly international work has suggested that alcohol abuse is linked to high levels of physical aggression toward one's partner and increased domestic violence.

Little is understood as yet about the links between substance abuse and violence in Johannesburg. However, given the easy access to alcohol in urban areas and the international trend toward increased alcohol consumption with urbanization, it is an issue that needs addressing.

CONCLUSIONS

Our discussions have suggested that the gap in wealth that characterizes the city and the differentials of gender, age, and nationality give rise to new and evermore violent forms of marginalization and social exclusion. In addition, current levels of crime cannot be divorced from the disruption of families and communities that resulted from apartheid social policies. Much of Zwane's (1997) description of the destruction of the social fabric in African township communities could well be applied in all parts of Johannesburg today:

> Hardly any family living in the black townships of South Africa has escaped the traumatic effects of violence in its many forms—political, criminal, familial and structural—either as direct or indirect victims. Structural violence has created barren communities with few, if any, self-sustaining resources. Political violence has swept through the townships, killing thousands of people and destroying homes, personal property and the minimal infrastructure that did exist.
>
> Familial violence has created physically and emotionally chaotic environments where children are not safe or nurtured. For many township children, their homes, which should be safe havens, are often the most dangerous places to be. Even in families that are not violent, parents are often absent

because they have to travel long distances to jobs in urban areas, where they work long hours for low wages. Young children are left to take care of themselves in violence-torn communities that do not provide any support services. (Zwane, 1997:1–2)

A new social fabric is needed to support the new political and cultural life of the city. Internationally, evidence is mounting that the most effective forms of crime prevention are early childhood development programs and efforts to improve parenting skills (ICPC, 1997). In Johannesburg, support for this type of intervention is almost nonexistent with the primary emphasis of local level prevention being on law enforcement. A range of agencies in the city must work together to provide the education, welfare, medical, nutritional, and mental health services needed to reverse the erosion of the city's social fabric.

NOTES

1. The online addresses are contained in the references.
2. This is sometimes referred to in the South African context as a "culture of violence."
3. In sum, official statistics are unreliable as regards crime trends. First, prior to 1994, the crime statistics did not include records kept in the former "black homelands." These areas were included in recordkeeping after 1994, which led to an "illusional" increase in crime statistics after that year. Truly "national" crime statistics have thus been available for only seven years. Second, the crime statistics released after 1994 lacked consistency in the way in which crimes were being categorized in different police stations. This was cited as the reason for an official moratorium on all SAPS crime statistics in 2001. After 2001, crime statistics were (according to the SAPS) more reliable. This means that there have been adequate official crime statistics recorded in South Africa for only one year. Third, SAPS and city boundaries do not coincide. The area that the SAPS defines as Johannesburg when recording crime data is, in fact, far smaller than the city of Johannesburg. Official SAPS statistics for crime in Johannesburg are, therefore, vastly inaccurate. Finally, as a result of the historical lack of service given by the SAPS to the "black" residents of South Africa, as well as the punitive and discriminatory apartheid laws that the SAPS enforced, there is a great deal of mistrust between the SAPS and communities in South Africa. This is likely to result in a great deal of underreporting of crime, particularly interpersonal crimes such as rape, assault, or domestic violence.
4. The bulk of the data supporting these findings came from the Johannesburg Medico-legal Clinic, which means that the findings are a reflection of rape in the Johannesburg magisterial district in particular, which covers a considerably smaller area than the metropolitan area of Johannesburg.

REFERENCES

Allan, A., M. C. Roberts, M. M. Allan, W. P. Pienaar, and D. J. Stein. 2001. "Intoxication, Criminal Offences and Suicide Attempts in a Group of South African Problem Drinkers." *South African Medical Journal* 91(2):145–150.

Altbeker, A. 1998. "Guns and Public Safety: Gun-Crime and Self-Defence in Alexandra and Bramley January–April 1997." Unpublished report.

Butchart, A. 1999. "A Profile of Fatal Injuries First Annual Report of the National Injury Mortality Surveillance System in South Africa." Cape Town: Unpublished report.

CIETafrica. 2000. "Beyond Victims and Villians: The Culture of Sexual Violence in South Johannesburg." Johannesburg: Unpublished report.

CSVR. 1998. *40 Schools Project Newsletter* 2(1).

Dladla, J. 2002. *Homelessness in Inner City Johannesburg.* Johannesburg: CSVR.

Goldberg, D. 1993. *Racist Culture.* Oxford: Blackwell.

Gun Control Alliance. 2001. [Online]: Available Internet: http://www.gca.org.za/facts/pamphlets/youth.htm.

Harris, B. 2001. *A Foreign Experience: Violence, Crime and Xenophobia during South Africa's Transition.* Violence and Transition Series 5. Johannesburg: CSVR.

International Centre for the Prevention of Crime (ICPC). 1997. *Crime Prevention Digest.* [Online] Available Internet: http://www.crime-prevention-intl.org/english/prevention/index/pres.html.

Louw, A., M. Shaw, L. Camerer, and R. Robertshaw. 1998. *Crime in Johannesburg: Results of a City Victim Survey.* ISS Monograph Series 8.

McKendrick, B., and W. Hoffman. 1990. "Towards the Reduction of Violence," in B. McKendrick and W. Hoffman, editors, *People and Violence in South Africa.* Cape Town: Oxford University Press, pp. 466–482.

Minaar, A., and M. Hough. 1996. *Causes, Extent and Impact of Clandestine Migration in Selected Southern African Countries with Specific Reference to South Africa.* Pretoria: Human Sciences Research Council.

Mokwena, S. 1991. "The Era of the Jackrollers: Contextualising the Rise of the Youth Gangs in Soweto." Paper presented at the Centre for the Study of Violence and Reconciliation, Seminar No. 7, 30 October.

Morris, A. 1998. " 'Our Fellow Africans Make Our Lives Hell': The Lives of Congolese and Nigerians Living in Johannesburg." *Ethnic and Racial Studies* 21(6):1116–1136.

Office of the President. 1995. *Promotion of National Unity and Reconciliation Act no. 34.* Pretoria: Government Printers.

Rauch, J. 1998. "Crime and Crime Prevention in Greater Johannesburg: The Views of Police Station Commissioners." Paper presented at the Centre for the Study of Violence and Reconciliation, Seminar No. 2, 12 March.

SALC. 2001. "A Compensation Scheme for Victims of Crime in South Africa." [Online] Available Internet: http://www.law.wits.ac.za/salc/discussn/html.

Segal, L., J. Pelo, and P. Rampa. 2001. "Into the Heart of Darkness: Journeys of the *Amagents* in Crime, Violence and Death," in J. Steinberg, editor, *Crime Wave: The South African Underworld and its Foes.* Johannesburg: Witwatersrand University Press, pp. 95–114.

Simpson, G. 1999. *Rebuilding Fractured Societies: Reconstruction and the Changing Nature of Violence—Some Self-critical Insights from Post Apartheid South Africa.* Research Report commissioned by the United Nations Development Programme (UNDP).

———. 2001. "Shock Troops and Bandits: Youth, Crime and Politics," in J. Steinberg, editor, *Crime Wave: The South African Underworld and its Foes.* Johannesburg: Witwatersrand University Press, pp. 115–128.

Sinclair, M. R. 1998. "Community, Identity and Gender in Migrant Societies of Southern Africa: Emerging Epistemological Challenges." *International Affairs* 74(2):339–353.

Van Onselen, C. 1982. *Studies in the Social and Economic History of the Witwatersrand 1886–1914. 1 New Babylon.* Johannesburg: Raven Press.

Vetten, L. 1996. " 'Man Shoots Wife': Intimate Femicide in Gauteng, South Africa." *Crime and Conflict* 6.

———. 1998. "Geography and Sexual Violence: Mapping Rape in Johannesburg." *Development Update* 2(2):11–17.

Vetten, L., and J. Dladla. 2000. "Women's Fear and Survival in Inner-City Johannesburg." *Agenda* (45):70–75.

World Bank. 2001. *What Is Social Capital.* [Online]: Available Internet: http://www.worldbank.org/poverty/scapital/index.htm.

Zulu, B. 2001. *An Exploratory Study of Sexual Violence and Coercion Among Soweto High Schools Youth.* Johannesburg: CSVR.

Zwane, W. 1997. "Restoring the Social Fabric." *Recovery* 2(12):6–11.

8

On Belonging and Becoming in African Cities

GRAEME GOTZ AND ABDOUMALIQ SIMONE

Municipal administrations are entrusted with ensuring order, equity, and the conditions for productive endeavour. They discharge this responsibility by providing some sense of coherence over who does what where. But exactly what "sense of coherence" is needed by those who inhabit the city? How precisely is this "coherence" structured?

In the past, coherence was based on clarity and certainty. Conventional approaches to urban governance emphasized the need for individuated, clearly identified units—be they citizens, households, communities, associations, or institutions—to be held in fixed relations to each other (Healey, 1997; Leftwich, 1994; Schmitz, 1995). This was in response to a general societal aversion to the insecurity of not being able to recognize the interests and actions of others on the urban landscape, and not being able to predict their response to one's actions. Even as governmentality has expanded to include complexities of all sorts, the role of government has remained that of occasioning moments in which clearly marked objects of government (property owners, say) respond in determinable ways to set stimuli (e.g., taxation, planning controls) (Dunsire, 1996; Kooiman, 1993). The traditional tools have been directed at tying identified actors to preferable behaviors in approved territories.

In some contexts, clarity and certainty are no longer a feasible basis for coherence. They are neither completely possible to effect by government, nor are they still desired by those subject to government. Many African cities are seeing long-valued frameworks of "social and spatial fix" gradually undermined. Displacement is accelerating and progressively eroding the conditions for clarity and certainty.

For our purposes two interwoven processes are important. Both of these processes represent new orientations to place-making. Each is launched by the deteriorating conditions for clear and predictable relations between defined social spaces and established interests and identities. And each, in their own way, reinforces the context that generated them.

The first process sees increasingly zealous attempts to claim rights to exclusive geographic territories within which to profess highly parochial identities. As historical connections to established social spaces erode, urban actors recoil from the felt loss of attachment and reassert a sense of connection in more defensive and particularized place-bound affiliations. The second, no less a response to a felt loss of membership, is the widening of social spaces to a multitude of new translocal connections and associations. This process reflects attempts to ensure the sustainability of what can be constructed in any specific place by multiplying the possibilities for defining and exploiting an array of non-place-bound identities. These two processes represent very different approaches to regrounding social relations. We call them "belonging" and "becoming."

This chapter is concerned with two questions. First, what do these two processes mean for the particular sense of coherence needed by African urban actors? How exactly do these actors come to terms with and live through the failure of traditional forms of social cohesion by embarking on each of these processes? What is to be valued or feared in each response? Second, how can government create coherence in the contexts shaped by these processes? What is the capacity of traditionally styled government institutions to take account of, negotiate, and utilize the processes of belonging and becoming? What eludes these institutions because of the irrelevance of old modes of knowing and acting, and what is "recontained" through the invention and elaboration of new techniques of government?

BELONGING AND BECOMING: MANIFESTATIONS AND EFFECTS

The erosion of long-accepted connections between place and identity is calling forth its inverse—an amplification of the need for social and political projects that enhance the sense of belonging. This process of asserting belonging in the face of its felt deficit is often experienced negatively. It manifests most frequently as passionate, even virulent, new struggles over place that center on which kinds of people belong to a particular place, as well as what kinds of places belong to particular people (Geschiere and Nyamnjoh, 1998; Osaghae, 1998; Zeleza and Kalipeni, 1999). Africa is currently seeing a proliferation of disputes over which identities have legitimate access to and rights over specific places and resources. They range from the contested citizenship of Kuanda in Zambia and Ouattara in Côte D'Ivoire aimed at eliminating their presidential candidacies to the expulsion of "migrants" in Gabon, to intensified ethnic claims of particular regions in Cameroon, to the fight over whether Shari'a belongs in Nigeria.

Ironically, struggles for belonging often erode the basis for social and spatial cohesion. Where social collaborations do arise, these are often highly fragmented and transitory clan or gang-based formations whose motivation for organization is the perception of encroaching threat or competition for a dwindling pool of resources and opportunities. Cooperation across these place-bound identities is rare. Struggles for belonging may therefore create deficits in the institutional mechanisms needed to plan and implement large-scale and long-term social and economic projects. Without these mechanisms, societies are unable to generate and capitalize on concentrations of resources, leading to further impetus to secure and consolidate particularistic place-bound identities. This may have profoundly destructive and restrictive effects, limiting the maneuverability and reach of urban residents.

Various African actors and social ensembles are not simply striving to reassert a sense of belonging. African identities also display a remarkable capacity *not to need fixed places.* Historically, African urban actors have revelled in the interstices of stability and instability, individuation and forms of social solidarity, rural and urban, colonial zones of domination and spaces of relative autonomy, the material and spiritual, home and nonhome (Apter, 1992; Coquery-Vidrovitch, 1991; Guyer and Eno Belinga, 1995; Hopkins, 1973; Martin, 1995; Robertson, 1997). In some instances, the tentativeness of established modes of belonging is calling forth an historical capacity to configure highly mobile social formations that focus on elaborating multiple possibilities. These formations embody a broad range of tactical abilities aimed at maximizing economic opportunities through transversal engagements across territories and disparate arrangements of power. Instead of focusing on the development and consolidation of specific places, this orientation emphasizes the construction of multiple spaces of operation through migration, specialization in trade, and the interpenetration of religion, politics, culture, and identity.

By way of example, urban quarters not only serve as platforms for popular initiatives around waste management, microenterprise development, or shelter provision. They are also sites in which local modes of sociality adapt to more regional and global frameworks. Here, religious fraternities, ethnically-based trading regimes, syndicates, and even community-based associations are configuring new divisions of labor and coordinating the cross-border small and medium-scale trade of individual entrepreneurs. They pool and reinvest the proceeds to access larger quantities of tradable goods, diversify collective holdings, and reach new markets.

Modes of becoming may also involve the forging of invisible spaces in which creative new intersections of forces, peoples, and economies can take place. African cities are replete with stories about geographies that are "off the map," demonstrated in popular descriptions of subterranean cities, spirit worlds, lucrative but remote frontiers, and underground highways along which pass enormous wealth. These geographies are not simply imaginary objects to which otherwise inexplicable events or woes can be attributed or impossible dreams attached. They are an enacted *terroir* being brought to life daily by unplanned, often indiscernible collaborations of actors from different walks of life, points of view, and positions.

On the face of it, these processes of becoming are far more creative and productive than the impetus toward belonging. But the mechanisms through which local affiliations expand in scale are often murky and prob- lematic and may exacerbate the conditions for disjuncture and displacement.

First, modes of becoming may entail highly tenuous and frequently clandestine articulations among religious networks, public officials operat- ing in private capacities, clientelist networks mobilizing cheap labor, foreign political parties, and large transnational corporations operating outside of conventional procedures. Multiplying possibilities also necessarily means "hedging one's bets," and this often requires deception, dissimulation, and circumvention. These practices thrive within, and work to produce, envi- ronments of intense doubt and unpredictability.

Second, urban residents often maintain multiple memberships in vari- ous associations, religious groups, clubs, and community organizations, si- multaneously participating in formal and informal economies, dispersing dependents across different localities, and spreading investments across a spectrum of different options. Such mobility and flexibility have dis- integrating effects. For example, because of the sheer labor intensity involved in maintaining multiple connections and the limited number of hours avail- able in a day, participation in any one institution can only be sporadic. Intermittent participation limits how effective or consistent institutions can be in managing the collective efforts necessary to attain long-term objec- tives. As with belonging, becoming may only further weaken customary modes of social affiliation.

NEW EXPERIMENTS IN GOVERNMENTALITY

What has all this meant for governmentality? What is now to be governed and how? How have the capacities of government to take in its objects changed as these objects shift into forms less obviously amenable to determinability? What do the acts and arts of government consist of when they can no longer be about ensuring coherence by ensuring economic and political equilibrium of determinable interests in fixed places?

As the deficit of belonging grows and new, highly problematic, opportunities for becoming multiply, government often fails to achieve the moments of fixity and clarity needed for its continued operation. Where true public space recedes in the face of the new more parochial and defensive forms of belonging, government may find it increasingly difficult to establish certainty and to determine in advance the identities, interests, and responses to policy of a diversifying set of actors. Similarly, since processes of becoming are often exercises in the deliberate manipulation of what is visible, or the construction of things invisible, they create blind spots for government. Almost always, as a condition for self-sustainability, collaborations that traverse fixed places and identities implicitly ascertain what the prevailing institutional powers are paying attention to and then locate themselves beyond their reach.

Where government has taken account of the logics of belonging and becoming, it has often been extremely clumsy. New governance arrangements, whether they be in the fields of urban development and settlement upgrading, planning, or participatory decision making, can potentially undermine the very mechanisms through which belonging and becoming were once naturally mediated and brought together. In many African cities, residents are increasingly compelled to speak about what they are doing and how they are doing it, as a precursor to constructing particular kinds of citizenship and as practices of incessant self-improvement and self-accountability (Rose, 1996). Government often encourages urban actors to navigate the intricacies of local social life through discourses of management and thereby to express approved identities that can be held accountable. But a heightened visibility and self-consciousness about local social dynamics may wear away more understated and less visible practices through which local residents are able to hold each other in view and assess and access other activities, collaborations, resources, and likely courses of action. Government control of residents may therefore intensify the sense of loss of belonging and call forth the reassertion of parochial identities.

Governments find the complex coexistence of belonging and becoming difficult to read and build on. But there are also a few instances of government responding creatively with new projects that work from the premise of the failure of belonging or toward the promise of fostering and sustaining the new possibilities of becoming. These seek to construct new senses of place via the deliberate opening of new avenues of becoming, by facilitating diverse interinstitutional interactions. Although embryonic, these projects establish new forms of coherence in the otherwise displaced social landscapes of African cities.

BELONGING AND THE JOHANNESBURG INNER CITY

The inner city of Johannesburg has changed more rapidly than perhaps any other inner city in modern history. For several decades, the race-based zoning of urban residential communities kept blacks out of the inner city. In the mid-1980s, accelerating white movement to the suburban areas coupled with economic recession pushed up vacancy rates in the neighborhoods of Hillbrow, Bertrams, Joubert Park, Berea, and Yeoville. Although officially still illegal until 1991, blacks began moving to what was known then as the city's gray areas (Tomlinson et al., 1995). This accelerated turnover of population provided a cover for the sizable immigration of foreign Africans to Johannesburg in the mid-1990s, a process that has profoundly reshaped inner city life and commerce and further contributed to its progressive internationalization (Bouillon, 1999). It also provided the impetus for new informal economies to flourish on the inner city's increasingly crowded streets. Because the inner city is one of the most circumscribed and densely populated urban spaces on the continent, with neighborhoods such as Hillbrow made up of row after row of high-rise apartment blocks, this social and economic reconfiguration has been largely invisible.

Today the inner city of Johannesburg is a cauldron of diverse peoples and agendas. Corporations, once accustomed to thinking that they owned the city, now agonize over whether to stay there and bet heavily on the chance that the inner city may be revitalized or to flee to the safer and greener suburbs. A minibus taxi industry provides the bulk of transportation. Over the years, it has become an intensely volatile and often violently competitive business that has steadfastly resisted public regulation. Vastly overcrowded residential areas of mass apartment blocks, many of which have not been serviced for years and whose occupants (many of them desperate immigrants forced to pay exorbitant rentals) are now accustomed to being always on the move. Street traders barely eke out a living selling farm produce and light consumer goods as their numbers have swelled by almost 500% during the past few years.

A VACUUM OF BELONGING

The inner city represents a veritable vacuum of belonging, where almost no one presently living there can claim an overarching sense of origin in this place or profess a real wish to stay. Why?

First, there has been precious little sense of community. Where the overwhelming majority of inhabitants of the inner city are recent arrivals to Greater Johannesburg, and so have no protracted history of settlement, there are few grounds for any established sense of social cohesiveness. Stable social institutions have never fully formed. Any that have emerged have been quickly displaced by the next wave of in-migration. This is true for both South African migrants and foreign immigrants.

The inner city has been seen as both a proper extension of the major black townships of Soweto and Alexandra and a place of escape from the townships. Former township residents, many with jobs in the central business district, view it as a more centrally located space of habitation. It is also a new zone of opportunity for the operation of township gangs and businesses. Many of the latter are trying to escape the oppressive and often implosive features of township politics with its incessant struggles over resources. The inner city thus becomes a place of relative freedom where migrants can live more independently and anonymously in large apartment blocks.

As black South Africans had almost no presence in the inner city prior to the late 1980s, the rapid influx of migrants far outpaced the development of local institutions. The demographic shift saw a vacating of critical social infrastructure. Although there has been a proliferation of fly-by-night academies and the development of several effective community associations and a broad range of schools, social service organizations, churches, and cultural and recreational centers, the mechanisms through which the traditional sense of belonging in urban areas is cultivated have been depleted. For years, therefore, children residing in the inner city continued to commute to Soweto for school and most of the trappings of community life remained located "back home."

The same is true for foreign immigrants. Foreign African residents have been driven to the inner city by the fact that life at home is simply untenable. This assessment has been ingrained into most immigrants. Young children in Senegal and Ghana prepare themselves to leave from an early age. Often this preparation entails prolonged periods of unpaid labor for relatives who will later supply airline tickets and other assistance. The absence of possibilities for real belonging at home—as represented by ownership of a plot of land or a house, the semblance of security embodied in a territorially rooted set of social connections, or the ability to marry and reproduce a lineage—is thus inculcated from nearly the start of an immigrant's life.

Although national associations, newsletters, and advocacy groups have appeared in recent years, there is little sense of immigrants trying to carve out a visible, long-term niche in the city. Unlike historic cities of African immigration elsewhere, the commercial activities usually associated with servicing a growing immigrant community (such as restaurants, record stores, beauty salons, telecenters, and nightclubs) reflect a tenuous history, and their numbers and vitality are not consonant to the numbers of foreign Africans actually living there.

Second, the sheer rapidity of demographic and economic changes has created uncertainty (Reitzes, 1999). Although fellow nationals or even immigrants of various nationalities may band together to share living expenses, information, and risk, the possibilities for corporate action are limited. Each is trying their best to make ends meet and deal with specific family, community, and political situations back home. Each is in some way a competitor, and cooperation is based only on self-interest, self-protection, and comraderie. Indeed, within the inner city where jobs are scarce, everyday life precarious, and the need to mobilize available social capital acute, the very act of counting upon others becomes a practice that leaves individuals vulnerable to further difficulties. A critical problem is the instability of household composition. Families that reside in an apartment for several months frequently then disperse, with new household arrangements then established elsewhere. This instability is directly related to the intensifying uncertainty permeating everyday kinship relations. If one interweaves the details of one's daily life too closely with those of family members, and if something goes wrong or a growing divide in economic capacity becomes apparent, one is then vulnerable to accusations of witchcraft or of selfishly putting personal needs above those of the family. The very process of mobilizing social capital to elaborate a viable sense of belonging becomes most difficult.

This instability and uncertainty have caused a huge reluctance to invest, and sudden and substantial divestitures of property and position occur at a cheap price. These divestitures exacerbate the sense of insecurity and the urge to move on and out. Consequently, even less interest exists in solidifying a long-term investment in Johannesburg (Morris, 1999). The actors that do inhabit the inner city do so only in the hope of leaving as quickly as possible with sufficient resources to have made the stay worthwhile.

The inner city therefore represents a process of "running away." Black South Africans are escaping the implosive sociality of township life. The townships had become "nowhere" places arbitrarily configured to be apart and to embody the essence of cultures long uprooted. Foreign Africans are running away from the impossibility of being at home, doing whatever is possible to maintain the sense (and often, the illusion) that they can have a home. But the inner city presents neither group with a real place to be "running to." Although most migrants might dream of a quick score that would enable them to return home with significantly enhanced prestige and purchasing power, many are acutely aware that this rarely happens. Instead, many years of toil in a series of low-paid jobs is the norm, with the bulk of the limited earnings remitted back home to support an array of family members. Additionally, bribes to police often must be paid, as well as unofficial surcharges to landlords. For traders at any scale, goods are often seized, lost, or stolen. For most residents, the Johannesburg inner city is neither the preferred nor the final destination (Sinclair, 1998).

HYPERREAL EXPRESSIONS OF PLACE AND POSSIBILITY IN THE DRUG TRADE

In response to the felt deficit of belonging, migrants to the inner city frequently exaggerate attachments of identities to particular places. For example, as Congolese networks began to extend themselves out of the foreign African ghetto of Ponte City—a large cylindrical apartment tower in Berea that dominates the skyline with its four hundred plus studios—into neighboring Yeoville during the early 1990s, the exodus took place in highly visible ways. Up Harley Street, Congolese drinking clubs, set up in the front rooms of dilapidated single-family households, would blare out Soukous music and the intonations of Lingala were loud on the street, conveying to all that a Congolese neighborhood was in the making. With their cultural traditions on public display and their exaggerated performance of confidence and capacity, Congolese immigrants conveyed the appearance of a developing community within the uncertain dynamics of the inner city.

The process can be fully explained only as a vigorous reassertion of parochial place-bound identities in the face of a felt insecurity. Immigrant communities in the inner city are subject to a large measure of xenophobia. Foreign Africans are blamed for an overcrowded informal trading sector, the growth of the narcotics trade, and the general deterioration of the inner city. Many South African residents believe that it is because of such a foreign presence that government authorities and the private sector are unwilling to make investments in upgrading and service provision (Mattes *et al.*, 2000). Most notably, South Africans view the apparent capacity of foreign Africans to elaborate a sense of supportive social connectedness as particularly threatening. Immigrants seem to have an undue advantage to thrive in this competitive urban environment. This is ironic. Although immigrant networks depend on always activating a sense of mutual cooperation and interdependency, such ties are often more apparent then real.

The Congolese's very public expressions of community on Harley Street were very likely not the product of a growing sense of connectedness to place. They were more probably attempts to symbolically colonize a space. They were an overexaggerated display of connectedness for the sake of amplifying a sense of inadequacy in local competition, and thereby effecting some sense, however momentary, of self-security. Far from establishing places, such occupations only bespeak a lack of placement and ultimately reveal and reinforce the extreme ghettoization felt by the immigrants.

The same exaggerated expressions of belonging are seen in the illicit drug industry. The drug trade is commonly seen as the purview of Igbo-dominated Nigerian networks. Although this might be generally true, such enterprises are by no means ethnically homogeneous, nor formed on the basis of national identity. Rather, in a business that has little recourse to law or official commercial standards, the appearance of ethnic or national homogeneity conveys a certain impenetrability from both external scrutiny, infiltration, and competition. Ironically, this allows the enterprise to incorporate the diversity of actors it requires to constantly change sources, supply routes, and markets.

Such enterprises act as a parody of belonging. In the commercial culture of the inner city narcotics economy, the discrete tasks of importation, circumvention of customs regulations, repackaging, local distribution, money laundering, relations with legal authorities, territorial control, market expansion, and plotting traffic routes are all complementary yet highly territorialized domains. Each domain is administered by discrete entities so that disruptions in one domain do not jeopardize the entirety of trade.

These units are highly spatialized. The modalities of operation of the drug business provincialize certain parts of the inner city, localizing it in terms of clearly marked territories and fiefdoms. Thus, most inner city residents know which hotels, residential buildings, and commercial enterprises belong to which syndicates and to which nationalities these syndicates belong. The inner city becomes a complex geography where residents must navigate according to a finely tuned series of movements and assumptions.

There are places where they know they must not go or be seen, producing a convoluted economy of safety. For example, Nigerian syndicates use the hotels in Hillbrow to accommodate a large transient population that in turn serves as a mask behind which to consolidate a steady clientele of drug users. The hotels become discrete localities, housing not only workers in the drug trade but also Nigerians working in a wide range of activities. Nigerians who are not involved in the drug economy are depended on to provide a semblance of internal diversity, even if they are often used and manipulated (Osita, 2000).

Despite this territorialization, the definitiveness of organizations and their spaces is often more a necessary performance than something descriptive of actual operational practices. The more entrenched and expansive the drug economy becomes, the more it must proliferate ambiguous interfaces.

Any particular narcotics enterprise handles only certain facets of the overall trade and leaves itself increasingly vulnerable if it expands in efforts to dominate more functions and more territory. Separate domains, then, must be maintained and these domains must be integrated such that complicity and cooperation prevail. Within each domain, each operator has a specific place and is expected to demonstrate unquestioning loyalty. At the same time, the illicit nature and the practical realities of the trade create an incessantly open space for participants to take their chances and seek greater profits and authority outside the hierarchies that each syndicate must attempt to rigidly enforce. Between the domains, spaces must also be maintained that clearly belong to no one. These spaces are often subject to the most vociferous claims of belonging. Conditions are wide open for the intersection of many groupings, and thus these spaces contain the most intense processes of becoming.

In these spaces, new connections are forged between supposedly discrete groupings, between illicit activity and legitimate investment, even between inner city Johannesburg as an increasingly well-known site of the drug economy and other more invisible and thus often more advantageous sites of operation. The drug economy with its hyperactive sensibilities and codes of belonging found a viable place to entrench itself in Hillbrow and Berea because a highly dense, highly urbanized area with massive infrastructure was being vacated. Yet, drug operations do not need the inner city either as market or center of operation. Already, several syndicates are seeking other locales. Thus, the sense that specific territories within the inner city have exclusively belonged to specific agents is revealed as arbitrary.

GOVERNMENTALITY IN THE INNER CITY

The inner city of Johannesburg presents a rapidly changing, highly fractured, and deeply contested urban milieu. Residents feel obliged to feign exaggerated and almost theatrical commitments to particular spaces and place-traversed connections and defend these against outsiders assumed to be laying claim to the resources and opportunities accumulated there. A more general commitment to the inner city as a community of neighbors, peers, potential partners in business ventures, or even compatriots in social action is absent. This is a place to pass through, to run from, not a place to collaboratively contest or coinvest in for the longer term.

In such an environment, government efforts to provide services, social development, or security in ways that assume a collective desire to stay and be together are not only tenuous. They may also be highly perilous. The dissipation (or the nonformation) of viable modes of solidarity, the uprooting of individuals from familiar domains, and the ghettoization of individuals within highly circumscribed identity enclaves constitute a potentially explosive mix. Attempts to configure viable domains in which discrepant identities and interests can belong together run the danger of miscarrying.

Conventional forms of governmentality aimed at producing clarity and certainty invariably invite their subjects to assume shared standards of behavior. They try to make it the responsibility of residents to embody and display the correct normative attitudes toward managing individual performances appropriate to their place. These efforts force urban actors into the cul-de-sac of having to confront their own ability to perform as self-responsible, accountable individuals. This expected self-reidentification via government may be too high a hurdle for many and so may simply make real a certain impossibility to secure themselves and perform well within any durable context.

The danger is obvious. By asking residents to disregard their traditional informal means of navigating urban complexity and uncertainty, these projects accentuate a lack of belonging and so invite pointed refusals in the form of reasserted parochialism. The result is reaffirmed commitments to the most exclusive bonds and particularistic connections. Government may thereby invoke the passionate reexpression of the identities of gangs, clans, associations, and cartels.

What is not needed is the proliferation of technical standards through which citizens' capacities can be compared and judged. In such a politics, everyone is to be found wanting, and vigorous group identities may all too easily be reaffirmed as both compensation and insulation from a newly felt loss of belonging. What is needed is mediation and dialogue among differing experiences, claims, and perspectives—a politics of becoming.

Has government in the inner city managed to negotiate this complexity? Has it provided viable new spaces for social collaboration, without inviting renewed commitments to aggressive or defensive bunker identities?

The answer is a mixed one. In some of its activities, government has indeed fallen into all the obvious traps and created conditions where the dynamite mix of contiguous parochialisms has exploded into urban conflict. In others, government has creatively provided just the right opportunities for the mediation of potentially discordant interests and identities. It has configured new arenas for urban actors to step outside the narrow confines of existing zones of control and paths of interaction. Once exclusive identities now intersect with each other in ways that do not increase the risks of being an individual. Moreover, they motivate a drastic "return to self" as a member of one or another group competing with others for the rights and benefits attached to their particular place. Two examples will illustrate.

THE INNER CITY OFFICE

During its brief existence between 1998 and 2001, The Inner City Office (ICO) of Johannesburg's Metropolitan Council was widely applauded as among the most innovative segments of South Africa's largest local government. The Office functioned as a project design and facilitation unit, structuring a range of urban environmental upgrade and social and economic development projects, most with various business and community partners. Over a period of three years, the ICO configured upward of forty medium-scale urban regeneration and development projects. In 2001, as part of a broader institutional restructuring of the Johannesburg Metropolitan Council, the Inner City Office was disbanded. Its responsibilities were reallocated to three new structures: the City's conventionally styled Planning Department, an Inner City Regional Administration responsible for overseeing urban service standards in the area, and the Johannesburg Development Agency focused on managing large-scale economic infrastructure investments in various parts of the city. Despite its closure, and the even higher profile of some of its successor structures, the ICO is still regarded as the generator of a new approach to the complex urban landscape of the inner city.

The Office deliberately chose a project approach. High profile political representatives argued that the challenge of urban management demanded a blunt reimposition of the rule of law. As they saw it, a prestige projects approach might appeal to the sensibilities of well-resourced investors who like the image impact of big urban upgrades, but targeted interventions aimed at making things happen in selected segments of the urban fabric invariably leave gaping holes. The ICO's approach, critics averred, did not impose a necessary gridwork of urban order, a matrix of properly formulated and rigorously applied codes that stops the misuse of cityscapes and buildings. For them, the ICO approach was doomed to fail because it did not clearly know its zone of application and could not apply the necessary prescriptions to bring this zone under control. In short, it could not remove Nigerian drug lords from Hillbrow flats for the simple reason that it did not know who really owned the buildings.

The Inner City Office disagreed. It argued that development projects are the only way to understand and manage the textures and nuances of the inner city. The ICO Manager, Graeme Reid, repeatedly made the point that one cannot intervene in the complex environment of the inner city with predetermined responses (personal communication). Development projects, needing to be designed with specific microcontexts and circumstances in mind, and negotiated with local stakeholders, are the only way to be sensitive to the complexities and dynamics of urban life. "Conventional service delivery management cannot deal with nuance," states Reid, "Designing projects means you have to be attuned to the fact that whereas the taxi associations are rigid and regimented, trader organizations are always weak and chaotic." More than this, he suggested, development projects provide concrete opportunities. They invite or expect beneficiaries to act in new ways for the sake of their own well-being, rather than directly proscribing how they must not act for the sake of the cleanliness and order of inner city spaces.

THE ROCKEY STREET MARKET

Since the early 1990s the Greater Johannesburg Metropolitan Council has sought to govern a burgeoning number of informal street traders in the inner city. Uncontrolled street trading is regarded as a major urban environmental and governance problem. From the vantage point of government, a large number of hawkers, selling all manner of goods, from shoes to fresh produce, most from makeshift wooden stalls, and without access to transport, cleansing, or storage facilities, presents various environmental risks and hazards. Street trading adds to the image of a city center supposedly beset by crime and grime, a malaise that has supposedly caused large numbers of formal businesses to flee to the pristine malls and office parks in the luxurious northern suburbs. The unavailability of transport and proper means of storage means that foodstuffs are kept in ways that violate local health by-laws. A large amount of cooking fat is washed into stormwater drains, causing blockages of pipe networks. Traders' stalls jut out dangerously into busy roads and sidewalks inhibiting pedestrians and passing traffic.

Excessive street trading is also perceived as a governance problem. Traders, most situated in front of the doors and shopfront windows of formal stores, block entrances and lines of sight and in many cases compete directly for the business of passing shoppers. Conflict between traders and formal shop owners is a daily occurrence. On the highly congested pavements, contestation over prime spots is also rife, spawning violence between individual traders, conflict between trader associations, and the growth of protection rackets. After one trader was knifed to death in a disputed claim over a particularly preferred trading site on one of downtown's busiest walkways, Graeme Reid was moved to remark, "The state was completely absent there. We totally failed those women."

Various measures have been introduced to manage the overtraded streets. In 1997, the City passed strict new street trading by-laws, most focused on addressing the public health and traffic obstruction risks posed by the traders. At the same time, portions of the inner city were declared restricted or prohibited zones in an attempt to thin out the traders. Later, a by-law enforcement unit was set up in the Inner City Office to more rigorously impose these regulations. The last few years, the City has also considered a licensing strategy, even extending this notion to a proposal to randomly assign each licensed vendor a marked space, a so-called "parking bay" on the sidewalk, from whence they would not be able to move without swapping their permit. These mechanisms, standard to governmentality, have had little effect. Against a 10,000 strong street trading population with no other means of livelihood and with limited resources, a governmentality meant to isolate, designate, assign, and individually permit or prohibit is neither viable nor appropriate.

In early 1998, the Inner City Office embarked on an ambitious process to develop a "trader markets" strategy for the downtown area of Johannesburg. The idea was that if a sufficient number of markets could be built and organized in a way that encouraged street traders to keep stalls within them, the prohibition on unorganized street trading proper could be enforced with more vigor. A two year research and design process ensued, starting with surveys of street traders to determine their profit thresholds, and, by extension, feasible rentals for market sites as the way to recover likely construction and operating costs. A planning firm mapped the busiest zones and, on an overlaying street plan of empty lots and convertible buildings, identified the best sites to build markets. A new special purpose vehicle, the Metropolitan Trading Company, was established by the Council. It would receive grants from donors and government and serve as a guarantee that the markets would be established and run on business principles with a view to gearing in investments from private sector partners. In late 1999, the first of eight planned markets was constructed with provincial government start-up funds on Rockey Street in Yeoville.

Despite the clear logic of a markets strategy, and a fairly elaborate process of consultation with affected traders in the run-up to its launch, the Rockey Street Market was a disaster on its opening. The problem turned on the issue of high rentals for stalls, but embedded in this was an invisible politics of street identities that the Inner City Office simply could not read.

The Rockey Street Market was built with R6.1 million of R8.3 million granted to the Metropolitan Trading Company (MTC) by the Provincial Government. Although it was never the intention to recover this investment, the designers of the Market were adamant that a profit margin be structured into its operation as proof to future private equity investors that the enterprise could realize a return. The MTC estimated operating costs of the market at some R95,000 a month for five internal management and administrative salaries, a cleansing and security contract, and miscellaneous items such as Council rates on the land. It further estimated that some management costs in the MTC would need to be recovered from the Council. It then pegged a required rate of return of some 25% (10% less than most large project finance lenders in Johannesburg would accept). On these calculations, the MTC set a required monthly income from traders rentals at R150,000. Shared between 300 sites, this meant a range of rentals of between R6.70 to R20 per day, depending on whether the stall was on the outer edge of the market and easily accessible to passing pedestrians or inside the market facing one of the courtyards.

These calculations led the MTC and Inner City Office to a difficult conclusion: the market was an extremely fragile project. On the estimated rentals, the MTC and ICO concluded, no survivalist traders, defined as vendors who were used to selling only sweets and cigarettes and other odd items for a few Rands a day, could possibly rent there. Because the establishment of the Rockey Street Market would go hand in hand with the prohibition of street trading in the broader Yeoville area, this meant that survivalist traders had to change their behavior if they were to survive. Only entrepreneurial traders would prosper, that is, vendors who were good at selecting the right products given the Market's pull-through of a particular profile of customers, who diversified their product range to minimize competition with adjacent stalls (a large number of street traders sold fruit and vegetables), and who became creative at sourcing their goods at cheaper rates. The designers realized that they had no other option. If the first market was to lead to a successful markets strategy across the inner city, they had to hold to the minimum rental thresholds and force traders to adapt. An estimated 30–40% of traders would probably not make it, the MTC believed.

Implicitly, the Rockey Street Market became a project of identity reconstruction through a reformatted community of self-responsible and accountable individuals expected to act in entrepreneurial ways. Only by shrugging off the subjectivities of vulnerable street traders who were struggling to get by and by becoming businesspeople could traders secure a long-term place in the market.

All of this caused considerable tensions. Two principal traders associations operated in Yeoville at the time of the market's construction. The Yeoville Traders Association was a local body made up principally of Yeoville residents used to hawking goods on the streets. The Gauteng Hawkers Association (GHA) was a much older structure with a murky history. ICO staff speculated that the GHA was set up by the apartheid security apparatus in the old predemocracy Johannesburg City Council as a "sweet-heart union" to contest the growing force represented by the once powerful African Chamber of Hawkers and Informal Business (ACHIB) led by a charismatic trader, Lawrence Mvundla. Though ACHIB's influence waned toward the end of the decade, the GHA stayed around, evolving, as the ICO tells the tale, into an archetypal street protection business. GHA leaders, as the Council staff understand it, had developed a vested interest in controlling the transport of a large number of hawkers and their wares to and from various sites in the inner city as well as the security of these traders. GHA leaders had no wish to see the development of markets that would provide on-site storage facilities, security through professional private security companies, and the transformation of previously dependent traders into stand-alone businesses.

In the run-up to the opening of the Rockey Street Market, all negotiations between Council and traders had occurred with the Yeoville Traders Association. The ICO and MTC believed that they had achieved a considerable level of buy-in to the extent that they had received over 700 applications for an available 300 sites (an oversubscription that forced them to redemarcate the market into 370 stalls just before opening) and up-front payment for stalls from some 65% of traders. On the weekend before the official opening, as the ICO came to understand it later, the Gauteng Hawkers Association mobilized against the market among Yeoville traders. The rallying cry was predictable: "The rentals are too high and have not been adequately communicated and consulted." The GHA instigation called forth an equally predictable response from the stronger Yeoville Traders Association. If it was not to be regarded among members as soft on traders' rights and compliant with Council plans, it had to take up GHA's call. The Market opened on a Monday morning with a full blown Yeoville Traders Association protest against the concurrent street trading ban and the Market's high rentals.

The sudden turn of events became a media fiasco for the ICO. It faced a heated attack from a range of quarters and was accused of acting against the interests of poor traders. Critics argued that the ICO was being authoritarian in forcing previously fluid street operators into a fixed and inflexible space and being insensitive to the conditions of people in poverty. Moreover, its cost recovery approach, critics continued, had blinded it to the need in Yeoville (and elsewhere) for a "social responsibility" project that protected a vulnerable group against crime and the elements. Some went so far as to argue that the City had concocted a mechanism for the sole purpose of making a profit off traders' meagre incomes.

The Inner City Office was quite eloquent about what it was trying to do via the MTC and the Rockey Street Market. Providing a market that obliged traders, on pain of failure and indigence, to become something different than they had traditionally been was the most socially responsible thing that could have been done, the ICO argued. On the streets of Johannesburg, it averred, all traders are trapped. All are vulnerable to various protection rackets that oblige them to make use of usually uneconomic means of transport, storage facilities, and bulk supply networks. And with the intense competition, virtually none can develop the market-share to accumulate enough capital to graduate to more productive and profitable ventures. Providing spaces within which survivalists are expected to, and can be enabled to, reidentify themselves as businesses is the only way to liberate many from the crippling and violent contestation and informal control of overtraded streets and launch them into less fragile enterprises.

Such arguments were lost on a media out for blood and traders associations enjoying both a public stage and a renewed popular legitimacy from which to reassert themselves. Faced with the prospects of individualization and reidentification of members, the associations reaffirmed, against exactly that which was offered as a means of escape, the identity of traders as vulnerable survivalists destined to have to vigorously defend their place against official prescriptions that would rob them of livelihood.

The MTC offered a two-month moratorium on rentals and then extended this again in an effort to allow traders to acclimatize to the market. Although market occupation rose gradually, protests and negotiations between ICO/MTC and the associations dragged on well into 2000. As the impasse continued, the Manager of the ICO began to explore drastic options to convince the Yeoville Hawker's Association to accept the inevitability of the MTC's approach. "Our biggest mistake," Graeme Reid remarked in frustration at one point, "Was not to offer the Associations the Market's R20 000 a month cleansing contract to keep them quiet. We would have been paying a premium to get the Market accepted, but at least it would have brought them in."

THE PARK CENTRAL TAXI RANK

One of Johannesburg's most deeply intractable urban problems has been continued warfare between the Province's major taxi transport associations. Competition over the right to operate the city's major routes, most running between the inner city and Johannesburg's major townships, Soweto and Alexandra, have frequently spilled over into violence. The taxi associations operate like mafia families, with invisible hierarchies, ruthless control over demarcated territories, and (many outsiders believe) an eye on using the "front" business of moving millions of people a day around South Africa's largest urban conglomeration as cover for the transport of illicit guns and drugs. Addressing the needs of this burgeoning, unregulated industry has traditionally caused major difficulties. The associations have a vested economic interest and hence good political motivations to run their operations (such as the ranking of vehicles) in separate locations in the city. It is also not surprising that they disagree with each other over proposed official plans meant to benefit the industry as a whole and fight to capture the economic benefits from urban upgrade interventions. This has made it extremely difficult to design taxi transport facilities to provide better access to passengers, structure contiguous economic opportunities such as markets, and address traffic bottlenecks from overcongestion.

In 1997, the Greater Johannesburg Metropolitan Council embarked on a major project to construct a modern taxi ranking facility at Jack Mincer Park on the southern edge of Hillbrow. The project was taken over by the

Inner City Office in 1998. The motivation for the initiative is summarized in a submission by the Inner City Office to Habitat's Best Practice database:

> Problems arising from the taxi industry are considered the biggest problem facing South Africa's inner cities. Nowhere is this more evident than in Johannesburg's Jack Mincer area where taxis invaded a 15 block area in the early 90's causing a rapid deterioration of the surrounding area. The infiltration of taxis in the area caused the roads to become nearly impassable and made proper sanitation impossible and crime inescapable. In the absence of formalized management, the central point, the Jack Mincer Park and parking garage deteriorated to a point where it became unsafe and a health hazard to the general public. (Inner City Office, 2000)

As the Inner City Office envisaged it, the area desperately needed a facility that could formalize the ranking of over 2,000 taxis belonging to some ten competing short-distance taxi associations. The ICO targeted an unused and partially flooded underground parking garage as the focal point of the intervention. In rehabilitating and substantially upgrading the garage, the ICO hoped that it could move a large number of minibus taxis off the overcrowded inner city streets and provide taxis with an economically beneficial place to which a greater number of users might be drawn.

The Inner City Office mobilized some R14 million for facility upgrading in two phases. In the first phase, the Council devoted R8 million to the reconstruction of the underground portion of the garage. R4 million worth of "free" construction services were secured from Grinkar Construction by offering the property company the transfer of an adjacent site and building for it to develop into a new shopping center. In the second phase, involving the addition of a rooftop level to the parkade to accommodate larger 35 seater taxis, the ICO geared in a further R6 million from the National Department of Transport and the Gauteng Provincial Department of Transport and Public Works.

With finances secured and construction plans drawn-up, the ICO was ready to build. The problem, however, was the taxi associations. How, the Inner City Office asked itself, could it possibly get 10 different associations, many of whom had frequently been at war with each other in the past, to work together to plan the detailed operation of a rank that would serve the industry's needs? Even more challenging, how could it convince them to overlook their differences and use the rank peacefully and cooperatively once the facility was built?

The answer was a new institutional space. The Inner City Office established a Rank Committee, representing all taxi associations, for the express purpose of guiding the development of the rank and overseeing the use of the facility by taxi drivers from all associations. Two members from each taxi association are represented on the Rank Committee. With the ICO and the rank developers, Grinkar Property Company, this Committee negotiated the various steps of the final design and construction. On completion of the rank, operating agreements were then struck with the same Rank Committee (now reduced by natural attrition from ten to just six associations). The key accords were a User's Agreement and a Management Agreement.

The User's Agreement structured the usage of the rank by some 2,000 drivers from the various associations and directed the financial flows from user fees and other incomes. It had a number of motivations. First, the ICO desperately needed active association involvement in the design of dropping off and picking up procedures, and the structuring of rights of use provisions with individual drivers. Over the ten years, the Jack Mincer area had developed its own informal ranking management system. It included invisible codes on when and where drivers could park and a unique, largely incomprehensible street sign language to control the movement of incoming and outgoing vehicles each trying to minimize the time needed to drop off and pick up passengers. The ICO found this system inscrutable, but was worried about trying to replace an operating system that had enabled the daily movement of some 2,000 taxis crowded in a few city blocks. "We just don't understand the informal system that makes this possible, but we want to re-demarcate and re-organise the rank to our specifications. Everything suggests that our plans simply won't work. If we don't try to mimic the system that already exists we will never be able to enforce," said Reid.

Second, only an agreement that formally recognized the *shared* use rights of separate associations—the Alexandra Taxi Association or the Diepmeadow Taxi Association or any other—could hope to obviate potential conflict from those maneuvering to capture the rank, or any part of it, as its exclusive territory. An Agreement was needed to make the associations collective partners, with the rights to decide how benefits would be distributed and the right to choose whether other associations might be allowed access to the rank.

The User's Agreement set out a number of taxi association responsibilities. First, it required the associations to reveal themselves, that is, to specify precisely who their member owners and drivers were and which vehicles belonging to the association would be utilizing the facility. The Agreement enjoined the associations to assist in issuing identification permits (in the form of windscreen decals) that named vehicle operator, association, registration, and route, and required them to guarantee that all operators were legally licensed and registered. In one remarkable clause in the Agreement, the Rank Committee was obligated to ensure "[t]hat all drivers have valid Public Driver Permits and valid Driver's Licences, and shall take all steps necessary to ensure that the drivers comply with all relevant laws and by-laws in the exercise of their functions and that they shall not engage in any unlawful activities" (Inner City Office, 1998). The Agreement further gave the associations, through the Rank Committee, the duty of sharing out holding and loading lanes and zones in the rank and organizing the movement of taxis within these agreed demarcations. Lastly, the Agreement made the associations responsible for collecting R20 a month (later changed to R1 per entrance) from each of their taxi drivers. These monies were to be handed over to the Council to be kept in a suspense account for repairs and maintenance to the facility via the services of a Management Company.

The Agreement automatically shifted the boundaries of spaces of belonging and, for this reason, could have called forth the aggressive reassertion of exclusive identities. Its very title spoke of common usage and it gave the signatories the collective right to decide whether other associations would be allowed access to special privileges. These are contentious issues.

Why did they not lead to conflict? One answer is that the arrangement offered an exciting new trajectory of becoming, *which in no way compromised the existing identities* of the signatories to the deal. In fact, precisely by being cosigned with each association separately and together, the new zone of becoming was *conditional* on each existing terrain of belonging remaining intact. Unlike the Rockey Street Market, where the traders associations were sidestepped in an attempt to provide a now atomized membership body with expectations of alternative ways of thinking about themselves and their operation, the User Agreement focused the offer of new possibilities on the taxi associations *as associations*. It multiplied the spaces of operation of these associations without negating their existing modalities of pursuing economic opportunity, and hence without deconstructing their existing place and connection bound identities.

The User's Agreement was supplemented with a Management Agreement that illustrates the point even more sharply and suggests that Reid's half-serious proposal for buying-off the traders associations by giving them a cleansing contract was precisely how a regressive logic of belonging should have been managed in the Rockey Street Market. The effort required to manage a taxi rank on an ongoing basis is considerable. The Inner City Office simply did not want the ongoing responsibility of maintaining the Park Central Taxi Rank after construction. Consequently, it put out for tender a Management Company contract that would provide for the daily management of the facility in exchange for a lump sum payment out of income from user fees as well as other revenues (for example, R1 million lease payments from a petrol company allowed to erect a station on the site, billboard advertising, and toilet facility use) being collected by the Rank Committee. At the same time, the ICO facilitated the formation of a closed corporation, *jointly owned by the taxi associations* represented on the Rank Committee, to bid for the contract. The taxi associations' new management company won the tender against a number of other bidders. Today, this new formation receives a large share of the money collected by the Rank Committee and handed over to the Council. The new company seals the rank collaboration by being one more space for becoming.

By all outward appearances, the ICO appears here to have successfully structured a project. It artfully mediated the logics of belonging and becoming in perhaps the most dangerous arenas of particularistic group identification and economic opportunity pursuit in the inner city. The ICO's endeavors suggest a dynamic, even if largely unconscious, process of experimentation with new ways to conceive the objects of urban management in complex and contested urban environments.

SUCCESSFUL EXPERIMENTS IN A NEW COHERENCE?

A recomposition of the relationship between space and social interests is causing a fundamental reformatting of urban identities. A deficit of belonging, wherein agents start to lose hold of traditional places in which they feel secure and known, is calling forth its logical inverse, that is, more frantic attempts to claim rights to places that embody connectedness, completeness, and coherence. Simultaneously, urban actors are seeking pure spaces of mobility in which ties to established territories of self-knowledge and collective recognition are neither compromised by a multiplication of the arenas of action and identification nor limited in the room for maneuver across multiple sites of opportunity. The Johannesburg inner city presents a context in which this pendulum between need for viable spaces of belonging and need for multiple spaces for becoming is beginning to swing ever faster.

The Inner City Office intuitively grasped that government simply cannot mean better and more policing. A government that works by identifying as objects of application the current or immanent transgressions of urban actors, which then require ever more finely tuned prohibitions, is a government destined for failure. Catalogues of government codes and prescriptions only sublimate the sense of chaos in government itself. They represent a predetermined response that does not bring government any closer to knowing what urban actors want, how they arrange themselves and move to get what they want, and how they can be induced to reconstitute their wants and ways of realizing them in ways less potentially damaging to the sustainability of a social body.

The ICO sought to intervene in the explosive milieu of the inner city through a series of intensely risky development projects. Because they aimed to configure institutional forms and arenas of interaction that necessarily required actors to reidentify themselves as more self-responsible and accountable, they implied a renegotiation of the current logics of belonging and becoming. Such negotiations and mediations, as the Rockey Street Market debacle showed, all too easily end up inviting the even harder reassertion of exactly those place-bound identifications that government wishes to supplant.

Where the Inner City Office *did* find success was by seeking to *replicate or build on the true logic of becoming.* Structuring a shared space for a transversal mode of interactions, which did not shut down or close off preexisting place- and connection-bound identities as the condition for opening new ones, was precisely what led the ICO to success in the Park Central Taxi Rank. The Rank Committee allowed the associations and their members to hedge their bets and to retain their existing identities, right down to the sign-languages of street ranking, while multiplying possibilities for the assembly of new interactions. The project suggests an approach to governmentality that respects and confirms multiple existing urban identifications in the process of elaborating new ones more appropriate to the exploitation of fluid new social, political, and economic opportunities.

In essence, the municipality worked well when it facilitated a process whereby different place-bound interests and identities could adapt themselves to each other without feeling that they had to give something up. Through the ICO, it sought to derive new regulatory frameworks from what was actually taking place and to circumvent unhealthy, illegal, or debilitating practices without altering their basic shape and dynamism. It tried to establish coherence not by effecting the clarity and certainty of what people were doing and where they were doing it, but by engineering opportunities for different actors to forge flexible, multifaceted, and multitiered relationships while maintaining their identities, interests, and sense of local authority. In this way, the Inner City Office conceived modes of becoming as its primary object and began to find success.

REFERENCES

Apter, A. 1992. *Black Critics and Kings: The Hermeneutics of Power in Yoruba Society*. Chicago: University of Chicago Press.

Bouillon, Antoine. 1999. "Transition et logiques territoriales en Afrique du Sud: 'Races,' (im)migration, territoires et réseaux." *L'Espace Géographique* 2.

Coquery-Vidrovitch, C. 1991. "The Process of Urbanization in Africa (from the Origins to the Beginning of Independence)." *African Studies Review* 34(1):1–98.

Dunsire, A. 1996. "Tipping the Balance: Autopoesis and Governance." *Administration and Society* 28(3):299–334.

Geschiere, P., and F. Nyamnjoh. 1998. "Witchcraft as an Issue in the 'Politics of Belonging': Democratization and Urban Migrants' Involvement with Home Village." *African Studies Review* 41(3):69–92.

Guyer, J., and S. Eno Belinga. 1995. "Wealth in People as Wealth in Knowledge: Accumulation and Composition in Equatorial Africa." *Journal of African History* 36:91–120.

Healey, P. 1997. *Collaborative Planning: Shaping Places in Fragmented Societies*. London: Macmillan.

Hopkins, A. G. 1973. *An Economic History of West Africa*. New York: Columbia University Press.

Inner City Office, Greater Johannesburg Metropolitan Council. July 1998. *Memorandum of an Agreement of Usage of a Common Area between Greater Johannesburg Metropolitan Council and Alexandra Taxi Association et al*, clause 6.5, p. 7.

———. 2000. *Park Central Taxi Rank*, submission for consideration by the Habitat Best Practices Technical Advisory Committee for the Dubai Best Practices Award.

Kooiman, J. 1993. "Socio-political Governance: Introduction," in J. Kooiman, editor, *Modern Governance*. London: Sage, pp. 1–9.

Leftwich, A. 1994. "Governance, the State and the Politics of Development." *Development and Change* 25(4):363–386.

Martin, P. 1995. *Leisure and Society in Colonial Brazzaville*. Cambridge and New York: Cambridge University Press.

Mattes, Robert, Donald Taylor, David McDonald, Abigail Poore, and Wayne Richmond. 2000. "Still Waiting for the Barbarians: South Africa's Attitudes to Immigrants and Immigration," in David McDonald, editor, *On Borders: Perspectives on Cross-Border Migration in Southern Africa*. Cape Town and New York: St. Martins Press.

Morris, A. 1999. *Bleakness and Light: Inner City Transition in Hillbrow*. Johannesburg: University of Witwatersrand Press.

Osaghae, V. 1998. "Managing Multiple Minority Problems in a Divided Society: The Nigerian Experience." *Journal of Modern African Studies* 36:1–24.

Osita, N. 2000. "The Ability to Squeeze Water from a Stone." *Daily Mail and Guardian,* November 8, 2000.

Reitzes, Maxine. 1999. "Patching the Fence: The White Paper on International Migration." Centre for Policy Studies, Johannesburg.

Robertson, C. C. 1997. *Trouble Showed the Way: Women, Men and Trade in the Nairobi Area 1890–1990*. Bloomington, IN: Indiana University Press.

Rose, N. 1996. "The Death of the Social? Refiguring the Territory of Government." *Economy and Society* 25:327–356.

Schmitz, G. J. 1995. "Democratization and Demystification: Deconstructing 'Governance' as Development Paradigm," in D. Moore and G. J. Schmitz, editors, *Debating Development Discourse*. London: Macmillan, pp. 54–89.

Sinclair, M. R. 1998. "Solidarity and Survival: Migrant Communities in South Africa." *Indicator South Africa* 15(1).

Tomlinson, Richard, Roland Hunter, M. Jonker, Chris Rogerson, and J. Rogerson. 1995. "Johannesburg Inner-City Strategic Development Framework: Economic Analysis." Greater Johannesburg Transitional Metropolitan Council.

Zeleza, P. T., and E. Kalipeni. 1999. "Rethinking Space, Politics and Society in Africa," in P. T. Zeleza and E. Kalipeni, editors, *Sacred Space and Public Quarrels; African Cultural and Economic Landscapes*. Trenton, NJ: Africa World Press.

Rodney Place and ZAR Works, Johannesburg

RETREKS, POST-CARDS (1999)

Post-CARDS were the first salvo in RETREKS—A Metro Allegory. Begun in 1998, RETREKS tracks the emergence of post-1994 Johannesburg through four social viewpoints: yuppie/buppie suburbanites; cowboy and girl calvinists; white tribes of KwaSandton; and new African immigrants. The project proposes—through a variety of mediums and actions, including urban design—a cultural approach to urbanism in South Africa, distinct from the continuing apartheid habit of overcapitalization and global postmodernism.

To date the on-going project has included: The Washing of the Soaps, a dance/performance/video piece; unTITled blURB, a 40,000 copy newspaper; Headline/Chorus line, a dance performance; mall distribution of unTITled blURB; unSUNg CITY, an urban opera staged in a disused nine-story car park in central Johannesburg, featuring a security company parade and sixty-five visual and performing artists and musicians; African Bowl, an urban design proposal for the phased, event-based construction of public space in Newtown, Johannesburg; and Bread City, Coming Trends in Interior Design and Street Wear, a video and fashion show of clothing designed on the proposition, Africa-meets-Infrastructure.

Figure 1

Umlungu! Scene from the blockbuster movie that chronicles the heroic struggle of the Umlungu people. Here the Umlungu take their final stand atop Sandton City, the fortress built on the pastures of Sandton, North of Johannesburg.

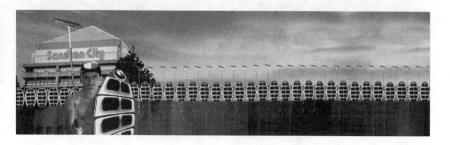

Figure 2

Traditional Life. Umlungu maidens pictured above the Valley of a Thousand Holes, KwaSandton. In an innovative policy, the South African government has established Umlungustans like KwaSandton where minority peoples can continue in their traditional ways and honor their ancient gods and goddesses.

Figure 3

The Blink of an Eye. New South African Television has piloted the use of advanced color reversal technology in programs to appeal to a wider audience.

Figure 4

Hot Pursuit. Kooiseun, West of Johannesburg.

Figure 5

Johannesburg—An African Metropolis. An African metropolis rises on the ruins of a deserted mining town.

Figure 6

Stealth. New livery for the Metro-Neighborhood-Watch Units of the South African Security Forces.

Section III

GOVERNING AND INSTITUTION BUILDING

Resistance to apartheid was resistance to the state. The demise of apartheid required universal citizenship—with voting rights a necessity—and a rethinking of the structure and function of national, provincial, and local governments. This rethinking has been going on for over a decade and has yet to be resolved. A new South Africa requires legitimate institutions.

Heller points out the tensions inherent to the quest for new local governments. Central to these tensions are the African National Congress and civic associations, whose role in the dismantling of apartheid is unquestioned. Civics have become sites of legitimacy for democratic practices and mechanisms for accountability of local government. They have not passed into history. Often in conflict with the ANC, they continue to reflect popular aspirations.

Governance is not only fraught with internal tensions. The HIV/AIDS epidemic poses numerous problems, as Thomas notes, ranging from health issues to housing and service provision. HIV/AIDS is severely altering the demography of the city and creating a large group of seriously ill people (not to mention orphans). Local government struggles to respond to backlogs still in place from the apartheid era even as it faces new demands.

Beall, Crankshaw, and Parnell continue the theme of local politics and its connection to housing and service delivery. They look at how increasing social differentiation in a former township in Soweto is creating new lines of cleavage: between homeowners and tenants and between elderly households and youth. These tensions are forcing local governments to pay more attention to a divided constituency.

In addition, local governments have to adjust to recent laws that give more rights to citizens. Emdon points out how the shift from a parliamentary democracy to a constitutional democracy has led not just to political rights but also legal rights that enable people to protect themselves from discrimination and the arbitrary behavior of landlords and government. Throughout its existence, the apartheid state had maintained a commitment to the law (even unjust ones). Today, those laws have been strengthened and augmented.

This section ends with a chapter on the history of the Johannesburg Art Gallery. Built early in the city's history, the Gallery was an institution white in its conception but never explicitly a part of the apartheid regime. Over many decades, it has served as a "cultural recorder and resource," as Carman describes in detail. To survive, however, it will have to foster strong relations with its immediate environment and the changing social mix of Johannesburg.

9

Reclaiming Democratic Spaces

CIVICS AND POLITICS IN POSTTRANSITION JOHANNESBURG

PATRICK HELLER

The posttransition period in Johannesburg has witnessed a dramatic recon-figuration in the relationship between political and civil society. The African National Congress (ANC) has rapidly consolidated its control by either incorporating or marginalizing the popular movements that brought it to power. In addition, the imperatives of a top-down vision and strategy of transformation have resulted in the increasing centralization and insula-tion of the institutions of municipal governance. A number of observers have read into these trends a decline in the autonomy and vitality of civil society. Indeed, it has become something of a cliché to claim that the civic movement in Johannesburg—celebrated not so long ago as the most vibrant urban social movement in the world—is in complete disarray.

Reports of the death of the civic movement in South Africa are pre-mature. First, as Cherry *et al.* (2000) argue, there has been a conflation of the demobilization of the civic movement with a generalized crisis of civics. Although it is true (and hardly surprising) that civics no longer engage in the kind of broad-based mass actions that marked the height of their power, ample evidence exists that civic associations continue to play an important role in the lives of the urban poor.

Second, many analysts have judged the civics movement against a nar-row and instrumentalist measure of how civil society organizations (CSOs) contribute to democracy; they focus on governance questions and the extent to which CSOs shape state policy and assist state intervention. But CSOs, and social movements in particular, often have their most lasting and democracy-enhancing effects in civil society by promoting horizontal (rather than vertical) ties of association and creating new spaces of voice and participation.[1] Independently of whether such efforts are successfully scaled up (that is, impact on the state), they have the valuable cumulative effect of enhancing citizen capacities and cultivating (or recultivating) solidarities (Cohen and Arato, 1995). The particularly strong brand of fiscal conservatism and technicism that has marked the transformation process in Johannesburg has seen the civic movement displaced from the center to the periphery of the organized political forces reconfiguring the city. Despite this political marginalization, and in part because of it, local civics continue to strengthen citizenship.

Third, social movements are, almost by definition, cyclical in nature. Their strength waxes and wanes in terms of their capacity to mobilize resources (internal and external) and with respect to the political opportunity structure.[2] In reviewing a number of cases of democratic transition, Hipsher (1998) found that in all those in which a dominant posttransition party emerged, urban social movements experienced rapid demobilization. Township civics that once made Johannesburg ungovernable have explicitly abandoned the politics of contention in deference to the authority and legitimacy of the ANC.

The consolidation of representative democracy has led many commentators, as well as the ANC, to question the very *raison d'être* of civic structures. But although the civic movement in Johannesburg is now ineffective and virtually invisible as a *corporate actor,* civics in Johannesburg at the branch level continue to play an important role in community life.[3] Based on research conducted between March and December 2000,[4] we found that a large number, and quite possibly a majority, of townships and informal settlements in and around Johannesburg have active civic branches, most of which are affiliated with the South African National Civic Organization (SANCO).

In this chapter we explore three factors that explain why civics in Johannesburg have persisted. First, civics represent an important means of bridging the gap between communities and the increasingly distant centers of authoritative decision making. Second, they provide important social protection functions in local economies that are exposed to the dislocating effects of liberalization and global integration. Third, they embody a powerful civic imaginary that represents an ongoing search and struggle to define *workable and virtuous* communities against a tide of socially disintegrative forces.

The political position of civics is fast undergoing significant changes. Throughout the transition period and through much of the democratic period, civics in Johannesburg affiliated with SANCO have enjoyed a close and almost symbiotic relationship with the ANC. This relationship reflects the strategic decision by SANCO to forgo the social movement politics of building autonomous civic structures in favor of a politics of incorporation. The latter has led to a unitary, hierarchical, and formal organizational structure designed to maximize SANCO's leverage in corporatist structures. Both politically and organizationally the strategy has failed. But somewhat paradoxically, the decline of SANCO has been accompanied by a revitalization of local civics. The most significant manifestation of this revitalization has been the increasing tension between local civics and the ANC. In many cases local conflict amounts to little more than intraelite struggles for political ascendancy or control over development resources. In most instances, though, this assertion of civic autonomy marks a revival of participatory democratic traditions as a reaction to the increasing centralization and insulation of representative structures. Specifically, this revitalization embraces the idea of a solidaristic, civic community in the face of the clientelization of politics and asserts a public moral economy in the face of the commodification of life chances.

A BRIEF HISTORY

There have been two peak moments for civics. During the mid-1980s the civics reached their height as a *movement*. They mobilized and contested state authority across hundreds of communities with the thinnest of organizational infrastructure. This was a movement in the classic sense: mobilizational, contestatory, and loosely and horizontally organized. Under apartheid, the civics initially arose in direct response to local grievances and functioned primarily as self-help organizations. As one Soweto civics leader explains, "We did not see civics as political structures—the majority of members were not ANC members, although many of the activists were. Organising was primarily around bread and butter issues like leaky roofs, water bills and rent."[5] But the civics soon became the fulcrum of an incipient urban revolt against the illegitimacy of Black Local Authorities.[6] As civics mushroomed across the country, regional civic structures—most notably Civic Association of the Southern Transval (CAST)—provided critical coordination functions. Direct, issue-based local protest actions were scaled-up into political actions that challenged apartheid directly. In this respect, this civics movement transformed local, immediate, and largely inchoate moments of protest and resistance into a cohesive, self-sustaining structure that produced its own distinct modes of contention (the boycotts) and its own ideology and vision of transformation.[7]

Organized movement capacity was rapidly brought into play with the political opening of 1990. The Soweto Accord of that year ended the civics-led boycott and resulted in the establishment of the Central Witswatersrand Metropolitan Chamber (CWM) "which aimed to set in place processes of remedying the apartheid city" (Tomlinson, 1999:8) and in many ways pre-figured national constitutional negotiations (Friedman, 2000). The Chamber became the model for the 1993 Local Government Transition Act, which set the stage for local negotiations based on principles of nonracialism, democracy, and a single tax base to establish new local government structures (Tomlinson, 1999). The civics were so central to this process that Chris Heymans could remark that it was "well-nigh impossible to discuss, plan or implement development in South Africa without engaging with, or at least having to take into account of, civic associations" (quoted in Seekings, 1997:10). However, tensions were already emerging between the ANC and the civics movement. As the role and legitimacy of the CWM expanded, so did apprehensions among ANC provincial leaders that the chamber was a "threat to their own desire to centralise political control of the transition" (Swilling and Boya, 1997:182).

This period marked the civic movement's second peak, one that was distinctly corporatist. With the opening in the political opportunity structure and the unbanning of resistance organizations, the power equation shifted from mobilization to negotiation. In 1992 the civics responded by creating SANCO, a unitary structure designed to centralize the civics movement into a corporatist interest group.[8] The immediate payoff was significant. SANCO was given the lead role in shaping the Local Government Transition Act and granted representation in the peak corporatist chamber, the National Economic Development and Labour Council (NEDLAC). SANCO authored key sections of the Reconstruction and Development Programme (RDP) (the new government's blueprint for economic and social transformation), notably the chapter on housing. Civics were assigned a direct and critical role in the transformation process.[9]

SANCO's corporatist moment was short-lived. Two factors undermined its efforts to institutionalize its influence. First, SANCO never had the organizational capacity to translate a conjunctural opportunity into a sustainable presence. Second, despite popular support, in the absence of a formal membership base SANCO could neither deliver nor withdraw support for government policy in a credible fashion. This problem was compounded by the rapid absorption of SANCO's leadership into ANC and government structures, blurring SANCO's identity and emasculating its independence. The 1995 local government elections dealt the civics a particularly harsh blow as most local councillors were plucked from the ranks of the civics movement. Not only did this virtually deplete the civics regional leadership, but by creating what were presumed to be robust and direct links between communities and local government the need to sustain independent civic structures outside of political society was all but obviated.[10] SANCO's corporatist stature and its ability to influence government policy were rapidly eviscerated, a turn of events most dramatically illustrated in SANCO's failure to stop the government's abandonment of the RDP's housing program in favor of a more market-and-bank friendly policy in 1995.[11] To make matters worse, the endemic violence of the transition period, in which civic leaders and structures were often directly targeted, saw many civic structures lapse into inactivity.

CRITICALLY ASSESSING THE CIVICS MOVEMENT

The civic movement's potential for deepening democracy in the posttransition period can be conceptualized along four dimensions. First, local civics provide a space in which ordinary residents of townships and informal settlements can associate and deliberate around community issues. Not only can common issues and needs be identified, but solidarities can be nurtured. Second, local civics can provide the resources and the framework for collective action, whether this involves self-help activities, various forms of social protection and development, or engaging the state. These roles have the potential to close the institutional and political gap that exists between the state and society and create modes and channels of participation outside formal political society. Third, civics can act as a "watchdog" by monitoring the actions of the state and holding public authorities accountable. Fourth, SANCO can proactively shape and influence policy as an organized interest group.

If the role of the civics movement in the antiapartheid struggle has generally been celebrated, its posttransition role has been the subject of controversy and criticism. Critics have focused on three legitimacy problems. As a peak organization, SANCO was created to scale-up the civics movement. In doing so it exposed itself to the classic dilemma faced by a maturing social movement: the need to manage the trade-offs between the two institutional goals of operational autonomy and political engagement.[12] On this score, the critics have been quite vociferous. Local level SANCO activists and leaders of breakaway civic movements argue that SANCO, as a national organization, has become so hierarchical and bureaucratized that internal democracy has become a sham and branches have lost their autonomy. SANCO has moreover been compromised by its close association with the ANC. In the words of former SANCO stalwart and president of a rival national civic organization, Mzwanele Mayekiso, the problem with SANCO has been "the introduction of a corporate culture into the civic movement, meaning both the imposition of a top down instructions instead of the bottom up approach associated with the participatory democratic culture of the civic movement as well as the introduction of business as an integral part of the organisation, namely SIH (SANCO investment holdings) with its projects that have commodified civic membership."[13]

A second critique concerns SANCO's representativeness and its claim to speak for the "community." Given that SANCO branch members represent only a small percentage of the community and that the communities for which it claims to speak are highly differentiated, with varied and often contradictory interests, SANCO's claims to be representative are specious at best and represent a usurpation of power at worst. The danger to civil society and democracy is significant. The claim to monopoly representation crowds out other interests and forms of representation, a threat made all the more serious by SANCO's close relationship to the party in power. In this role SANCO becomes little more than an instrument for the hegemonic colonization of civil society by the ANC (Friedman, 1992).

A third critique, and the one that has received the most media attention, sees SANCO as little more than a vehicle for local factional or elite interests. Acting as the gatekeeper between the state and the community, SANCO becomes a platform for opportunistic actors to build local power bases and to position themselves for future political or government careers.

THE CIVICS MOVEMENT TODAY

In evaluating the state of the civics, the first observation is that in areas such as Johannesburg where civics have a long history, SANCO is not a civic movement but a *movement of civics.* The character of local civics—that is, branches—has less to do with the formal unitary structures and the chain of command laid down in SANCO's constitution than with local dynamics and configurations. Mayekiso's assertion that the movement has become overly bureaucratized is accurate, but only with respect to the strategic intent of the higher leadership. In practice most SANCO branches operate quite independently of higher structures and have maintained a strong sense of local identity. In Johannesburg, SANCO is often little more than the title taken by existing and very rooted civics (mostly in townships). In other instances it is a useful framing logic and structure for constituting a new local civic. This is especially true in informal settlements.

In comparison to the early 1990s, civics have lost much of their clout. They are no longer capable of coordinated action beyond the local civic. In April 2000 Johannesburg civics did organize protest marches against the banking council to protest conservative bank lending practices, and in November 2000 a number of Alexandra civics marched on Pretoria to protest housing policy. Such extralocal actions are rare. At the local level there are occasional protests, but these receive little media attention and have little political effect. Civics are not in a mobilizational phase. Demobilization, though, should not, as Cherry *et al.* (2000) have argued for the Western and Eastern Cape, be confused with the demise of civics. In fact, Cherry *et al.* found that current levels of support for and engagement with civics are as high today as they were in the past.

Interviews with regional and branch level officials in Johannesburg and outlying areas paint a picture of a precipitous decline of civic activity in 1994–1997, followed by a modest but significant revitalization of civic structures since 1998. Of the six SANCO regions in Gauteng, all except the Vaal have registered a significant increase in the number of active branches over the past two to three years. Johannesburg—the largest of the regions—currently has twenty-eight active branches (SANCO, 2001). Most of the growth has come in informal settlements. Dissatisfaction with SANCO and its close ties to the ANC have also fueled the creation or revitalization of independent civics. After bitter and debilitating internal struggles, the Alexandra Civic Organisation (ACO) has reemerged alongside the SANCO-affiliated civic and at least three other civics. SANCO dissidents—most notably Ale Tleane of the Tembisa Residents Association, Maynard Menu of the Soweto Civic Organisation, and Mzwanele Mayekiso (ACO)—launched a Gauteng chapter of their breakaway National Association of Resident and Civic Organisations (NARCO) in late 2000. To this can be added the mushrooming of concerned resident associations, derisively referred to by established civics as "popcorn" civics.

Many have seen in this proliferation of civic organizations the fragmentation of a once powerful and unified civic movement at the hands of ambitious political operators. Though this is certainly part of the story, it obscures two equally important points. First, this proliferation of civic organizations can also be interpreted as a healthy pluralization of civil society that was the inevitable result of the decline of the unifying logic of liberation politics. Second, although many civic associations may indeed be little more than vehicles for the personal ambitious of local powerbrokers, the most active and visible ones are governed by robust democratic practices and enjoy high levels of community support.

A number of case studies have documented the elite capture of civics. The Community Agency for Social Inquiry (1997) found that in the Soweto community of Tladi-Moletsane the local civic only represents homeowners. Shack dwellers have been left to depend on "a mysterious Mr. K to act as their benefactor." The study concluded that an observed rise in xenophobia and assertions of ethnic identity are the result of a "retreat of some community groupings either into organisations that represent their parochial interests or into clientelist relations with local power-brokers" (cited in Beall *et al.*, 2000:34). Everatt (1999) similarly argues that civics are controlled by petty bourgeois professionals and have been used primarily to defend middle-class homeowners against perceived threats of informal settlement encroachment. Some NGOs, such as Planact, have reduced their involvement with civics because internal fighting, such as in Alexandra, made cooperation impossible. In some communities, SANCO branches continue to stake a monopoly claim. In the Soweto community of Diepsloot, for example, efforts by Planact to build a multistakeholder community development forum were actively and often violently resisted by the local SANCO branch, which saw the forum as a threat to its control.[14] For at least one former civic activist who is now in city government, "SANCO still has the attitude that it is 'the' voice of the community and that if you don't go through them, they won't co-operate."[15] SANCO officials themselves note that in some civics the chairperson behaves "like a chief" and one SANCO organizational document noted that in many branches, "Members form consortia with unscrupulous developers for personal rapid upward mobility and delivery [sic] substandard housing products" (SANCO, 2000a:10).

These sobering assessments represent an important corrective to the often romanticized assessments of the liberation struggle. Not only did many commentators exaggerate the democratic character of civic structures, but they often took at face value civic leaders' claims that they represented the "community." Yet recent assessments also suffer from reductionism. Most notable is the tendency to take well-documented cases of elite capture and gatekeeping and extrapolate them to the whole of the civics movement. The resulting canvass depicts a Hobbesian world in which anomie, violence, distrust, self-interest, and fragmentation predominate. It is a world of clients, not citizens; of strongmen, not democrats. It is a world in which formal institutions (the powers of the state) and informal institutions (norms and values) have all but collapsed. Survival becomes a matter of investing in exclusive, interpersonal, and often extralegal networks of protection and patronage. This marked trend in urban South Africa can hardly be contested.

The other trend is one in which civics (as well as other civil society organizations) are actively resisting the pulverization of civil society (to borrow O'Donnell's [1993] term) by reconstituting communities through democratic and participatory structures. One such example is Beall's (2000) case study of a successful environmental movement in Meadowlands. She found that SANCO was the most prominent community-based organization (CBO) in Meadowlands and that it represented the "more marginalized members of the community." Respondents expressed faith in SANCO as a watchdog and in its ability to "represent their interests in relation to the local councillors and to ensure the latter delivered on their promises" . . . "SANCO was seen as both more accessible and more accountable to the community than local politicians" (SANCO, 2000a:20).

Given the above, there is only one safe generalization: the state of civics today is symptomatic of the state of civil society in Johannesburg—uneven, changing, diverse, and highly differentiated. Moreover, not all forms of associational life promote democracy. Associational ties can be based on exclusive and clientelist exchanges rooted in unequal relations that ultimately promote narrow and parochial interests, just as much as they can be based on more horizontal forms of interaction rooted in non-hierarchical forms of communication and trust and geared toward securing public goods.

CIVICS AND DEMOCRATIC PARTICIPATION

Under apartheid, state repression made it virtually impossible, outside the union movement, to build transparent democratic organizations. Accordingly, despite broad-based popular support, civics never had formal memberships. This posed problems of representiveness and made formal democratic practices difficult (White, 1995). Most civics today, and especially SANCO civics, are membership-based organizations governed by a constitution and formal democratic practices of elected representation and accountability. Office holders are elected at every level of the organization and branch level elections are held every year. Regional officials preside over branch elections and branch officials over zonal, street, or area elections. None of this guarantees that officials will be accountable; it does provide the rank and file with significant leverage and moral authority.

The higher one goes up SANCO's organizational structure, the more influential the ANC. Though the ANC does not intervene directly in elections, candidates for key positions at the regional and provincial level are often "tipped" or selected by ANC structures. The Chairperson of the Gauteng Province, Richard Mdakane, is also the ANC Gauteng Legislator's Chief Whip. Most of the Johannesburg leadership also has close ties to the ANC. The disciplinary powers of the ANC at this level are directly felt. SANCO leaders such as Ali Tleane and Mzwanele Mayekiso who have defied the ANC have been quickly expelled.[16] But at the branch level, leaders seek and sustain their legitimacy by accounting to the community, not to the party. The operational autonomy of SANCO branches is also quite substantial. With the exception of making political endorsements, branches enjoy full discretion of action, including organizing protests. Criticisms of upper structure leadership moreover are fully aired at regional and provincial council meetings.

Based on attendance at ten branch meetings (all in Gauteng and half in Johannesburg), including branch executive meetings, branch councils (in which substructures attend), and mass meetings (open to all community members), local-level democratic culture appears to be robust. Each branch has its own mix of executive, substructure, and open meetings, but on average Branch Executive Committees (BEC) and substructure meetings are held weekly and councils are held fortnightly.[17] In some of the larger branches, heads of department from different substructures also meet routinely. Branch conferences in which the leadership is elected are held yearly. Meeting procedures are fairly uniform, with the reading of agendas, review of attendance, presentation of minutes, and discussion of agenda items as the basic format. Attendance at meetings is very uneven. Weekly branch meetings in Diepkloof, Wattville, Winnie Mandela Park, and the Joe Modise area in Alexandra attract anywhere from 30 to 100 people.

However, community support for civics extends well beyond regular members. When the SANCO branch in Wattville held a mass meeting (the third that year) in November 2000, more than 700 residents attended and sat patiently through three hours of report-backs that included an hour long presentation (and critique) of the government's housing policies. A protest march on council in May attracted over 2,000 participants. Protests in Finetown and Winnie Mandela Park attracted similar numbers. The responses from the focus groups we conducted were especially revealing.[18] Though we selected only residents of townships and informal settlements who were not members of SANCO, virtually all claimed to support SANCO. Though only a small number admitted to having attended a meeting, a majority could provide examples of instances in which they had approached a SANCO official for assistance.[19] When we visited SANCO offices in Diepkloof and Vosloorus we found long lines of local residents seeking help with bills, bond payments, and other problems.

The participatory and deliberative quality of branch meetings is high. Though we have witnessed rather autocratic styles of leadership at higher levels, presiding officers at the branch level are made to answer to fairly robust and clearly highly valued rules of order and accountability. Participation from the floor is built into every agenda item and discussions are often quite animated. Officials who monopolize or divert the discussion are often called to order. Discussions are conducted across the full range of African languages spoken in Gauteng, and we have yet to observe any patterns or language indicative of ethnic affiliations or alliances. The membership and leadership, as far as we can tell, reflect Gauteng's African ethnic plurality.

Most of the branches we visited have fairly homogeneous constituencies, being predominantly either townships or informal settlements. Of the two mixed branches we examined—Alexandra and Wattville—there is ample evidence that local civics represent a cross section of residents. In Alexandra the SANCO branch has significant representation in squatter areas and hostels. Alexandra moreover has witnessed the multiplication of civic structures, with no fewer than four civics currently active. In Wattville—which is mostly a township but also has a significant squatter population—the leadership comes entirely from the homeowning strata. The squatter camp, Harry Gwala, does however have a SANCO subbranch. Its leaders regularly attend branch meetings and are extremely vocal. The branch executive has been actively engaged in pressuring the Benoni city council to provide property titles to shack dwellers and adamantly opposed the demarcation board's decision to locate another informal settlement under a different jurisdiction.

The most observable barrier to equal participation is gender. Although it is widely accepted that women form a disproportionate percentage of SANCO's membership, they are dramatically underrepresented in elected positions. This becomes less true at the branch level where women constitute roughly 30% of elected leaders. Women's participation in branch meetings is quite high, representing at least 60% of those in attendance.

ENGAGING THE STATE

In the early 1990s liberation politics were animated by a hegemonic impulse: "the central problem was that the unity of the 'people' tended to be conceived in terms of an abstract and monolithic 'general will' . . . embodied in a single movement . . . there was a tendency for 'unity' to be imposed from above."[20] In 1990, for example, some civic activists from Johannesburg insisted "that there was no need for local elections because civics already constituted a democratic form of local government" (Friedman and Reitzes, 1995). This hegemonic impulse posed two intertwined threats to civil society. On the one hand, political elites claimed for themselves the right to interpret community needs and, on the other hand, they elevated the strategic and organizational imperatives of resisting the state above the principle of nurturing associational life. The politics of hegemony demanded that the communicative rationality of civil society (deliberation and pluralism) be subordinated to the instrumental rationality of capturing state power. As Shubane (1992:37) has argued, there "are characteristics inherent to liberation movements that militate against the emergence of civil society. This arises fundamentally from the structural limitations imposed by colonial domination and the exclusion of the dominated from the state."

More problematic for the civic movement, at least in its expressed aspiration to be an autonomous organ of people's power, is that this ideological and strategic reflex carried over into the postapartheid period. As early as 1992 Friedman could detect the emergence of a "new hegemony" in which a civics movement aligned to the ANC would act as "a hegemonic power annexing civil society on behalf of the movement, not as a guarantee of its independence" (Friedman, 1992:88). The formation of SANCO in 1992 can be interpreted as an effort to rein in the centrifugal tendencies of independent civics. The ANC itself has never been apologetic about its determination to control the civics movement. In a 1991 discussion paper, the ANC demanded that civics recognize their role as leaders of the liberation movement, asserted its primacy in all matters of political concern, and argued that civics "in a democratic South Africa need not remain as 'watchdog' members of civil society."[21] SANCO quickly and rather quietly accepted these terms in exchange for inclusion in the state. Thus, despite the fact that a large number of civics continued to be involved in the direct negotiations with white municipalities in the CWM, SANCO supported the ANC's call that local government negotiations take place within the process of national transformation.

SANCO's decision to forgo a politics of contention for a politics of incorporation was a reasonable strategic calculation. Social movements are notoriously difficult to sustain. With the demise of the apartheid state, the civic movement lost the unifying and mobilizing frames of resistance to an oppressive, racist state. The civics' intimate ties to the ANC and a transformation project that envisaged a central role for community structures promised to give the movement influence and resources, and hence the means to secure new sources of legitimacy. A movement can engage the state, even to the point of incorporation, without compromising its autonomy.

To do so successfully requires that the movement maintain a credible exit option—the operational capacity to disengage from the state when inclusion is no longer meeting movement objectives. This in turn requires sustaining an independent support base and an independent set of goals and commitments. On both counts SANCO's position has rapidly deteriorated. The problem stems from the complete organizational disconnect between SANCO's higher structures, which are close to the ANC and primarily concerned with exerting political control, and it grassroots branches, many of which have significant support and are willing to engage in contentious actions. The result is a catch-22. Branch-led efforts to organize citywide actions have been actively discouraged and even sabotaged by the region and the province in the name of alliance discipline. At the national level, SANCO has been unable credibly to threaten mass action since 1992 when it threatened a bond boycott.[22] This "fear of rocking the boat" (as SANCO's president puts it) has undermined the credibility of SANCO's claims to mass support and hence its bargaining leverage with the ANC.

To compensate for the loss of corporatist influence, SANCO has relied increasingly on interpersonal ties and a form of elite pact making. This has involved deploying large numbers of SANCO officials to government and extending direct political support to the ANC. When SANCO's National Conference decided in 1997 to allow its officials to simultaneously hold positions in government, it was banking on the advantages of being an insider. It assumed that its deployees could be held accountable to SANCO. Instead the lure of the ANC's internal labor market proved far more powerful. Having forgone the bargaining leverage associated with mobilization, and with it any autonomous support base, SANCO had become almost exclusively concerned with securing its place in political society. In doing so it compromised its historical role and its core capacity *as a movement*. More than anything this explains the precipitous decline of civics in Johannesburg in the 1994–1997 period.

On issues that directly impact the urban poor, rather than mobilizing support through public actions, SANCO has opted to work through its "channels of influence," that is discretely and without embarrassing the government. Not only has this strategy proved futile, but it has preempted efforts to do what regional coordinating structures are supposed to do, i.e., provide conduits for aggregating and framing local sources of moral outrage and protest.[23] Thus, if at the local level the problems of bond and rate payments and the nonperformance of local government remain (along with crime) the most important sources of popular outrage, their political articulation remains inchoate. As a result, at the branch level, SANCO is engaged in a firefighting action. Extraordinary energy and commitment are put into defending households threatened with evictions or cut-offs, but in the absence of regional leadership there is little strategic or programmatic engagement of the issues. Most notably, whereas local branch leaders despair about the lack of access to and accountability of councillors and bemoan the disappearance of local development forums, with rare exceptions none of the leaders we interviewed expressed strong opinions on local government transformation. When asked about the transformation of Johannesburg into a unicity, the standard response was to endorse the "one city, one tax base" principle.[24]

The most serious consequence of this failure to interrogate the role of the state has been the erosion of the institutional infrastructure of participatory democracy. By lending almost unequivocal support to the ANC's transformation agenda, SANCO has endorsed not only the *substance* of transformation (where there is much common ground) but also the *modalities* of transformation. These modalities have increasingly been marked by an emphasis on neoliberal and neomanagerial criteria of delivery, including a heavy reliance on technocratic forms of decision making. As one commentator has noted, "The push towards technicism has resulted in struggle NGOs and mass-based organisations losing their previously clear political direction" (Meer, 1999:112). In a context of fiscal constraint and tighter regulatory frameworks, outsourcing, and increasing technocratic dominance, community-driven and politically negotiated initiatives have been marginalized (Khan, 1998).

Similarly, by privileging the role of a ruling political party SANCO has invested in the media of authority and discipline (the modalities of state power) rather than in the media of communication and contestation, which are the powers (of persuasion) of civil society. Contentious politics have given way to the politics of bargaining and lobbying. SANCO's pact with the state has been to deliver, not to build democracy. Although SANCO branches retain a strong and potentially democratizing presence in civil society, the institutional terrain through which participation was to be given substance is rapidly shrinking.

THE EROSION OF PARTICIPATORY SPACES

Between 1991 and 1994, the CWM was described as "the most consultative policy development process for a single metropolitan area that has ever been conducted in South Africa" and was applauded by the World Bank as unique in the developing world (Swilling and Boya, 1997:182). But as the city has progressed though its multiple rounds of reorganization, the decision-making process has become increasingly insular, driven by small committees of high-level technocrats and private sector consultants. SANCO denounced the Transformation Lekgotla (or meeting) appointed in 1999 as an unconstitutional structure that reduced elected representatives to mere spectators. The formulation of Johannesburg's overall strategy (*iGoli 2002*) for long-term development has also been contested. Though peak-level consultations were initiated, public employee unions and SANCO withdrew in protest over the Municipality's unilateral decision to privatize public services. Repeated protests from SANCO's Gauteng office against credit control measures, the failure to introduce an effective system for registering indigents, and tariff increases for water, sewage, and electricity have fallen on deaf ears. The regional secretary of Johannesburg even goes so far as to compare the city's decision-making style with "the old system."[25]

The city has opted for a transformation trajectory that is essentially a neo-Thatcherite model calling for reducing the municipality's role to "core" functions. In parallel will occur the expansion of private sector provision based on a "purchaser–provider" contract management model "that rests purely on the assumption that everything can be managed by contracts, financial controls and performance management" (Swilling, as quoted in Tomlinson, 1999:27). The downsizing of the state to its neoclassical incarnation as nightwatchman marks a clear rupture with the vision of integrated local development laid out in the Constitution and the Local Government White Paper. Technical and economic merits aside, the model's political impetus is of a classically high modernist inspiration (Scott, 1998). In the high modernist worldview, state managers have an unbounded faith in the ability of experts to apprehend and transform the world.[26] In this light, the primary effect of reducing local government to a contractor of services and making fiscal principles of cost recovery the key measures of good governance are political: citizens are reduced to customers, and democratic principles of accountability (including participation) are replaced with market signals.

If SANCO has been marginalized from the decision-making process at the city level, the situation on the ground has been even more dramatic. Local governments in South Africa are mandated by law to engage in two separate but overlapping planning exercises: the preparation of Land Development Objectives (LDOs) under the Development Facilitation Act and the preparation of Integrated Development Plans (IDPs) under the Local Government Transition Act. Both processes were designed to be consultative and to engage civil society. In Johannesburg, the process has been a largely top-down affair dominated by technocrats that has afforded few, if any, opportunities for meaningful community participation. When local councils submitted their LDOs in 1997, those that had adopted a technocratic approach were accepted whereas those that had relied on more participatory approaches were rejected (Bremner, 1998). Bremner (1998) concludes that "because [the process] is seen as imposed by both political and administrative officials [it] has been left to the planning departments of the metropolitan and local councils to co-ordinate and manage (118)."

Interviews with civic leaders in Johannesburg paint a picture of increasing exclusion from the planning process whether in its intensive IDP phase or in various ongoing fora and steering committees. Not a single civic official claimed to have made effective use of participatory structures. The problems ranged from criticisms of the top-down and technocratic nature of the process to conflicts between civics and local government and politicians that resulted in outright exclusion.

In one case civic organizations were invited to participate, but largely to review plans that had already been elaborated by line departments and consultants. The problem was only exacerbated by open tensions between the SANCO branch and ANC councillors. As the SANCO branch chair explained:

> At first we did participate in Community Development Forums (CDFs) at the ward level which were supposed to submit to the LDO. But the CDFs were controlled by ANC, and individuals who wanted to enrich themselves. The councillors did not follow the recommendations. When SANCO became more involved, the ANC scuttled them [the CDFs].

Even where ANC–SANCO relations are good, the LDO process has been a dismal failure. Diepkloof was the first community in Soweto to establish a civic. Today it remains a very active branch with a permanent office and an estimated membership of 11,000. Its chair, Vuyisile Moedi, a young, energetic, and extremely well-informed activist, was nominated as ward candidate for the ANC in the December 2000 elections. He won with 94% of the vote, the third highest winning margin in Johannesburg. Yet when asked his experience with the LDO he explained:

> The process was hijacked by consultants. Consultants were called in to make things easier. But it became a nightmare. The community submitted a list of priorities, but it was not taken into consideration. In the end the LDOs went their own way. . . . We were so frustrated by experience that we stopped attending meetings and wrote letters of protest. The consultants betrayed the aspirations of the people.

A regional official echoed this view:

> You go to meetings and the well paid consultants have done all the research and have all the answers. There is no place for the community leader to intervene. The problem is that councillors, officials and consultants are all more capacitated than community leaders. So in the end you get a document that has a list of participants, but if you go through the document you won't see where there has been any real community input.

Given the virtual collapse of formal participatory spaces, SANCO's ability to influence local government has been limited to working through representative structures. Yet even these channels of influence have become increasingly subject to centralized authority, especially under the new executive mayor structure. The medium of influence in Johannesburg has shifted decisively in favor of structural power. One has but to consider the immediate effects of a white residents' rate boycott, redlining by banks, threats of capital flight, or the emigration of professionals to underscore the weak bargaining position of SANCO and other black civil society organizations.

RECLAIMING DEMOCRATIC SPACES

The failure of SANCO's politics of incorporation coupled with the shrinking of institutionalized participatory spaces has made effective engagement of the state increasingly difficult for the civics movement. Yet civics continue to play an important role in many communities. They do so neither through control or delivery of development nor through large-scale mobilizations, but through far more prosaic, even mundane interventions. Most of these activities can be conceived as efforts to bridge the gap between community needs and state action. Broadly speaking, SANCO's local level activities can be grouped into three clusters: providing brokerage services, advocacy and watchdog functions, and conflict mediation.

Across a wide range of issues, SANCO officials and volunteers provide assistance and guidance to all residents in addressing individual and community complaints. These brokerage functions are provided either as direct assistance or advice to an individual or group or by engaging the relevant authorities on behalf of aggrieved parties. The most common services provided in Johannesburg are assistance in dealing with billing problems and cut-off threats, renegotiating bonds, and bringing (and following up) cases to the police. In informal settlements, civics assist residents with securing title deeds and registering for government benefits.

The representation of community interests can also assume a more contestatory character. Civics often get involved in bargaining for services or exposing the poor performance and corruption of elected representatives and government officials. These activities involve competition with other interest groups for scarce resources and often involve challenging the authority and probity of local government officials.

The most successful civic-led interventions that we documented in Gauteng were instances of responding to state failures. In Winnie Mandela Park and Finetown (both informal settlements), SANCO branches identified the need for a secondary school (in each case schoolchildren were traveling long distances to school and were often the victims of crime) and lobbied the Department of Education. When these appeals fell on deaf ears ("We got the duck and dive treatment"), the SANCO branches raised money and labor to construct or rent facilities and then identified volunteer teachers and principals to staff the schools. The schools are now operating and SANCO continues to lobby the department for support.

In addition to these brokerage and advocacy functions, many SANCO branches, and most of its subbranches, also take on significant conflict mediation functions. These are not the people's courts of the 1980s.[27] Officials do not sit as judges, instead they simply hear disputes brought to them by complainants and limit their interventions to providing advice or referrals. The vast majority of the disputes involve domestic issues or minor conflicts between neighbors and are handled by subbranch structures. These meetings are particularly well attended and point to the fact that civic leaders enjoy significant legitimacy and respect in the community.[28]

Many complainants are actually referred to SANCO by the police. Focus group respondents repeatedly expressed a preference for "first trying to solve the problem as a community" and "resolving disputes as neighbors" rather than taking matters directly to the police. One mother noted that when it came to dealing with children involved in petty crimes, "I prefer SANCO because it disciplines in a parental manner unlike the law." Another township resident explained: "SANCO . . . helps us with our youth when they've been involved in crime. Instead of getting them arrested we take the matter to SANCO . . . we meet together and the youth get a scolding."

Brokerage and conflict resolution functions have long been the bread and butter of the civics movement. That this role has persisted, despite the ebbs and flows of the civic movement, attests to the degree to which civics have gained a significant institutional presence in many communities.[29] It also points to the extent to which the postapartheid state has failed to bridge the gap between communities and government. SANCO's brokerage role remains important because of the distance and insulation of local government and the difficulties ordinary residents have in interfacing with local authorities. Government bureaucracies and service providers (e.g., of electricity) are distant and often user hostile, and many residents have neither the skills nor the resources to effectively engage them. As one shack dweller succinctly noted, "They (SANCO branch officials) are our mouthpiece to government." Another said that "SANCO negotiates better for us than if we go there personally. If SANCO goes as SANCO we get quick responses."

Support for civics is clearly tied to a critique of the effectiveness of formal representative structures. When we asked our focus groups why they would turn to SANCO rather than their councillors, one participant responded:

> We don't even know where he (our councillor) stays (laughter). When you go to look for him you are told he stays in the suburbs. With SANCO it is better because we live with these people in the informal settlements.

Another added:

> It's different with SANCO officials because they are easily accessible. Your presence applies pressure whereas councillors don't feel the pressure because we do not see them. With SANCO, if they've made a promise and don't report, you are able to call them to a meeting after a week and ask for feedback. If you've reported something to him [the SANCO official] he will see that you are desperate, he must feel the pressure because he is your next door neighbour. Every time he steps out of his door he remembers that this person has made this request.

Although residents and SANCO leaders remain overwhelmingly supportive of the ANC, they have quite clearly become increasingly critical of the logic of party politics. Councillors, they argue, are primarily interested in advancing their careers and are as such more accountable to the ANC than to the community. As for the ANC they see it as more concerned with rewarding its supporters (examples of patronage and nepotism are readily given) than with promoting development. That access to power breeds self-interest ("from the struggle to the gravy-train" in the words of one local activist) is hardly a new idea. What is new is that communities or, more accurately, those outside of political society have become increasingly skeptical of the liberation-era claim that emancipation and redistribution can be achieved through the instrumentalities of political power. This is reflected in the view (forcefully expressed in focus groups) that what defines civics is their nonpartisanship and that civics should not, accordingly, play a role in electoral politics.[30] One might, then, cautiously point to the emergence of an *autonomous politics of civil society,* characterized by a rejection in practice (if not yet in officially stated positions) of ANC hegemony and the assertion of the necessity of more participatory forms of democracy.

Concretely, the "spaces" that such a politics has created are reflected in two strong empirical trends. The first is the pluralization of civil society. The weakening of the ANC's ideological hegemony coupled with accelerated socioeconomic differentiation has given rise to new interests, identities, and demands and spawned the creation of a range of new associations, including special interest groups, concerned resident associations, and non-SANCO civics. SANCO's once facile claim to representing "the community" is now hotly contested. Not a single respondent claimed that SANCO should have exclusive representation in development fora or exclusive bargaining powers for the community. Most readily endorsed the principle of multiple-stakeholder representation and we were provided with multiple examples of SANCO's branches working closely with other local associations. A regional official even remarked that "Popcorn civics are right to criticise SANCO for its lack of autonomy. . . . In the East Rand having opposition is healthy. It has kept SANCO on its toes."

The second development has been an increasing rift between the ANC and SANCO at the branch level. In some cases the rift has been occasioned by little more than power struggles, with local faction leaders using SANCO branches to leverage their position with the ANC. These are clearly cases in which communities are pawns in a struggle for political power. Predictably, this trend accelerated during the run up to the local government elections. In most cases, the tension arises from the frustration related to the state's disengagement and the ANC's failure to take up local grievances. In the informal settlement of Winnie Mandela Park, SANCO officials have documented corruption by local ANC branch officials who have been taking bribes for allocating serviced plots. When repeated demands to the Provincial ANC office and government officials for an investigation were ignored, the civic organized protests and even investigated the possibility of obtaining a court order to stop the development project. In the informal settlement of Ruth First, the local SANCO branch has challenged the dominant and exclusionary role of a private developer and the local ward councillor in supporting a housing project for the community. In both cases, the civics effectively expanded the democratic space by challenging the ANC's gatekeeping and the top-down logic. In almost every branch we have investigated, including Alexandra, which is the most pro-ANC branch in Johannesburg, the ANC's hegemony is being openly challenged.

Even when SANCO is politically inseparable from the ANC, the sheer centrifugal pull of local grievances and dissatisfaction with the performance of local government have created new sources of contention and new opportunities for autonomous action. Civics that are well organized and have community support are doing much more than providing stop-gap services. By facilitating citizen participation and deliberation they are constituting a *civic community*—that is, a shared imaginary of the *virtuous* community nurtured through horizontal forms of association and communication (Chipkin, 2000).

Our focus groups revealed the persistence of a powerful civic imaginary and, specifically, a manifest desire to identify and address common interests against a backdrop of increasing social disintegration and political fragmentation. Respondents spontaneously equated the "civic" with the "community" and indeed repeatedly noted that civics were an active force "in uniting the community." This is reminiscent of liberation politics, with a key difference.[31] Whereas in the past being a member of the community automatically meant being a member of the civic (at least in the later stages of the struggle) and given that the binary categories of "us" and "them" implied a necessary and inescapable political affiliation, there is no such conflation today. Though none of our focus group respondents was a SANCO member, almost all expressed support for "the civic" and emphasized that assistance from the civic did not require membership. The civic is valued because it is "a home to all" and residents hold very strong opinions about insulating civic affairs from political interests.

Much of the popular support for civics and much of the commitment of its activists flow from moral outrage at the wrenching effects the market economy is having on the urban poor. Civics resist the transformation of citizens into clients and consumers by providing a modicum of protection to socially and economically vulnerable categories—pensioners, the unemployed, those without land rights—who are exposed to the vicissitudes of the market. Bargaining for those who cannot pay their bonds, arguing for fair tariffs, demanding housing for the needy, and providing assistance to a family in crisis are all elements of defending a public moral economy. As communities have found themselves increasingly excluded from formal politics and authoritative decision making and increasingly subject to state predations in the form of cost–recovery measures, forced displacements, and rent seeking, the locus of political activity has shifted from representative structures to direct democracy. Of course, in the absence of more robust and inclusionary institutions of state–society engagement, these forms of public politics are difficult to scale-up and generally have little impact on public policy. But this should not blind us to the important role that participatory spaces can have in empowering residents *as citizens* and mending (through community-embedded mediation) relations *between* residents.

This is reflected in the fact that residents are more likely to attend civic meetings than ANC meetings. Residents point out that ANC branch meetings address only "party matters" and that executive members "don't want to address community issues because they might have to criticise the local ANC council." In contrast, SANCO meetings serve two dominant purposes. First, branch officials provide a wide range of information to residents, including both general information about government policies and specific feedback on SANCO activities. Second, the meetings serve as sounding boards and rallying points for popular grievances. At a meeting we attended in Winnie Mandela Park, every zone was given an opportunity to report. What followed was a litany of complaints, including accusations of bribe taking by community liaison officers, broken water pipes, the blockage of sewage systems, and confusion about the allocation of toilets. Various SANCO officials were then mandated to take up these issues.

The conflict mediation role of SANCO also fills a critical gap in this respect. Given the crowded conditions in townships and informal settlements, the unevenness of property rights, the continuous in-flow of new residents, and the pressures on common and public resources, petty conflict (not to mention criminality) is endemic. Few, if any, states have the capacity to effectively regulate or manage such fluid conditions. Community-based mechanisms of mediation thus remain critical and represent an important countervailing force to atomized or clientelized modes of intermediation. Much the same is true of SANCO's brokerage functions. They are a direct response to state failures and strengthen the collective rights of residents.

As long as civics are subject to community-based accountability, do not enjoy state-sanctioned advantages in representing community claims, and do not resort to extrademocratic means in competing with other associations for support, they clearly enhance democratic associational life. The intermediation functions of civics, moreover, have two specific democracy-enhancing effects.

First, by mediating conflicts within the community and serving as public spaces for the assertion of community values, civics contribute to producing what Chipkin (2000:14) calls a *virtuous* community. Communities are reclaiming for themselves a vision of the good society that challenges the crime and conflict within their midst and their transformation into atomized clients and consumers. The referent of "the civic" is critical; it invokes the legitimating principle of modern citizenship, rather than traditional or charismatic authority.

Second, civic activity is an alternative to clientelization. State disengagement from civil society provides powerful intermediaries with opportunities for securing client access to government or scarce resources in exchange for political loyalty (Bratton, 1994). Within South Africa, continued support for chiefs in rural areas and shacklords in informal urban areas is a perfect example. Such vertically organized forms of brokerage compromise associational autonomy and undermine civic life (Fox, 1994). In this respect, the fact that the brokerage and mediation services that civics provide are available to all residents, irrespective of political affiliations, and are as such public goods, takes on new meaning. To the extent that civics provide these services without demanding political or organizational loyalty in exchange, they are creating an alternative to clientelism and expanding the scope of associational life.

TOWARD A NEW POLITICS?

More than ever the ANC subscribes to a hegemonic view of civil society. Governance, not democracy, is the challenge, and the role of civil society organizations is to provide support to the government's transformation project. Delivery, not deliberation, is the order of the day. As former Soweto civic activist and now mayor of Johannesburg Amos Masondo explains:

> The watchdog idea is still very strong in the civic movement. But it does not work. You can criticise from the sidelines, but in the end the community will judge you on the strength of what you have delivered. Of course there is also a role for civics in promoting democracy, but what matters is what is practical.[32]

Politically the ANC's hegemonizing impulse is reflected in its equation of state power with people's power and its strategy of subordinating independent arenas of popular action to political control. A key ANC Gauteng official recently deplored the "dichotomy between political and civic matters" that is implicit in the very existence of SANCO and called for ANC branch committees to supplant SANCO by engaging directly in civic activities (Makura, 1999:17). A 1999 ANC Gauteng document is even more explicit in its colonizing logic: "There are regions and specific localities in our country in which, for instance, the ANC does not enjoy hegemony. In cases like this, politically non-aligned civics might be appropriate organisational forums alongside of our branch structures . . . [But] unless we convert ANC branches into [a] more civic type function . . . we will have produced a generation of 'cadres' whose experience and understanding of 'politics' will be very alien to our Congress tradition" (ANC, 1999:10).

Many ANC leaders have repudiated this position, but if the list nomination process that preceded the December 2000 local government elections is any indication, vanguardism is alive and well in the party. In principle, ward councillors were to be nominated at branch level meetings. Though the distribution of delegates heavily favored the ANC, alliance partners were given representation (10% for SANCO, SACP, and COSATU each). Given that many ANC members are also SANCO members, there was a real possibility that SANCO could muster majorities. In many branches throughout Gauteng this was indeed what happened. Yet in the end, SANCO nominations were rarely accepted. In some cases, ward conferences were hijacked by the ANC through procedural manipulation. In most cases, the decision was simply made at a higher level. Most analyses of the criteria ultimately used range from outright nepotism and favoritism to a systematic process of selecting only councillors with the appropriate credentials and capacities.[33] Local popularity was not an important consideration, a point that even a key ANC Provincial official subsequently acknowledged and deplored.

Civic leaders in Johannesburg are quite bitter about their treatment by the ANC. A standard refrain is that "at the local level, there is no alliance." Many branch leaders in Gauteng were so indignant about the autocratic manner in which the ANC nominated candidates that they openly challenged SANCO's National Executive Committee call for supporting ANC candidates.[34] The Johannesburg regional leadership however remained steadfast in its support for the ANC. In contrast, the East Rand region complained publicly to the Provincial list committee and demanded an investigation into blatant irregularities. The chair of the East Rand region explained that if SANCO were to campaign for ward nominees that do "not have the support of the community" it would lose its credibility. "A lot of dead horses have been nominated. I am not going to campaign meetings to support dead horses."

The cement of SANCO's close relationship with the ANC has always been interpersonal networks. The credibility of a hegemonic front and the strategic rationale for the politics of incorporation were secured through the deployment of SANCO's leadership into ANC and government positions. In many areas there have been concrete personal and political payoffs for working closely with the ANC.[35] These network ties, though, can no longer support the contradictions of the partnership. The most obvious problem is that the internal labor market has reached a point of saturation. The reduction in the number of councillors with the redrawing of local government boundaries only aggravated an already precarious situation. One solution to this problem has been to sponsor independent candidates, as the Eastern Cape did in the December 2000 elections. Even in an election marked by a decline in support for the ANC, machine politics still guaranteed the defeat of all candidates challenging for the African vote. Not a single SANCO-supported candidate in the Eastern Cape was elected.

A second response has been to propose that SANCO withdraw entirely from competing for political positions. The rewards of ANC loyalty are increasingly viewed as incompatible with accountability to SANCO. The idea that SANCO deployees can effectively represent the organization's goals within government—the "two hats" position—is now widely disparaged: "how can you bite the hand that feeds you?" is the common refrain. This has occasioned the emergence of what might be called an "autonomy" faction within the Gauteng leadership.

The Gauteng leadership first took a public stance when it endorsed a document drafted by SANCO's president. Entitled a "Strategy Discussion Document to Radically Re-Shape the Vision and Role of SANCO," the document (which SANCO's national executive committee had tried to suppress) rejects a politics of incorporation and argues that movements should "not have aspirations to be in power." Although movements and the state can complement each other, "more importantly, the power of the people and the power of the state must also contradict each other, so that the balance of forces is tilted towards the people" (SANCO, 2000b:15). The document cautiously argues that the ANC does not have a monopoly over the

National Democratic Revolution and that social movements must have an independent political role: "The lack of appropriate opposition to the ANC is a clear indication that the real political opposition can be found in the grassroots mass movements of South Africa" (SANCO, 2000b:15).

The Gauteng commission applauded the document for being the first "written by SANCO" and not the usual documents "released by the alliance, which proffers (sic) to offer advice to SANCO." At the urging of the commission, the Gauteng Provincial Conference in March 2000 passed resolutions calling for constitutional changes that would bar SANCO officials from holding government and ANC positions. (The position was fully endorsed at SANCO's April 2001 national conference, though the actual constitutional amendments are still pending.) With hindsight, the civics movements' decision to embrace the ANC's hegemonic view of transformation is now being openly questioned. As one of the most prominent leaders of the Alexandra Civic, and current chairperson of SANCO's Gauteng Province candidly explained,

> When we came to power we ignored social movements because we assumed that state would deliver. The state had a lot of legitimacy, so there was no need for social movements . . . [But] it was a mistake to disband local civics. People were attached to names—TRA, ACO, etc. . . .—and disbanding them created a lot of confusion. By creating SANCO we opened a vacuum for people to set up their own structures. We were too hasty. We should have gone with a federal, not unitary structure.[36]

As a movement, SANCO is clearly caught between significant but uncoordinated assertions of autonomy and a legacy of engagement with the ANC. Defining a more autonomous position would mean a significant rupture of networks ties and ideological affinities and require articulating a clear and focused alternative to the politics of incorporation. This is precisely what SANCO's principal competitor, NARCO, has done. SANCO, however, remains far too entangled, politically and ideologically, with ANC structures for such a clean break to occur.

If we take movements seriously, the importance of such strategic shifts in direction must not be given too much weight. To Lenin's question of "What is to be done?" the answer is (to borrow from Ferguson, 1994) that it is already being done. As we have seen, because of the degree to which they remain deeply embedded in communities, and in many cases have nurtured community self-representation and action, civics are already carving out autonomous political spaces.

CONCLUSION

The history and current state of the civics movement in Johannesburg are testaments to how rapidly powerful urban social movements can be demobilized in the aftermath of negotiated transitions to democracy. Despite the fact that the transition was the first in history in which negotiations started at the level of local government (Swilling and Boya, 1997), the popular sectors, and most notably civics, have subsequently been excluded from Johannesburg's transformation process. One must not confuse what has transpired in political society with the state of civil society, however. Civics in and around Johannesburg still enjoy a high degree of legitimacy, both as an incarnation of popular aspirations for a *virtuous* community and as a structure of democratic participation.

Because Johannesburg is the standard bearer in the ANC's modernist aspirations of global competitiveness and designated would-be "world class city," the impulse to implement a top-down technocratic transformation insulated from politics has been especially acute. The resulting efforts to exert political control over autonomous civil society organizations and the collapse of institutions of state–society engagement have threatened the very existence of civics. That they have survived underscores three points. First, given the unevenness of local government capacity and the considerable difficulties that ordinary citizens have in engaging democratic authorities, civics still have a critical role to play as brokers and interlocutors. Second, the existence of large numbers of hard working and committed activists is a testament to the fact that though the political context has changed dramatically, the voluntarism and sense of political engagement born of the years of struggle remain important motive forces. Third, the very idea of the civic is powerful in the popular imagination as an expression of solidarity and self-help under trying and desolidarizing circumstances. It remains powerful because it resonates with popular aspirations for an inclusionary and participatory democracy.

NOTES

1. For an extended discussion with reference to the Indian case see Heller (2000).
2. Tarrow defines the political opportunity structure as the "consistent—but not necessarily formal or permanent—dimensions of the political environment that provide incentives for people to undertake collective action by affecting their expectations for success or failure" (Tarrow, 1994:85).
3. Cherry et al. arrive at a similar conclusion. On the strength of surveys conducted in an Eastern Cape and Western Cape township, they conclude that though the level of mass mobilization has predictably declined, "there continue to be high levels of popular engagement with self-governing civic structures at the local level" (Cherry et al., 2000:1). Working in a Soweto community, Beall (2000) also found high levels of support for the civic.
4. The research on which this chapter is based was conducted for the Centre for Policy Studies and was carried out with the indispensable assistance of Libhongo Ntlokonkulu. We conducted inter-

views with national and provincial level South African National Civic Organizations (SANCO) officials, government officials who have interacted with SANCO, and international donors. We examined three Johannesburg branches in depth and conducted interviews with officials from six other branches, including two non-SANCO civics. We also attended SANCO meetings at every level of the Gauteng organization—provincial, regional, and branch—including executive meetings and public meetings. For a full report, see Heller and Ntlokonkulu (2001).

5. Interview with Amos Masondo, July 24, 2000.

6. The strength of the civic movement is most dramatically reflected in the wide-based support for rent and service boycotts. In Soweto, for example, 80% of formal rent-paying households withheld payments for four years (Swilling and Boya, 1997:181).

7. For the most extended treatments see Seekings (1997) and Zuern (2000). For one of the most insightful case studies written by a civic activist, see Mayekiso (1996).

8. The circumstances of SANCO's formation were highly contested. Many local civics were concerned that subordinating themselves to a national organization and a single constitution would compromise their autonomy. Some civics, especially those with long histories and strong bases of support, were specifically concerned that SANCO's ties to the ANC would compromise the civic and specifically nonpartisan character of the movement. The Alexandra Civic Organisation in fact moved that SANCO should be a federal structure and that individual civics be allowed to keep their own constitutions and raise their own funds (Mayekiso, 1996).

9. The RDP promised that "Social Movements and Community-Based Organisations are a major asset in the effort to democratise and develop our society. Attention must be given to enhancing the capacity of such formations to adapt to practically changed roles. Attention must also be given to extending social-movement and CBO structures into areas and sectors where they are weak or non-existent" (quoted in Bond, 2000:95).

10. SANCO Gauteng Chair Richard Mdakane estimates that 75% of elected ANC councillors in 1995 had civic backgrounds (Interview). In her exhaustive study, Zuern (2000) puts the figure at 80%.

11. For a detailed account, see Bond (2000:Chapter 4).

12. Foweraker, as cited in Lanegran (1996).

13. Personal communication, October 24, 2000.

14. Interview with Hassen Mohamed, April 18, 2000.

15. Interview with Laurence Boya, October 30, 2000.

16. Mayekiso was expelled in part for publicly criticizing SANCO's decision to endorse Mbeki as President even before the ANC had officially nominated him (Zuern, 2000:223) and Tleane was expelled when he sided with the Tembisa Residents Association (TRA) against the ANC council in supporting a flat rate for tariffs.

17. The Johannesburg branches in which we attended meetings were Alexandra, Mdeni (Soweto), Winnie Mandela Park, Ruth First, and Finetown.

18. Each group consisted of nine to twelve same-sex participants, ages 25–40, and all met in November 2000. All participants were Johannesburg residents (predominantly from Soweto) who were selected on two criteria: they knew of a SANCO branch in their community and they were not SANCO members. Two of the focus groups were drawn from informal settlements (shack dwellers) and two from townships (house dwellers).

19. In their survey research, Cherry et al. (2000) also found high levels of support for civics despite lack of formal membership. In the Cape Town township of Guguletu only 14% of residents said they were members, but twice as many had attended branch meetings recently and 58% had attended street committee meetings.

20. R. Fine quoted in Friedman and Reitzes (1995:6).

21. As interpreted by Lanegran (1996:114).

22. The threat was withdrawn under pressure from Nelson Mandela. A call by the SACP in late 2000 for protests against the failure of banks to extend housing credit to low-income families was endorsed by SANCO, but participation was limited to a few Provincial officials.

23. When we asked a Diepkloof branch official what support the regional structure provided he gave a fairly typical response, "Don't make me laugh. We don't get any support. We don't even get correspondence from the region."

24. SANCO's current president has articulated an important critique of current tendencies toward political centralization and technocratic domination, but these views have not been diffused to the media, and have not been communicated successfully to lower structures.

25. Interview with Siphewe Tusi, October 23, 2000.

26. For an extended discussion see Heller (2001). As applied specifically to Johannesburg, see Friedman (2000).

27. As Zuern has pointed out, the people's courts of the 1980s were established by civics in response to the vigilante courts set up by gangs and kangaroo courts set up by councillors. The state tried to demonize people's courts, but all the evidence shows that violent, summary justice came from vigilantes and councillors, not from the civics (Zuern, 2000:110).

28. In the Joe Modise area of Alexandra conflict resolution meetings are held weekly and attended by seventy to eighty persons. Similar weekly meetings in Wattville attract fifty to sixty people. Most of the other branches we visited tend to hear cases during regular open SANCO meetings. Among the cases we heard was an elderly man complaining about the drug abuse of a nephew and his threatening behavior; a young, married couple who were fighting over the fact that he had moved into his sister's household; and a family of sisters who were disputing their deceased mother's inheritance amid accusations of witchcraft.

29. Zuern (2000) makes a similar point.

30. SANCO's leadership, having long espoused the politics of inclusion, has predictably taken much longer to come around to this view. Many branch leaders we interviewed insisted that they would not provide active support to ANC candidates who did not have community support. Nonetheless, most of the SANCO leadership continues to espouse the contradictory view that SANCO can be a "home to all" and at the same time provide electoral support to the ANC.

31. I owe this observation to Ivor Chipkin.

32. Interview with Amos Masondo, October 16, 2000.

33. The ANC has defended this position on the grounds of needing to improve the caliber and performance of councillors. That it also serves as a means of exerting more centralized control over the party and shutting out challengers with independent bases of support need hardly be emphasized. When a very popular SANCO leader from an informal settlement was rejected as the ward nominee by the provincial list committee, a SANCO official explained that the "ANC doesn't like populists who can stand up and challenge the leadership. They did not like this guy because he is too consultative—he takes everything back to meetings."

34. Information based on attendance at Gauteng General Council meeting, September 17, 2000.

35. In the Pretoria region for example, a factional spit within the ANC created a vacuum that has been filled by SANCO. Among other gains, this gave SANCO the power to call for and secure a Credit Control Summit in which local government's credit control measures were significantly renegotiated. In the December 2000 local government elections SANCO more or less took charge of the nomination process, and a large number of elected councillors are from SANCO.

36. Interview with Richard Mdakane, October 3, 2000.

REFERENCES

ANC. 1999. Gauteng Province: State of the Alliance—Unity in Action.

Beall, Jo. 2000. "Where the Dust Settles: Limits to Participation and Pro-Poor Urban Governance in Post-apartheid Meadowlands, Johannesburg." Unpublished manuscript.

Beall, Jo, Owen Crankshaw, and Susan Parnell. 2000. *Urban Governance, Partnership and Poverty: Johannesburg.* Working Paper 12. University of Birmingham.

Bond, Patrick. 2000. *Elite Transition: From Apartheid to Neoliberalism in South Africa.* London: Pluto Press.

Bratton, Michael. 1994. "Peasant-State Relations in Postcolonial Africa," in Joel Migdal, Atul Kohli, and Vivienne Shue, editors, *State Power and Social Forces: Domination and Transformation in the Third World.* Cambridge: Cambridge University Press, pp. 231–254.

Bremner, Lindsay. 1998. "Participatory Planning: Models of Urban Governance: Porto Alegre and Greater Johannesburg." *Urban Forum* 9:1.

Cherry, Janet, Kris Jones, and Jeremy Seekings. 2000. "Democratisation and Urban Politics in South African Townships." Paper presented at the Urban Futures Conference, Johannesburg (July).

Chipkin, Ivor. 2000. "Area-Based Management and the Production of the Public Domain." Unpublished manuscript, Centre for Policy Studies, Johannesburg.

Cohen, Jean, and Andrew Arato. 1995. *Civil Society and Political Theory.* Cambridge, MA: MIT Press, 1995.

Everatt, David. 1999. "Yet Another Transition? Urbanization, Class Formation, and the End of the National Liberation Struggle in South Africa." Strategy and Tactics (Research Consultancy), Johannesburg, June.

Ferguson, James. 1994. *The Anti-Politics Machine: "Development," Depoliticization and Bureaucratic Power in Lesotho.* New York: Cambridge University Press.

Fox, Jonathan. 1994. "The Difficult Transition From Clientelism to Citizenship." *World Politics* 46(2):151–184.

Friedman, Steven. 1992. "Bonaparte at the Barrricades: The Colonisation of Civil Society." *Theoria,* University of Natal, 79.

———. 2000. "A Quest for Control: High Modernism and its Discontents in Johannesburg, South Africa." Unpublished manuscript, Centre for Policy Studies, Johannesburg.

Friedman, Steven, and Maxine Reitzes. 1995. *Democratic Selections?: State and Civil Society in Post-Settlement South Africa.* Midrand: Development Bank of South Africa.

Heller, Patrick. 2000. "Degrees of Democracy: Some Comparative Lessons from India." *World Politics* 52:484–519.

———. 2001. "Moving the State: The Politics of Decentralization in Kerala, South Africa and Porto Alegre." *Politics and Society* 29(1):131–163.

Heller, Patrick, and Libhongo Ntlokonkulu. 2001. "A Civic Movement or a Movement of Crisis." Research Paper No. 84, Centre for Policy Studies, Johannesburg, June.

Hipsher, Patricia. 1998. "Democratic Transitions as Protest Cycles: Social Movement, Dynamics in Democratizing Latin America," in Sidney Tarrow and David Meyer, editors, *The Social Movement Society.* New York: Rowan & Littlefield, pp. 152–172.

Khan, Firoz. 1998. "A Commentary on Dark Roast Occasional Paper No. 1: 'Developmental Local Government: The Second Wave of Post-Apartheid Urban Reconstruction.' " Cape Town: Isandla Institute.

Lanegran, Kimberly. 1996. "South Africa's Civic Association Movement: ANC's Ally or Society's "Watchdog"? Shifting Social Movement-Political Party Relations." *Critical Sociology* 22(3):113–134.

Makura, David. 1999. "The MDM, Civil Society and Social Transformation." *Umrabulo,* no. 7 (3rd Quarter): http://www.anc.org.36/amcdpc/pubs/umrabulo.

Mayekiso, Mzwanele. 1996. *Township Politics: Civic Struggles for a New South Africa.* New York: Monthly Review Press.

Meer, Shamin. 1999. "The Demobilisation of Civil Society: Struggling with New Questions." *Development Update* 3(1):109–118.

O'Donnell, Guillermo. 1993. "On the State, Democratization and Some Conceptual Problems: A Latin American View with Glances at Some Postcommunist Countries." *World Development* 21(8):1355–1359.

SANCO, 2000a. Gauteng Provincial Organisational Report (March).

———. 2000b. "Strategy Discussion Document to Radically Re-Shape the Vision and Role of SANCO." Presented in April 2000 at the 3rd National Conference.

———. 2001. Organisational Report. Third National Conference, Mogale City, Johannesburg, April 19–22.

Scott, James. 1998. *Seeing Like a State: How Certain Schemes to Improve the Human Condition Have Failed.* New Haven: Yale University Press.

Seekings, Jeremy. 1997. "Sanco: Strategic Dilemmas in a Democratic South Africa." *Transformation* 1–30, 34.

———. 2000. *The UDF: A History of the United Democratic Front in South Africa 1983–1991.* Cape Town: David Philip.

Shubane, Khehla. 1992. "Civil Society in Apartheid and Post-Apartheid South Africa." *Theoria,* University of Natal, 79.

Swilling, Mark, and Boya, Laurence. 1997. "Local Governance in Transition," in Patrick Fitzgerald, Anne McLennan, and Barry Munslow, editors, *Managing Sustainable Development in South Africa.* Cape Town: Oxford University Press, pp. 165–191.

Tarrow, Sidney. 1994. *Power in Movement.* Cambridge: Cambridge University Press.

Tomlinson, Richard. 1999. "Ten Years in the Making: A History of the Evolution of Metropolitan Government in Johannesburg." *Urban Forum* 10:1–39.

White, Caroline. 1995. "Democratic Societies? Voluntary Association and Democratic Culture in a South African Township." Centre for Policy Studies, Research Report No. 40 (June).

Zuern, Elke. 2000. "Democracy from the Grassroots? Civic Participation and the Decline of Participatory Democracy in South Africa's Transformation Process, 1979–1999." Ph.D. Dissertation, Department of Political Science, Columbia University.

10
HIV/AIDS

IMPLICATIONS FOR LOCAL GOVERNANCE, HOUSING, AND THE DELIVERY OF SERVICES[1]

ELIZABETH THOMAS

South Africa was once notorious for apartheid. Today, it is the acquired immunodeficiency syndrome (AIDS) epidemic. In 2000, President Thabo Mbeki made international headlines with his claim that AIDS was not caused by the human immunodeficiency virus (HIV) and his reluctance to implement the health responses that were unanimously adopted by the worldwide medical community. Over 4.7 million South Africans are living with HIV and AIDS-related deaths are estimated to make up to 31% of all deaths, increasing the mortality rate by 45%.[2] By 2010, when such deaths are projected to peak and to comprise 67% of all deaths, mortality will increase by 205%.

HIV/AIDS is a serious impediment to the development problems that challenge the country and its cities. Johannesburg is estimated to have over 287,000 people living with HIV, making up 10.4% of the population. As the city attempts to solve its HIV/AIDS health crisis, it finds that the crisis is not easily confined by public health measures.[3] HIV/AIDS has specific repercussions for housing, the delivery of public services, employment, and social services. Johannesburg is a city struggling with government reorganization, political integration, increasing economic inequalities, lingering racism, poverty, and economic restructuring. On top of this, the AIDS epidemic weighs heavily.

Not surprisingly, local authorities have experienced increasing demand for health and welfare services arising from the increasing incidence of HIV/AIDS. However, aside from the health sector, there has been a very limited response by local authorities largely due to a lack of awareness of the broad developmental impacts of the epidemic. At the local level, the non-governmental sector has addressed HIV/AIDS through HIV/AIDS awareness, caring for orphans, and operating hospices. However, these efforts have been constrained by limited resources.

Using Johannesburg as a case study, this chapter explores the development impact of HIV/AIDS as well as the city's existing policy responses and possible future policies. The chapter is organized into six parts: a brief review of HIV/AIDS data and projections for South Africa and Johannesburg, the identification of the development impact relevant to local government's role in delivering housing and services, Johannesburg's initial response to HIV/AIDS, Johannesburg's subsequent attempt in the *iGoli 2010* strategic planning exercise to formulate a coherent response to HIV/AIDS, and a further discussion of housing and services.

Housing and services deserve particular attention. HIV/AIDS impacts occur primarily at the household level and increasing poverty due to HIV/AIDS exacerbates limited access to basic services. The ability of households to cope with the care of people with HIV/AIDS is strongly dependent on the quality of housing, sanitation, and water supplies (HIV Management Services, 1997). At the same time, local governments struggle with the delivery of water and sanitation, roads and storm-water, waste removal and electricity. These services consume an average 64% of their annual expenditures.

POLITICS AND DEMOGRAPHY
THE POLITICS

For South Africans, the growing rate of HIV in the country is of major concern; those not *infected* are certainly *affected* by the pandemic. Increasingly the impact on a range of sectors such as education, health care, the economy, and especially rising levels of unemployment and poverty is alarming. Behind the "facts" are the escalating numbers of heart-rending stories related to HIV/AIDS. These include an ongoing stream of HIV-related personal tragedies such as abandoned babies, increasing numbers of street children, stigma for those infected and their families, breaches in confidentiality, rapes, suicides, and evictions.

This escalating human tragedy takes place against a backdrop of indifference and denial by President Mbeki. The dogged refusal to make anti-retroviral drugs available though the health system, on which 85% of the population is dependent, has considerably tainted the public's respect for him. His stance regarding the "questionable" link between HIV and AIDS has also undermined the "safe sex" messages targeted at the youth.

There is a strong resistance to Mbeki's position on access to drugs. Some provincial governments have made antiretroviral drugs available and there is a growing number of voices in opposition to the "official Mbeki's" drug policy. In atypical style, Nelson Mandela has spoken out critically (in February 2002) regarding the government's policy and praised those who have widened access to treatment though public hospitals. Pilot studies to limit mother-to-child transmission through drug therapy are being unofficially expanded. The Treatment Action Campaign, moreover, has challenged the government's drug policy in the Courts.

Access to drugs can potentially limit new infections in a third of all babies born to HIV-positive mothers and also help maintain the health of 4.7 million infected people. However, the key action must be prevention, specifically targeting young people. Recent studies have confirmed the importance of hope and optimism in affecting safer sex choices of young people. HIV/AIDS is clearly an issue impacting and impacted on by the broader socioeconomic and political environment. The following statistics paint a clearer picture.

DEMOGRAPHY

HIV/AIDS data and projections for South Africa vary considerably, depending on the source. This chapter errs on the conservative side, primarily relying on the national projections of Whiteside and Sunter (2000) and the Johannesburg projections of Van der Heever (2000).

Three categories of data are referred to below: the prevalence of HIV/AIDS in the general population, the prevalence of HIV/AIDS in adults, and the prevalence of HIV/AIDS among women attending antenatal clinics.[4] Antenatal data collected annually from government health facilities is the only comprehensive (though incomplete) HIV data available. The other two data sets are extrapolated from the antenatal data based on comparative experience and assumptions.

The epidemic in South Africa is best understood within the context of southern Africa. Despite containing 10% of the world's population, 70% of the world's HIV/AIDS-infected population is to be found in the region (Whiteside and Sunter, 2000:44). There are, however, regional variations in regard to the prevalence of HIV/AIDS. In 1998 one in four adults in Zimbabwe and Botswana (25.8% and 25.1%, respectively) was infected, whereas the South African prevalence rate at the time was 12.9% (Whiteside and Sunter, 2000:54).

The lower prevalence rate in South Africa results from the epidemic being less advanced. There are, however, certain areas of the country that have much higher rates (for example, KwaZulu-Natal and Mpumalanga), whereas others, such as the Western Cape, are further behind in the stage of the epidemic. The South African HIV prevalence rate among sexually active women, based on antenatal data, was estimated to be 0.8% in 1990 (Whiteside and Sunter, 2000:51), rising to 22.4% in 1999 and 24.5% in 2000 (Department of Health, 2000a). This rapid increase has given South Africa the notorious status of having the fastest growing HIV/AIDS epidemic in the world. The epidemic is expected to peak near 2010 with an adult HIV prevalence rate of 21.7% and with 6.2 million people infected with the virus (Whiteside and Sunter, 2000:69).

In 1999, Gauteng Province had an HIV infection rate of 23.8%, 1.4% above the national average (Department of Health, 2000a). In the same year, the antenatal data for the Central Wits (health) region showed a HIV-positive prevalence rate of 26% (see Table 1), which is higher than the prevalence for Gauteng as a whole and for most of South Africa.[5] When these prevalence rates are extrapolated to the population of the metropolitan area, the HIV-positive prevalence in the area was estimated to be 10.4% in 2000. This translates to over 286,000 cases (Van der Heever, 2000, Table 4.1).

The Soweto magisterial district, which has 42% of the city's total population, has the highest HIV infection rate in the metropolitan area. In addition, the incidence of HIV/AIDS has been found to differ between urban formal and informal areas. Those living in private housing have a lower rate than those living in informal urban communities. This intraurban variation is likely to be indicative of the varying sociodemographic profiles and the differing stages of the epidemic across the city.

Returning to the Central Wits area and looking beyond 2000, HIV-positive persons are projected to make up 8.6% (251,728) of the popula-

Table 1

Central Wits HIV Seroprevalence by Age Group in the 1999 Antenatal Sample

Age Group (Females)	HIV-Positive Prevalence Rate (%)
Central Wits antenatal total	26.1
Gauteng	23.8
South Africa	22.4

Source: Department of Health (2000a).

tion in 2010, with slightly more women than men likely to be infected throughout the next decade. The projections show the number of deaths from AIDS declining from 31,000 to 26,000 per annum between 2000 and 2010, but remaining higher than normal mortality, that is above 21,000 per annum (Van der Heever, 2000). What is most noticeable "is the marked increase in the death rate of the younger generation of people between 20 and 50" (Whiteside and Sunter, 2000).

Children will bear the brunt of the impact as a result of the sickness and the death of parents. There will consequently be an increasing burden on older caregivers and the community at large. AIDS orphans in Johannesburg are expected to increase from a total of 77,000 in 2000 to 139,000 in 2010, with approximately 9,000 children being orphaned each year during the next decade in the Central Wits area. In the period 2000 to 2010, the greatest number of orphans will occur in the Soweto (32,000–55,000) and Johannesburg (26,000–43,000) magisterial districts.[6]

The increasing numbers of young people dying has a marked impact on life expectancy. In the Central Wits area, it is projected to decline substantially from 61 to 48 years by 2010. Between 2000 and 2010 the life expectancy of African males is anticipated to drop 13 years from 60 years to 47 years. For African females, the drop is projected to be by 19 years from 64 to 45 years.[7]

DEVELOPMENT IMPACTS OF HIV/AIDS

Two groups are most at risk: those who are infected with the virus and the affected population who experience the ripple effects of the epidemic. The affected group is much broader and includes families, extended families, communities, and society at large. The economic impact includes the loss of income as infected persons take increasing breaks from work and then cease to work altogether. Other family members, some economically active, are drawn in to care for the sick and dying. Likewise, children are taken out of school to care for the dying and to earn money, since an ever-greater proportion of the household income goes toward health care. Household savings and ultimately the income and savings of the extended family and community are thus drawn on to care for the dying and to pay for burials. This leads to diminished resources available for nutrition, education, clothing, and basic services for the remaining household members.

Evidence from research elsewhere suggests that households dealing with chronic illness and financial consequences adopt multiple coping strategies ranging from the sale of assets and the securing of loans to changes in consumption patterns and even crime (Goudge and Govender, 2000). Likewise, Foster (1990) points out that paying for the treatment of a terminally ill patient can set off a downward financial spiral from which the family may not recover for generations. In the context of housing, these are likely to include overcrowding as tenants are taken in, lack of maintenance of structures, and defaulting on payments for housing and service charges. Eventually, there is likely to be an unraveling of the social fabric that will be most evident in the psychosocial impacts on children and inadequate guidance and support for child-headed households and orphans.

As household incomes spiral downward and as the demand for goods and services declines and shifts toward health care, local manufacturers and merchants, already struggling with the illness of employees, may have to cut back production, increasing unemployment and reducing household incomes still further. According to the World Bank, these and other impacts of HIV/AIDS will result in South Africa's GDP being 17% less in 2010 than it would have been were it not for HIV/AIDS. In this view, despite the high death rate, the unemployment level is not expected to drop (Lewis, 2000). Others have a less pessimistic economic view, anticipating the impact on the national economy to be minimal considering that the largest share of those infected are low-income earners and the unemployed, and that national (health) treatment expenditures are being curtailed.

The consequent and immediate implications for local government involve the delivery of housing and services and the reduced resources available for doing so. As Crewe (2000a:1) has written, the "failure to pay rents, rates and taxes . . . will be dramatic—the street children problem will be exacerbated and the delivery of services will be affected both by the lack of payment but also by the infection of the workforce. Issues such as burial space, crime and family and community disintegration will have a major impact." These difficulties will be exacerbated by the fact that national government's grant funding for housing and infrastructure is directed toward households, whereas for many individuals, including orphans, the need will be for publicly available shelter, a roof and a bed, and access to clean water and sanitation (Tomlinson, 2001).

THE ESSELEN STREET CLINIC

Johannesburg's initial response to HIV/AIDS did not take account of the development impact. Nonetheless, from a health care point of view, it was dynamic and creative. Unfortunately, it was not sustained either politically or financially (Crewe, 2000a). In her review of Johannesburg's initial response to HIV/AIDS during the late 1980s and 1990s, Crewe (2000a:70–72) observes that

> The [old] Johannesburg City [Council] health department developed a dynamic and creative AIDS programme that was run from the center in Esselen Street simply known as Esselen Street. With the change in local authority structures this programme became the Greater Johannesburg Metropolitan Council (GJMC) AIDS programme. It developed from being a vibrant and creative programme which was well funded by the local authority with a small supplement from the province, attracting many people from inner city Johannesburg and all surrounding areas, to one which, while serving a growing number of people in the late 1990s, was under-resourced and underfunded. The political commitment to the programme collapsed and the infighting between the province and the district for the control of the AIDS programme had a debilitating effect.
>
> [By 1999] the programme was forced to shrink, posts were frozen, the media budget dried up and there was no encouragement for training and new projects. The integration with the other services was complicated by the wider integration of the various local authorities that made up the greater metropolitan council.
>
> There was an initiative to integrate HIV and AIDS work into the library services and into the sports and recreation centres. Staff from these departments was trained, but was not given the leeway to integrate programmes. Other departments were involved more through training than through new initiatives.

On reflection, the key challenge is to convince "local authorities to recognize how dramatically this epidemic will impact on their services and their ability to deliver services as well as on the population they serve." In response to the question of what should be done differently, it was proposed that "the politicians and bureaucrats" should be made to be "far more accountable and give far greater education and responsibility to the ratepayers" (Crewe, 2000a).

This health focus was not unique to Johannesburg. From a survey of Johannesburg and many other local authorities, the key theme that emerged was the extent to which HIV/AIDS was seen as a health issue. Those interviewed saw the key local authority challenges as being prevention and developing appropriate responses to mitigate the impact of the epidemic (Thomas, 2000).[8]

iGOLI 2010

The current mayor of Johannesburg, Amos Masondo, has recognized the severity of the impact of the HIV/AIDS pandemic on the city, which will be in his words an "enormous threat to the development of Johannesburg into a world class city" (Cox, 2001). Guided by the *iGoli 2010* plan, a number of practical strategies have been undertaken. One of the key decisions has been the launch of a Metro AIDS Council, a cross-sectoral advisory body for the mayor.

Johannesburg revised its HIV/AIDS strategy contemporaneously with the formulation of its *iGoli 2010* strategic planning initiative. The *iGoli 2010* draft strategy acknowledges the enormous challenges facing the city: poverty and slowing economic growth in the context of lower life expectancy and slowed population growth resulting from HIV/AIDS. It states that these issues "need to be faced and addressed if they are to turn the tide away from the current crisis to achieve the aim of being a world-class city."[9] In contrast to the past approach to HIV/AIDS, which was to parcel the epidemic as a health concern, the analysis recognizes a broad range of sectors and functions that will be impacted by HIV/AIDS. HIV/AIDS impacts are expected to contribute to an increasing burden on current services in terms of capacity and appropriateness. In response to the challenges, *iGoli 2010* proposes a number of key actions:

- HIV/AIDS programs–antiretroviral drugs, treatment, counseling, vocational skills training, and the improvement in economic opportunities for low-income women and girls,
- Intersectoral investments for health focusing on women's empowerment, and
- Improved preventative care (greater access, affordability, appropriateness and availability of preventative services) (GJMC, 2000b).

The analysis acknowledges that the HIV/AIDS pandemic will impose increasing income insecurity and additional burdens on households as well as further extending the deteriorating social fabric.[10] It also acknowledges that the province has taken a curative approach to health and social welfare services rather than a developmental and preventative approach. This has resulted in inadequate funding for the latter.[11] One of the key strategic actions proposed for the Human Development strategic action agenda is that of Targeted High Impact Intervention on HIV/AIDS, although how this is to be achieved is not discussed.[12] One of the identified actions is to work with the provincial government to develop a "high impact sustained HIV/AIDS programme and implement sexually transmitted disease treatment, counseling and vertical transmission prevention."[13]

Short-term actions also identify targeted HIV/AIDS interventions under the broad heading of Housing Infrastructure and Services, however no mention is made of HIV/AIDS in the service delivery section.[14] Because the delivery of water, sanitation, roads, electricity, and waste removal consumes nearly two-thirds of local government's operating expenditure, this is a significant oversight.

HOUSING DELIVERY IN A CONTEXT OF HIV/AIDS

The national Department of Housing is responsible for housing policy and the management of the housing subsidy, a capital grant used for delivering a top structure and internal services such as water and sanitation and roads. Housing subsidies are available to households having an income below R3,500 that have not previously benefited from a housing subsidy. The national Department of Provincial and Local Government is responsible for policy and for managing infrastructure grants for bulk and connector services. The availability of infrastructure grants is calculated on the basis of household income. In Johannesburg the grants have gone entirely toward supporting low-income housing projects. Counterpart provincial departments are responsible for channeling the funding to local authorities and, in the case of Gauteng, allocating housing funds to specific housing programs in specific local authorities.

In the cases of services, this means that local authorities apply for infrastructure grants and implement the relevant projects. In the case of housing, local authorities act as developers of housing projects and use the private sector as contractors. In effect, within the presently constraining policy framework, local authorities are responsible for delivering housing and services to those with HIV/AIDS. The significance of the last point is especially apparent as a result of *iGoli 2010*'s assertion that provincial government has responsibility for housing.[15]

Tomlinson (2001) argues that this strategy is premised on two key assumptions: a "functioning household able to be the recipient of a housing subsidy and an ability and willingness to invest in housing." Tomlinson further notes that HIV/AIDS calls these assumptions into question due to the breakdown of families and the pressures on household income that cause households to minimize their housing expenditures. In practical terms, this means that Johannesburg should strive to deliver housing to still-functioning families and, preferably via community-based and nongovernment organizations, shelter and services to those displaced by HIV/AIDS.

Government has been slow to come to terms with such changes in the housing and services policy environment. The national Department of Housing has still to launch an investigation into the impact of HIV/AIDS on housing policy and the application of the housing subsidy. The Gauteng Department of Housing did so in 2000, but this has still to come to fruition. The Johannesburg housing strategy (April 2000) encompasses a range of delivery options and none of these directly address HIV/AIDS.[16] Although reference is made to HIV having been taken into account in terms of calculating the demand for housing units over the 2000–2010 period, the projections are based on the continued existence of functioning households. The draft sustainable housing strategy for Johannesburg (revised June 12, 2001) acknowledges the importance of considering the impact of HIV/AIDS, but fails to identify the issues to be considered or how HIV/AIDS is likely to have an impact on the housing demand, policy, and delivery. HIV/AIDS is considered a narrowly defined housing issue rather than in terms of sociodemographic impacts and their housing policy and delivery implications.

The significance of this erroneous policy framework arises from the incredibly complex array of housing needs. Not just the terminally ill and dying need care. What of street children, not all of whom will be AIDS orphans?[17] What of family members expelled for being HIV positive? What of child-headed households who cannot enter into contracts for service delivery and whose inheritance of a house will be coveted by members of the extended family and others in the community? What will be done to help the AIDS sick who are too poor to cover the costs of housing repayments and service charges? How will the creation of ring-fenced utilities for service delivery and the outsourcing to the private sector of the management of the utilities gel with the need to provide water and sanitation to make home-based care a cost-effective option for the care of the terminally?

In terms of the South African Constitution, housing is a basic right and like other sectors, housing policymakers have as failed to respond to the challenge of HIV/AIDS in any meaningful way. The fundamental assumptions of the current housing policy are challenged by the increasing levels of poverty and the growing need for access to housing by people who are not in the traditional family structures on which current policy is based. HIV/AIDS further focuses attention on the need for revising housing opportunities and reassessing infrastructure policy so as to provide for excluded single mothers, child-headed households, orphans, the terminally ill, and hospices.[18]

CONCLUSION

Within the HIV/AIDS policy framework, creative responses must be mounted at the local level to address the multidimensional consequences of this disease and the multitude of needs experienced by HIV-infected and affected persons and households. HIV/AIDS is not only a health problem. Its detrimental consequences reverberate across the social life of Johannesburg and thus across policy areas. Of particular policy significance are housing and services. There, the incidence and destruction of HIV/AIDS are likely to exacerbate shelter and services needs and undermine current policy. Moreover, as the employment consequences of HIV/AIDS spread, local governments will find it increasingly difficult to collect revenues and to deliver basic services. Without decent housing and basic services, HIV/AIDS will become an even greater burden on Johannesburg's future.

NOTES

1. I would like to thank Richard Tomlinson for the help he provided with the drafting of this chapter.
2. The Medical Research Council's 2001 study into HIV/AIDS mortality estimates that AIDS accounted for 25% of all deaths in the year 2000 and 40% of all deaths of those aged between 15 and 49 years (Dorrington et al., 2001).
3. For a discussion of the National HIV/AIDS and Sexually Transmitted Disease Strategy, see Department of Health (2000b).
4. Prevalence refers to the existing level of infection from a particular disease in a given year.
5. The region corresponds closely to metropolitan Johannesburg prior to the 2001 elections.
6. See Table 4.1 in Van der Heever (2000). A study in KwaZulu-Natal suggests that for youth living in settings characterized by political violence and high crime levels, HIV has become accepted as a new and inevitable part of growing up (Lecrerc-Madlala, 1997).
7. The quoted figures are from a different source resulting in slightly different figures for 2000. However, the data in Table 1 reflect the difference between women in different groups/contexts whereas the data in the text refer to the change in life expectancy over time. Monitor (2000) is the source of data in the text.
8. A full reporting of the results is to be found in Thomas and Crewe (2000).
9. GJMC (2000b:43).
10. GJMC (2000c:74).
11. GJMC (2000c:77).
12. GJMC (2000d:130).
13. Ibid., p. 145.
14. Ibid., p. 160.
15. GJMC (2000d:162).
16. GJMC (2000a).
17. For a provocative discussion on care of orphans, see Crewe (2000b).
18. UNAIDS, http://www.und.ac.za/und/heard/publications/publications/htm.

REFERENCES

Cox, A. 2001. "Jo'burg to Set up Urgent Aids Council." *The Star,* March 10, 2001.
Crewe, M. 2000a. "Johannesburg Local Authority Response to HIV/AIDS." *Urban Health and Development Bulletin* 3(2):9–22.
———. 2000b. "AIDs Orphans: The Urban Impact." Paper delivered at the workshop on HIV/AIDS orphans: Building an Urban Response to Protect Africa's Future, Woodrow Wilson Seminar, Johannesburg, Centre for Policy Studies, July 21, 2000.

Department of Health. 2000a. *Summary Report. National HIV Sero-Prevalence Survey of Women Attending Public Antenatal Clinics in South Africa in 1999.* Pretoria: Department of Health.

———. 2000b. *HIV/AIDS and STD Strategic Plan for South Africa, 2000–2005.* Pretoria: Department of Health.

Dorrington, R., Bourne, D., Bradshaw, D., *et al.* 2001. "The Impact of HIV/AIDS on Adult Mortality in South Africa." Technical report of the Burden of Diseases Research Unit, Medical Research Council, South Africa. September 2001. http://www.mrc.ac.za/bod.

Foster, S. D. 1990. "Affordable Clinical Care for HIV Related Illness in Developing Countries." *Tropical Diseases Bulletin* 87(11):121–129.

Goudge, J., and Govender, V. 2000. "A Review of Experience Concerning Household Ability to Cope with the Resource Demands of Ill Health and Health Care Utilisation." *Equinet Policy Series,* No 3.

Greater Johannesburg Metropolitan Council (GJMC). 2000a. Housing Strategy (April 2000), compiled by Spadework, Johannesburg, unpublished.

———. 2000b. "iGoli 2010 Economic Growth-04-11-2000KN," unpublished.

———. 2000c. "iGoli 2010–Recommended Vision and Strategic Agenda," 2000. Information Package for the Executive Committee of the Council of the City of Johannesburg, Final Draft, December.

———. 2000d. "iGoli 2010 JHB-CTY-Package for Mayoral Committee—Integrated Story—December 2000-HH," unpublished.

HIV Management Services. 1997. "Assessment of the Impact of the HIV/AIDS Epidemic on the Gauteng Provincial Government," unpublished.

Lecrerc-Madlala, S. 1997. "Infect One Infect All: Zulu Youth Response to the AIDS Epidemic in South Africa." *Medical Anthropologist* 17(4):363–380.

Lewis, J. D. 2000. "The Macro Implications of HIV/AIDS in Southern Africa: A Preliminary Assessment." World Bank, Africa Region Working Paper Series No. 9. Washington: World Bank.

Monitor. 2000. "iGoli 2010 Demographics Report," November 4, 2000.

Thomas, E. P. 2000. "HIV and Integrated Development Planning." Study commissioned by the Department of Provincial and Local Government and GTZ, Pretoria, unpublished.

Thomas, E. P., and M. Crewe. 2000. "Local Authority Responses to HIV/AIDS: An Overview of a Few Key Cities." *Urban Health and Development Bulletin* 3(2). http://www.mrc.ac.za/UHDbulletin/june2000/contents.

Tomlinson, R. 2001. "Housing Policy in a Context of HIV/AIDS and Globalisation." *International Journal of Urban and Regional Research* 25(3):649–657.

Van der Heever, A. 2000. "iGoli, HIV/AIDS Impact and Intervention Analysis: Report on the Greater Johannesburg Metropolitan Region," November 2000.

Whiteside, A., and C. Sunter. 2000. *AIDS: The Challenge for South Africa.* Cape Town: Human & Rousseau.

11

Social Differentiation and Urban Governance in Greater Soweto

A CASE STUDY OF POSTAPARTHEID MEADOWLANDS

JO BEALL, OWEN CRANKSHAW, AND SUSAN PARNELL

When Meadowlands, a township in Greater Soweto, was first established during the 1950s, the population was fairly homogeneous. As a result of the sustained economic upswing during the 1960s, a number, but not all, township residents experienced upward occupational mobility. Social differentiation increased even further over the 1980s and early 1990s. Some residents benefited from the drive toward encouraging homeownership, whereas many others simply fell foul of persistent structural poverty and rising unemployment levels. Social differentiation looks to continue with some Meadowlands's residents benefiting from the deracialization processes accompanying the end of apartheid. Many others, by contrast, constitute the "new poor," victims of economic reform measures and adverse changes in labor market opportunities. According to the 1996 Census, 70% of Meadowlands's population over 15 years of age earn either nothing at all or under R500 a month; only 12% have incomes of more than R3,500 per month. Although a small number of professionals are service workers (such as teachers and nurses), only 49% of the economically active population is formally employed. Of these (see Table 1), the largest numbers are in crafts and trades and elementary occupations.

Social differentiation in Meadowlands, as in Greater Soweto more generally, is characterized not only by the differential access of individuals to employment and income but also by differential access to housing and basic services, such as water supply and sanitation, refuse removal, and electricity. In this chapter we trace social differentiation along the axis of different

Table 1

Occupations of Employed Residents, Meadowlands

	Number	Percent
Legislators, senior officials, and managers	682	2
Professionals	1,960	7
Technicians and associate professionals	1,788	6
Clerks	3,340	12
Service workers, shop and market sales workers	4,114	15
Skilled agricultural and fishery workers	129	0
Craft and related trades workers	5,576	20
Plant and machine operators and assemblers	3,683	13
Elementary occupations	6,348	23
Total	27,620	100

Source: 1996 Population Census.

housing types: homeowners, tenants in backyard shacks, and hostel dwellers. Social relations based on assets (the proxy for which is housing type), gender, and generation combine with ethnic and political conflict to constitute a crucial dimension of urban governance in Meadowlands.[1]

THE SOCIAL ORIGINS AND CHARACTER OF MEADOWLANDS

The origins and character of Meadowlands are fairly typical of African townships within Greater Johannesburg. Meadowlands was built during the 1950s, a period during which most of the existing housing stock in Johannesburg's African townships was constructed.[2] Consequently, all formal family housing there takes the form of the well-known "matchbox" house.[3] Meadowlands also has a hostel for rural migrants, another typical feature of African townships built during this period.[4] Moreover, along with other areas of Greater Soweto, in the 1980s Meadowlands saw a proliferation of backyard dwellings built in the face of government restrictions on the supply of family housing provision in urban areas and a growing urban population. Since the late 1980s many squatter camps were established on vacant land within Greater Soweto and on its periphery. No such squatter settlements were established within Meadowlands or on its northern boundary.

The social origins of Meadowlands are somewhat unusual in at least one respect. Whereas many of Greater Soweto's original residents were squatters who had invaded land in southwestern Johannesburg, the original residents of Meadowlands had been forcibly removed from within central Johannesburg itself, from areas such as Sophiatown and Western Native Township.[5] Meadowlands was established when, in accordance with the Native Resettlement Act of 1952, the apartheid government forced the Johannesburg authorities to remove African tenants and subtenants from multiracial neighborhoods as part of the Western Areas slum clearance schemes.[6] Orchestrated by the newly established Urban Resettlement Board, the forced removals to Meadowlands began on February 10, 1955.[7] Within ten years Sophiatown ceased to exist and was replaced by a white residential area, tactlessly and cruelly named Triomf (Triumph). People were resettled according to their ethnic group, a decision that was to mark the character and development of the area for years to come.[8] Ethnic separation can be identified through street names in the different zones and the fact that a range of languages constituted the vernacular in schools. Today most of the population of Meadowlands speak IsiZulu (33.5%) and Setswana (27%) with the younger generation opting for the lingua franca of the townships, sometimes known as *tsotsi taal* (gangster speak).

The area that today comprises Meadowlands has a long history of urban settlement, predating even the creation of Meadowlands itself. However, almost half (46%) of all residents not born in Meadowlands moved to their present home in the late 1950s and early 1960s when most of the township was built. Another 39% moved from 1966 to 1980. Of all residents, estimated at 127,568 people, 55% were born in the house they currently occupy.[9] These findings reveal a striking lack of mobility that was probably brought about by a combination of influx control laws that prevented Africans with urban rights from moving about within urban areas and a shortage of housing. This is significant; for urban Africans security of tenure and housing type are important indicators of social mobility and of social polarization.

Since the late 1950s, the development of Meadowlands has, with some exceptions, followed a similar trajectory to other townships in Greater Soweto. This trajectory entailed a steady movement away from the provision of standardized low-cost family housing. In the 1950s, the vast bulk of Sowetans had no choice but to accept the standard "matchbox" house. By the 1990s, however, state reforms that privatized the provision of housing meant that Sowetans lived under increasingly differentiated housing conditions. On the one hand, reforms that introduced homeownership offered the wealthy few the opportunity to purchase housing of a relatively high standard. On the other hand, the withdrawal of the state from low-cost housing provision led to overcrowding of formal houses and the proliferation of shacks in backyards on open land.

Social differentiation in Greater Soweto can be traced through its housing. The present day landscape of Meadowlands reveals housing of varied colors, styles, alterations, additions, and accoutrements, although there remains ample evidence of the 1950s housing stock. The original "matchbox" houses were built in three slightly different designs. The most common type, measuring 40 m², is the "51/6"[10] and comprises two bedrooms, a kitchen, and a living room. A later design, the "51/9," incorporates a small bathroom (with basin, bath, and lavatory) and is slightly larger at 44 m². The 51/6 design was often built as a semidetached unit. Rudimentary standards were employed with only earthen or ash floors and without internal doors and ceilings. Water was piped to a tap outside the kitchen door and an outside toilet was connected to a waterborne sewerage system. Most stands in Soweto are 260 m² in size. Soweto houses had no electricity until the late 1970s, but by 1988 all formal houses were supplied with electricity.[11]

Toward the end of the 1960s, the apartheid government began to channel funds for housing from townships such as Meadowlands to townships in the so-called "homelands." This meant an effective freeze on family housing provision. As a result of urban in-migration and the natural growth of the urban population, a chronic shortage developed. As early as the late 1970s, this shortage of housing manifested itself as residents were forced to overcrowd their standardized four-roomed houses and to build temporary shacks in their backyards to accommodate their adult offspring. By 1979, the official waiting list for housing in Soweto was about 14,000 persons. Because this list excluded those who did not have rights to live permanently in Soweto, it is an underestimate of the real extent of homelessness, which was calculated at about 173,000 people. Estimates of the shortage of houses in 1979 ranged from 25,000 to 32,000 units.[12]

A survey of Soweto households was conducted in early 1978. By then, 7% of houses in Soweto had subtenants, most of whom occupied a room in the main house.[13] In addition, in 49% of houses, residents used either the living room or the kitchen for sleeping.[14] The low rate of backyard subtenants, which was only about 1%, was probably because it was illegal and closely policed by the authorities. However, after the state lost control of the townships during the political rebellions of the 1980s, the housing shortage produced a rise in the number of backyard shacks and the proliferation of squatter settlements by homeless residents.[15] By 1987, 40% of all formal houses in Johannesburg's African townships had at least one backyard shack and 23% had a formally built "garage" that was inhabited by subtenants. Five percent of houses had subtenants occupying a portion of the house.[16]

By 1997, the number of backyard shacks (about 121,000) in Greater Soweto almost exceeded the number of formal houses, providing homes for about 20% of the population. In addition, there were approximately 18,000 shacks in 27 squatter settlements and they housed about 6% of the population.[17] Although Meadowlands has its fair share of backyard shacks, no large squatter settlements exist within the township or on its immediate boundaries. The 1996 Population Census estimates that 22% of households in Meadowlands live in backyard accommodation, half of which are formal rooms and half of which are shacks.[18] By contrast, only 2% of residents live in shacks outside of backyards.

The provision of services and the quality of accommodation available to the residents of backyards are significantly better than those available to the squatter camps. Although almost half of all backyard structures in Greater Soweto are crudely built wood and corrugated iron shacks, a majority (54%) are formal structures made from bricks and cement. What varies rather little is the number of rooms. Virtually all backyard accommodation, whether a formal structure (89%) or a shack (93%), has only one room. The quality of the shacks does vary however: some are built with old and rusty corrugated iron sheets (with many holes from previous constructions) whereas others are built with new sheets. Some shacks lack windows, whereas others have them. What they all lack is any form of insulation because it is considered a fire hazard and ants soon make nests in the gap between the outer wall and the insulation. Although very few backyard households have a legal connection to an electricity meter (5%), almost all backyard households (92%) have some access to electricity. Most get their electricity through an illegal extension cable from the main house (86%). Similarly, because backyard taps were originally fixed to an outside wall of the council houses, almost every backyard household (99%) has access to water. With few exceptions, neither residents of the main structure nor backyard tenants have water in the house unless they have had it piped in at their own expense. In addition, because almost all council stands were provided with outside toilets, all backyard tenants have access to a flush toilet.

Given the limited size of backyard accommodation there is a great deal of overcrowding in Greater Soweto. Although most (55%) backyard structures house either one or two people, 22% have three occupants and

23% have four or more. There are no significant differences between the occupancy rates of backyard shacks and formal rooms; severe overcrowding is common in both shacks and formal structures. If the houses are frequently overcrowded, there are also many people living on the stand. The average number of occupants on council house stands is 7.4 and almost one in five council house stands contains more than ten people. Where backyard accommodation exists, the stand accommodates an average of four backyard residents. Meadowlands itself has a significant population of backyard tenants.

Although one should not exaggerate the differences between backyard tenants and their owner–occupier landlords, some significant social differences exist between these two groups with respect to urbanization, age, tenure, class, and social status.[19] The most important difference relates to the fact that backyard tenants are more recent arrivals to Greater Soweto. Approximately two-thirds of backyard household heads arrived in Soweto after 1975, whereas two-thirds of landlords in "matchbox" housing arrived between 1946 and 1965. Second, backyard tenants are significantly younger than their landlords. Whereas the average age of the heads of backyard tenant households is 36 years, the average age of the heads of landlord households is 56 years. This age difference has a number of direct demographic consequences. Compared with the heads of backyard tenant households, landlords are five times more likely to have children who have left school and five times less likely to have a child who is not yet attending school. Landlords are ten times more likely to be retired than the respondent in the backyard and are twice as likely to be married. Finally, backyard tenants are more likely to be foreign immigrants than their landlords. About 16% of backyard tenants who were not born in Johannesburg are foreign immigrants. This is substantially higher than the overall rate (5%) for the whole of Greater Soweto.

These social differences between backyard tenants and landlords translate into important differences in political involvement. Everett's study of social divisions in Soweto's townships of Tladi and Moletsane has shown that backyard tenants are reluctant to participate in local politics for fear of reprisal from their landlords. Curiously, this fear is not because landlords and backyard tenants support different political parties. They both support the African National Congress. Instead, within the context of Congress politics, landlords dominate the structures and processes of local governance in order to ensure that their interests take precedence over those of their tenants and squatters.[20]

MEADOWLANDS AND THE STRUGGLE FOR DEMOCRACY

During the transition from apartheid, Meadowlands experienced periodic episodes of violent conflict growing out of the generalized opposition to the apartheid state and its institutions, but also turned inward, with the primary fissure being between those living in houses and those living in hostels. To understand this and subsequent axes of social differentiation within Meadowlands, it is important to trace the role played by successive policy interventions and community level reactions in relation to housing and urban services.

HOMEOWNERSHIP

In 1986, after various attempts at reforming existing leasehold arrangements, freehold ownership was introduced to African townships.[21] The government also introduced a once-off discount scheme, which effectively allowed council tenants to take ownership of the houses they occupied without any cost. Thus, by 1996 most Sowetans owned their homes. In Meadowlands, 92% of formal houses were owned.[22] The reforms that reintroduced homeownership allowed wealthier residents either to renovate their state-built homes or to raise the capital to buy new homes in suburban developments in other parts of Greater Soweto. In both cases, the new homes were a great improvement on the standardized, low-cost units that were built in the 1950s and 1960s. Consequently, housing standards in Soweto became increasingly differentiated according to the social class and income of the residents. By 1997, there were some 20,000 new houses built by the private sector for homeowners in greenfield developments.[23]

These new developments did not take place within Meadowlands. Instead, Meadowlands is characterized by renovations to the existing state-built housing stock. In many cases, these alterations are extensive; the original "matchbox" house is no longer recognizable beneath new rooflines, additional rooms, and perimeter walls. Nevertheless, by 1996 only 16% of all formal houses in Meadowlands had more than the original four rooms, suggesting that although much more differentiated than in the past, Meadowlands is probably less differentiated than some other Soweto townships.[24] The reason is that renovated and extended houses are usually scattered among the original "matchbox" houses and have not given rise to separate middle-class areas. Further explanation lies in the lack of private sector housing developments and the fact that Meadowlands's residents have prevented the establishment of squatter settlements within its boundaries.

HOSTEL ACCOMMODATION

Hostels in Johannesburg were not only for miners. From before World War I, the City housed its unskilled workers in downtown hostels. Private companies also used the accessible multiethnic, single-sex institutions to accommodate male migrant workers. At the height of the apartheid period, the government wished to racially segregate the inner city and sought to relocate hostels from within "white" Johannesburg to the townships. The government hoped that hostels would become an increasingly important component of housing in African townships. In fact, the government actually tried to turn the township of Alexandra, to the north of Johannesburg, into a township solely for the accommodation of migrant workers, with family housing being demolished and replaced with hostels throughout the 1960s and 1970s.[25] As part of the apartheid "total segregation" plan, a large hostel was built in Meadowlands. Ironically the location of the hostel in the midst of family housing made it an important site in the township uprisings that underpinned the demise of apartheid, albeit with an unusual twist.

Clashes between hostel and township residents became a common feature of the unrest of the 1980s. As a consequence of a violent clash between the migrant residents of Meadowlands hostel and township youth, the hostel was temporarily vacated in late 1976. Then, in early 1977, following the severe flooding of the Klipspruit River valley, some 1,200 homeless families were temporarily accommodated there. Because no alternative accommodation subsequently became available, they have occupied one section of the hostel ever since. Later, other homeless families from Soweto joined them.[26] This section of the hostel became known as *Mzimhlope,* or "Transit Camp."[27] It housed about 1,000 men, 1,120 women, and 3,500 children in very overcrowded conditions in what the Diepmeadow Housing Director called "Soweto's No. 1 slum."[28] Perhaps as a consequence of the *de facto* presence of women and children, it was targeted for upgrading into family housing as early as 1980. Today, substantial upgrading into family accommodation has taken place. Nevertheless, many of its residents remain socially disadvantaged and excluded.

Even though Meadowlands's hostel population has a unique family component, the usual social divisions between urbanites in family housing and rural migrants in hostels still obtain. Moreover, social fissures were widened by the violence that characterized Greater Soweto in the decades immediately prior to the democratic elections of 1994. The hostels, although linguistically mixed, had become the residence for many Zulu-speaking migrants from rural KwaZulu-Natal. They were sympathetic to

Mongosuthu Buthelezi's ethnically mobilized Inkatha Freedom Party (IFP)[29] and during the 1980s and early 1990s the IFP waged an aggressive campaign to expand its support among hostel residents in Gauteng. Although Meadowlands is often characterized as having been "backwards in coming forwards" in terms of the national liberation struggle, it has had its fair share of violence, serving to politicize existing social fractures and deeply implicating both sides in the conflict between IFP and African National Congress (ANC) supporters.[30]

LOCAL GOVERNMENT AND SERVICE DELIVERY

Hostilities in Meadowlands were not confined to those between township and hostel residents. Conflict also characterized the relationship between residents and local authorities. Although part of Greater Soweto, the administrative separateness and differences in Meadowlands' history have meant that the political scene there was also idiosyncratic. Following the upheaval of resettlement in the 1950s, Meadowlands was politically quiescent during the 1960s and 1970s.[31] Even during the Soweto student uprisings of 1976 Meadowlands was on the periphery of events. Marks sees it as a signifier of a tradition of political action evident in the early days of Sophiatown that was more closely linked to civic action such as boycotts rather than mass political mobilization.[32] In the 1980s, Meadowlands residents participated in the Soweto "rent boycotts" whereby they refused to pay council rent and service charges.[33] They argued that the services provided were far from satisfactory, the local authorities were corrupt, and the local government structure was illegitimate. The act of nonpayment was also a symbolic protest against the apartheid system. The rent boycott together with the changing political situation led the Transvaal Provincial Administration to enter into negotiations with the Soweto Civic Association. This culminated in the signing of the Greater Soweto Accord in September 1990, which made provision for the writing off of rent and service arrears and laid the foundations for the creation of a democratic metropolitan authority for the city.

BUILDING UNITY IN THE COMMUNITY

In late 1990s, the Greater Johannesburg Metropolitan Council (GJMC) officially united the 13 local government structures that had existed in Greater Johannesburg under apartheid. Although Soweto fell under the Southern Municipal Local Council (SMLC), Meadowlands fell under the Western Municipal Local Council (WMLC). (These were two of the four municipal substructures that remained part of the Greater Johannesburg

Metropolitan Council until the formation of a single metropolitan structure in 2000.) Like other substructures, the WMLC had to balance demands from privileged residents of previously white areas such as Roodepoort and those of the historically disadvantaged such as Meadowlands. Meadowlands contained nearly 47% of the municipality's total population though it occupied a relatively small proportion of the land area.

In the early postapartheid years, the WMLC made a small but visible impact on Meadowlands with some of the improvements being tarred roads, the construction of two clinics, a multipurpose hall, and a comprehensive welfare center. Partly as a result of these initiatives there has been an increase in the payment of service charges. According to the Chairperson of the Western Municipal Local Council, the level of payment at the time of the research was between 30 and 40%, a significant improvement on the very poor recovery rates that characterized the boycott.

The successful inroads made by the WMLC in terms of service delivery were a response to community priorities articulated through the Land Development Objectives (LDOs). The LDO consultation process, which took place in 1996/1997, asked communities to identify their priorities for local government attention. In terms of both housing and urban services, Meadowlands started off in a better position than many other urban settlements and although the progress made by the WMLC has been modest, it was reasonably responsive. Many of the improvements made by local government were more likely to benefit better-off homeowners who, having obtained water supply, sanitation, and electricity in the past, could now focus on improved infrastructure, garbage collection, and social services. For the poorest residents, for example the backyard tenants, improved services served to raise rents and other charges by landlords.

WHAT PEOPLE WANT FROM LOCAL GOVERNMENT

In assessing what people expected from the new dispensation, we found that the problem of unemployment was the chief concern. Informants and discussants generally believed that insufficient attention had been paid to job creation by postapartheid governments. Pleasure was expressed over central government's promotion of business development as well as efforts by the local authority to develop commercial and shopping areas. Unemployed youth pointed out that although the development of their local shopping center had not led to access to more formal jobs, there had been positive effects related to the growth of options for informal income-generating activities, e.g., the opportunity to set up a busy car washing service serving local residents as well as taxi firms operating both within Greater Soweto and between Soweto and Johannesburg.

All the respondents were concerned with housing supply. Homeowners (many of whom were also pensioners) and pensioners were concerned with increased housing provision for their extended and growing families. Women complained bitterly of the various stresses, financial and otherwise, associated with having to provide shelter for adult children and more distant relatives. They saw a direct link between having so many family members dependent on them and their own impoverishment and vulnerability. The extent of household survival strategies employed by women was evident when a group of elderly women was approached and asked by the researchers to participate in a focus group. They were found to be returning from the *veld* where they had been gathering *morogo* (wild spinach). Although some were homeowners and all were pensioners, they had many dependants and no income and were seeking ways in which to provide food for their families.

Backyard tenants and the youth also wanted affordable accommodation independent of parents, relatives, or landlords. They identified tensions between landlords and tenants as being a major source of discord and also complained of communication breakdowns between younger and older family members forced to live in close proximity as a result of persistent unemployment. Similarly, hostel dwellers were keen to have access to new housing. Apart from a small number of houses built under the Reconstruction and Development Programme (RDP) and through various low-income housing schemes, postapartheid housing delivery had been largely confined to turning the hostels into family units. Progress was slow. Consequently, a number of the hostel dwellers were anxious to be rehoused. This was particularly the case among younger hostel dwellers and among the families who had been relocated to the hostel after the Klipspruit flooding in the 1970s. The older migrants, particularly those from KwaZulu-Natal, preferred to be shielded by the relative privacy and safety afforded them by the hostels, in what remains a fragile and tense social climate.

Infrastructure and services were also of fairly universal concern. Revealing of their own lives and frustrations, the youth said that poor people were those without anywhere to go or anything to do and without prioritized sport and recreational facilities. Lacking in education, skills, and income-earning opportunities they reflected that young people with time on their hands were tempted to crime. Older people, many of whom owned their home but were income poor and could not afford to pay their service bills, focused on affordability. One elderly woman showed us her bills for rates and services and for electricity. She owed a total of R7,834. All the discussants confirmed that their own arrears were in excess of R7,000, despite the fact that these women regularly paid R300 toward the electricity account and R90 toward the rates and services account each month out of a monthly pension of R520. The group also reported that they were regularly cut off if they failed to pay. Their arrears never seemed to be reduced and there was a high level of mistrust toward the utility companies. As one of the discussants commented:

> As an old person I find it strange that I only receive R520 and my municipality bills range between R700 and R800 monthly. What is left for me to buy food with? I have never once seen any official of the municipality who comes to my house to record the water meter readings, so where do they get these figures?

Others concurred with another woman saying "we spend sleepless nights thinking how and when we are going to settle these huge bills."

The local council acknowledged these problems and the fact that poverty is rife in Meadowlands. This is explained in terms of the changing nature of the labor market. A further explanation both for extensive poverty within a seemingly working class area and for problems associated with people meeting their utility bills was offered by the Chairperson of the WMLC. He said that "Meadowlands is primarily a community of the aged who rely on pension grants for their survival" which makes issues of affordability paramount for the residents but which also renders it difficult for the cash-strapped local authority to deliver and maintain adequate services. The WMLC has responded by implementing the Indigence Policy developed by the GJMC. This is intended to ensure that punitive action is not applied to people who are unable to pay rates and utility bills due to circumstances beyond their control. However, the WMLC also acknowledges that the policy is ineffective.

Along with older people, backyard tenants were the group most exercised over service affordability issues. The anger of older homeowners on pension was directed at the local council and utility companies for exploitative bills and a lack of accountability in their operation and maintenance. The gripe of backyard tenants was with their landlords. The better-off homeowners who rent out backyard shacks pass on the cost of water supply, sewerage, and electricity to their tenants by insisting that backyard dwellers pay their share of the bills, with considerable dispute as to what constitutes a fair share.

In terms of community-level services such as tarred roads, storm water drains, solid waste management, and leisure and recreational facilities, these preoccupied the better-off informants and homeowning residents. For them, basic needs have already been met. They are an upwardly mobile group with a strong commitment to their neighborhoods and a more finely honed sense of civic pride. For example, Meadowlands was recently the center of an extraordinarily successful environmental campaign that not only took on the local authority in terms of improving refuse removal and solid waste management in Meadowlands, but engaged in (and won) a protracted environmental campaign with the Durban Roodepoort Deep Goldmine. The struggle was that of a small community-based organization with the support of a local NGO against the company's refusal to take responsibility for the dust emissions from a disused mine dump.

A number of social issues are also of deep concern to the community. Respondents readily raised the crucial issue of HIV/AIDS. This can be attributed in part to the fact that NGOs have been recently active in advocacy work around this issue. For older people, HIV/AIDS was seen as divisive, leading to shame and secrecy, whereas younger people saw the social mobilization associated with it as unifying. The youth were on the whole more open and AIDS-related deaths in the area are usually reported among the youth.

TOWARD AN ENGAGED LOCAL POLITICS

The first democratic elections in 1994 were an unprecedented thrill for most residents and it was generally agreed that the local elections of 1997 further entrenched democracy. Our field research also found an emerging malaise in respect to local politics. On the one hand, there is a level of burnout after the long years of antiapartheid activism and political violence that accompanied the transition. On the other hand, a sense of disappointment with postapartheid delivery focuses on local government politics and its perceived inefficiency and lack of accountability. The same political disillusionment does not appear to have attached to higher levels of government. For example, a number of discussants expressed the wish to by-pass local councilors in their negotiations for housing and services and to engage directly with national or provincial level politicians. As one informant said, "Mandela is the only politician I trust."

Despite a generalized disaffection with local politics, the ANC continues to enjoy significant support in Meadowlands. It is the only political organization that has a branch office in the area and that has organized zonal structures. Furthermore, there was widespread respect for some members of the local level leadership, especially those who came from Meadowlands and who had a history in the struggle. On the part of the more politically literate, there was also an understanding that councilors were bound by party lines and resource constraints and, as such, were not making decisions solely on the basis of personal predilection. People recognized that the political problems faced by councilors related to their limited control and jurisdiction over sectors, services, and budgets. Councilors themselves felt frustrated by being hamstrung in this way and by their own difficulties in balancing their obligation to conform to party (ANC) lines and address the immediate and vocal demands of their impatient ward constituents.

The dominance of the ANC can be attributed in part to the general popularity of the party as part of the liberation movement. In addition, most of the older members of Meadowlands society were members of the ANC in Sophiatown and were involved in the ANC's protest politics of the 1950s. Not surprisingly, the ANC Veterans' League is the most organized part of the ANC in Meadowlands. This also explains why community organization is the terrain of the more established, respectable older working class, most of whom are now homeowners. Strongly supportive of the ANC and closely allied to local councilors and local government officials, this group is most deeply involved in community-level collective action. The backyard tenants have been systematically excluded from local political arenas and are looking elsewhere for a vehicle through which to express their grievances and engage in public action. The youth also voiced frustration at not being taken seriously by the older homeowning activists. A much larger disaffected group of alienated and unemployed youth, a silent majority, eschews civic public action of any sort.

The most prominent civil society organization active around issues of service delivery is the local branch of the South African National Civic Organisation (SANCO). Notable among the issues raised by SANCO on behalf of their constituency in Meadowlands is the matter of jobs and affordability of services. For example, SANCO engaged the local authorities on behalf of residents on the strong-arm tactics used on the part of service payment defaulters as well as environmental issues and the repair and maintenance of services. This has won them widespread support. The organization has been less successful in attracting engaged support from the better-off members of the community and sees its greatest appeal among the more marginalized.

The youth and the backyard renters have allied themselves with SANCO, arguing that they have most faith in this organization as a "watchdog" for the community. For them SANCO is more accessible and more accountable than local politicians, even though most of the ward councilors belong to SANCO. Although the organization has a cordial and working relationship with the ANC and the councilors, the alliance is strained, with many SANCO members feeling that government decisions are not properly informed by consultation or adequate understanding of the real conditions faced by people. As such, SANCO is attempting to rejuvenate itself as a radical civic organization and has a growing constituency among younger people and backyard shack dwellers. It is carving out an increasingly significant position as an intermediary and broker, particularly in relation to the delivery of housing.

By virtue of its remit and competencies, local government is particularly concerned with the older more established homeowning citizens, notably around issues of service delivery. Many more issues remain to be addressed by local government that could target younger residents, notably the construction and maintenance of recreational and sporting facilities. However, many issues are beyond its competencies, scope, and resources. For example, the most pressing issue for informants and discussants of any age or asset base was that of crime, something about which the local authorities have done very little. At the time of the research, policing was not a local government competency but well maintained street lighting was. Such a relatively simple and low-cost deterrent against crime and gender violence is an issue that concerns people across a variety of social divides and, yet, has been neglected.

CONCLUSION

This chapter has sought to understand and explain a continuing tradition of public action in Meadowlands against a background of horrific community-level violence, persistent local-level mistrust, increasing social differentiation, and a pervasive disappointment with postapartheid local government. Political disillusionment persists despite relatively impressive service delivery since the 1994 elections and the fact that the majority of people interviewed cited Mandela's release from jail or voting for the first time as the most critical episode in their own recent history. A deep history and tradition of antiapartheid struggle and a context of continued and heartfelt loyalty to the ANC overrode social differentiation and tensions based on class, age, and gender.

Nevertheless, increasingly evident is a growing weariness with political conflict at the local level and a deepening resentment over poor housing and service delivery. This disaffection is occurring at a time when the face of civil society in Meadowlands is diversifying and growing in significance. Citizen demands may become less coherent as a variety of different and sometimes competing interest groups assert their own priorities and as NGOs and other advocacy organizations help vent and direct their demands. If and when this occurs, mass public action is likely to be more difficult to orchestrate and collective action will become more fragmented and potentially more conflictual.

In the case of Meadowlands, at least one reason for this can be found in the political fallout of an increasingly socially differentiated community in which, despite the fact that widespread unemployment is affecting all households, a large proportion of older residents have become relatively asset rich through coming to own their council homes during the death throes of the apartheid era. In this context, mainstream civic engagement at the local level has become the preserve of a middle-aged working class, often publicly represented by older women. They are primarily concerned with improving the quality of urban services and their built environment. This in turn has been reinforced by local government policy that has linked urban poverty reduction to the extension and improvement of urban infrastructure and services in historically disadvantaged areas and by higher tiers of government placing similar emphasis on housing delivery.

By contrast, the vast majority of younger people are asset poor, having no homes and relying on renting backyard shacks and paying their landlords disproportionately for services. Moreover, economic differentiation has to be understood also in relation to access to income. For National and Provincial government competencies, income streams such as pensions are as vital as housing. For local government it suggests that local economic development and job creation are as important as urban services. That they have received second-class treatment to date in postapartheid South Africa explains why younger city dwellers in Meadowlands are increasingly disenchanted with local ANC representatives while retaining a loyalty to the Party more broadly. In the multifaceted nature of social disadvantage in cities, the fault line of social differentiation lies as much across gender and generation as it does across class. This has potential political fallout, particularly when public action is viewed over the long term.

NOTES

1. For a fuller account of the findings of this research see Beall, J., O. Crankshaw, and S. Parnell, *Uniting a Divided City: Governance and Social Exclusion in Johannesburg* (London: Earthscan, 2002). The field research was conducted in 1999 and 2000 and included both contemporary and historical documentary evidence, as well as the perceptions and experiences of a cross section of people living and working in Meadowlands that were gathered from interviews.

2. Parnell, S., and D. Hart. "Self-Help Housing as a Flexible Instrument of State Control in Twentieth Century South Africa." *Housing Studies* 1999, 14:367–386.

3. Morris, P. *Soweto: A Review of Existing Conditions and Some Guidelines for Change* (Johannesburg: Urban Foundation, 1980), pp. 142–143.

4. Meadowlands hostel was built in 1957 and housed some 4,500 residents.

5. It was mainly the backyard shack residents and tenants who were destined for Meadowlands. People resettled from Martindale, Albertsville, Western Native Township, New Clare, Vredesdorp, Alexandra Township, and George Goch, and the neighboring white suburbs later joined them.

6. The fact that Meadowlands residents were removed under antislum legislation and not, as is often erroneously reported, under the Group Areas Act of 1950 means that they have no rights under the postapartheid land restitution acts to reclaim Sophiatown properties. It may also explain their obvious commitment to Meadowlands.

7. Morris, P. *A History of Black Housing in South Africa* (Johannesburg: South Africa Foundation, 1981), pp. 56 and 60.

8. Pirie, G. H. "Letter, Words, Worlds: The Naming of Soweto." *African Studies* 1984, 16:43–51. The Sotho group comprised Tswana, North Sotho, and South Sotho people, while the Nguni group consisted of Zulu, Xhosa, Ndebele, Venda, and Tsonga.

9. Marks, M. "Organisation, Identity and Violence Amongst Activist Diepsloot Youth, 1984–1993." M.A. Dissertation, Faculty of Arts, University of the Witwatersrand, 1993, p. 99.

10. So named after the date (1951) and the number (6) assigned to the prototype; Calderwood, D. 1953. "Native Housing in South Africa." Unpublished Ph.D., University of the Witwatersrand, Johannesburg.

11. Mashabela, H. *Townships of the PWV* (Johannesburg: South African Institute of Race Relations, 1988), p. 149.

12. Morris, P. *Soweto,* p. 149.

13. Swart, C. *Swartbehuising Deel I: Gesinsbehuising in Soweto* (Johannesburg: Randse Afrikaanse Universiteit, 1979), p. 89.

14. Swart, C., p. 90; see also the case study of Alice Makuma, whose two sons slept in the living room of their 51/6 house in Naledi: Ginsberg, R. "Now I Stay in a House: Renovating the Matchbox in Apartheid-Era Soweto." *African Studies* 1996, 55(2):130.

15. Mashabela, H. *Mekhuku: Urban African Cities of the Future* (Johannesburg: South African Institute of Race Relations, 1990).

16. Frankel, P. *Urbanisation and Informal Settlement in the PWV Complex,* Volume 2. Department of Political Studies, University of the Witwatersrand, Johannesburg, 1988, Appendix 6.

17. Crankshaw, O., A. Gilbert, and A. Morris, "Backyard Soweto," *International Journal of Urban and Regional Research 2000,* 24:841–857; Morris, A. (editor), B. Bozzoli, J. Cock, O. Crankshaw, L. Gilbert, L. Lehutso-Phooko, D. Posel, Z. Tshandu, and E. van Huysteen, "Change and Continuity: A Survey of Soweto in the Late 1990s." Department of Sociology, University of the Witwatersrand, 1999, pp. 5 and 70.

18. The Population Census's estimate of the percentage of households in backyard shacks is probably an underestimate.

19. This paragraph is drawn from Crankshaw, Gilbert, and Morris, op. cit.

20. Everett, D. "Yet Another Transition? Urbanisation, Class Formation and the End of National Liberation Struggle in South Africa." Comparative Urban Studies, Occasional Paper Series, No. 24. (Washington, D.C.: Woodrow Wilson International Centre for Scholars, 1999), pp. 18–20.

21. *Race Relations Survey 1986: Part 1* (Johannesburg: South African Institute of Race Relations, 1987), p. 349.

22. 1996 Population Census.

23. Morris, A. (editor), *et al.,* p. 73.

24. These figures are questionable because the Census reports that 29% of formal houses had only three rooms.

25. Sarakinsky, M. *Alexandra: From 'Freehold' to 'Model' Township* (Johannesburg: Development Studies Group, University of the Witwatersrand, 1984), pp. 50–51.

26. Grinker, D. *Inside Soweto: The Inside Story of the Background to the Unrest* (Johannesburg: Eastern Enterprises, 1986), p. 45.

27. "Mzimhlope" is the popular name for Meadowlands hostel, being named after the nearest railway station.

28. Grinker, p. 45; Morris, *Soweto,* p. 148.

29. Inkatha was formerly a tribal organization formed in 1975 with an exclusively Zulu membership. It subsequently transformed itself into a national political party, the Inkatha Freedom Party, at a launch, which (significantly) took place in Sebokeng in present day Gauteng rather than in its earlier base in KwaZulu-Natal.

30. Garson, P. "The Killing Fields." *Africa Report* 1990, 35(5):46–49; Ki, A., and A. Minnaar, "Figuring Out the Problem: Overview of the PWV Conflict from 1990–1993." *Indicator South Africa* 11(2), *Conflict Supplement* 1, 1994, 25–28.

31. Steve Lebelo ascribes this to the fact that the population were subtenants who had much to gain from their removal from Sophiatown backyards to their own housing in Meadowlands.

32. Marks, op. cit., p. 113.

33. "Soweto Rent Boycott," *Indicator South Africa* 1997, 5(50).

12

The Limits of Law

SOCIAL RIGHTS AND URBAN DEVELOPMENT

ERICA EMDON

O ver the past decade, South Africa has undergone a transition from parliamentary sovereignty to constitutionalism. This transition has involved the extension of the three components of citizenship—civil, political, and social—to the previously disenfranchised black majority. The civil component refers to the rights necessary for individual freedom: free speech, freedom of faith, property rights, the right to contract, and the right to fair justice. The political component involves the right to participate in political power. The social component includes rights to economic welfare, education, and social services (Marshall, 1964:71).

T. H. Marshall argues that these components of citizenship have evolved over different historical periods with civil rights obtained in the eighteenth century, political rights in the nineteenth century, and social rights in the twentieth century. Citizenship and social class, though, have been at war (Marshall, 1964:84). Nonetheless, twentieth-century capitalist economies have experienced a diminution in social inequality as governments have guaranteed certain minimum social services to which everyone is entitled.

In South Africa, citizenship was extended to all and constitutionalism was established in 1994. The civil and political components of citizenship are now secure. However, the social component of citizenship, that which ensures greater social equality, has not materialized so as to transform social relations in a meaningful fashion. The new constitution has extended formal rights to certain social goods, but the prerogatives of capital continue to ensure that real social equality is a dream for most people. In the arenas of housing, neighborhood life, urban form, and development, South African society is characterized by private control over capital investment.

This chapter looks at whether the shift to constitutionalism and changes in the law relating to urban governance have made a significant impact on peoples' lives. It does so by drawing examples from the inner city of Johannesburg. The chapter begins by tracing changes in the development of law and the emergence of constitutionalism, turns then to the consequences in street trading, transportation, and housing in inner city Johannesburg, and ends with a reflection on the role of the local council in development planning.

APARTHEID LAW

Prior to 1994, law and legislation reflected the prevailing ideology of the nationalist government and was an expression of the needs of the apartheid and capitalist state. Partial civil, political, and social rights were extended to the white ruling elite only and the law embodied, at the ideological level, the fundamental economic and social needs of the ruling class (Lloyd, 1964:204–205). The entire spatial nature of South African society and the manner in which development took place reflected the apartheid policies of the white ruling elite. Urban areas were characterized by highly developed cities and towns, the former "white" areas. On the outskirts of these areas were impoverished ghettoes, the former "black" areas. Rural areas were made up of large irrigated commercial farms managed by a white farming class and the former "bantustan areas," overcrowded pockets of land in which blacks were allowed to farm and live (see Native Land Act 27 of 1913 and the Native Trust and Land Act 18 of 1936).

Although many jurists during the apartheid period could be characterized as more "reform minded" than their political leaders, they were unwilling to change the law and unsettle the legal edifice of apartheid. A firm ideological commitment to the law characterized the legislature. The role of jurists was to carry out the law as faithfully as possible (Du Plessis, 1999:64–70). The doctrine of parliamentary sovereignty, holding that Parliament may pass any law but that no person or institution can challenge the law, also contributed to an environment in which laws went unchallenged. Parliament derived its power from the electorate, which, once it had passed a law, had to abide by its decisions. Any act of Parliament could be set aside only if it had not been passed in accordance with the appropriate procedures. Most jurists, being in the main legal positivists, were reluctant to question this (De Waal et al., 2000). They saw no fundamental problem with the fact that only whites had voted for Parliament.

Laws that divided the country along racial grounds (i.e., that disallowed people of certain racial groups to live and work in certain areas or that propped up the state by disallowing protest) were never struck off the statute book and very rarely queried by the courts. At the same time, laws were passed that curtailed civil and political rights so that over time, even for whites, these rights ceased to exist. With a few exceptions, jurists tended to support the status quo, merely interpreting the laws that were in place.

As the edifice of the apartheid state came under increasing attack, the law began to shift and reflect less boldly the full apartheid state machinery. It began to play the role of mediating competing needs and claims. During social transformation, perhaps more than at any other time, the law's hegemony began to crumble. The law became more fluid and less clearly reflective of one particular set of interests (Lloyd, 1964:208).

REFORM AND REPRESSION

In South Africa, this fluidity was manifest prior to the inception of the constitutionalism that fundamentally changed the nature of law and legislation. As early as the late 1970s, the law began to shift in response to the changing nature of urban development. For instance, the entire edifice of the apartheid state was premised on the notion that black people were temporary sojourners in the urban areas and their political and social needs were realized in designated "bantustan" areas. To support this political ideology, a number of laws, collectively known as "pass laws," made a person's legal status in the urban areas highly tenuous. [The Black (Urban Areas) Consolidation Act 25 of 1945 was one of the most significant.] Well before the political transformation of 1994 and the first democratic election, the pass laws and the Group Areas Act 41 of 1950, a law that restricted movement of people into former white areas and that disallowed people from living and working in areas that were protected for particular race groups, were repealed. The removal of these legal cornerstones of apartheid reflected the growing recognition that urbanization was inevitable and could not be reversed (Hindson, 1987:88–103). The ruling elite believed that by giving people certain civil and social rights, such as the right to live and work where they choose, the inevitable cry for political rights, particularly the right to vote, would be muffled.

As political turbulence increased, particularly after the Soweto student uprising in 1976, voices among the white, liberal establishment called for more dramatic reforms that would make land and the ownership of land available to black people in urban areas (Bernstein, 1991:322–333). This, it was believed, would create more social stability and diffuse the political demands that were being articulated by black political leaders.

From the late 1970s, laws were passed that allowed blacks to gain, for the first time, tenure rights. In 1978, special regulations made it possible for blacks to hold 99-year leasehold rights in urban areas. [These regulations were subsumed under the Black (Urban Areas) Consolidation Act 25 of 1945.] This early leasehold was not registerable in the main deeds registry offices, nor was it the same as freehold ownership (the ownership form available to whites). In 1984, the Black Communities Development Act 4 of 1984 created a stronger version of the 99-year leasehold. In particular, the leasehold was made registerable in deeds registry offices and the possibility of full freehold ownership rights was broached. Once a township register was opened, the 99-year leasehold right would be converted to freehold ownership. Finally, in 1991, the Upgrading of Land Tenure Rights Act 112 created mechanisms to upgrade a number of "lesser" forms of tenure to full freehold ownership (Emdon, 1993:4–5).

This period has been characterized as one of reform and liberalisation. Yet, it was accompanied by an increase in repressive security law. As political opposition to the apartheid state escalated, state response, at the level of law and security, increased dramatically. At the same time, laws that made up the very essence of the apartheid state were being repealed.

CONSTITUTIONALISM: A NEW BEGINNING

The full transformation to a constitutional democracy, the extension at a formal level of civil and political rights to all, and the coming into being of the Constitution changed the nature of the South African legal system (Constitution Act 108 of 1996). (The interim Constitution came into force on April 27, 1994 and the final Constitution came into force on February 4, 1997.) One of its most significant consequences was to replace the doctrine of parliamentary sovereignty by the doctrine of constitutional sovereignty.

Constitutional sovereignty changes the status of law. The Constitution shapes ordinary law, determines the way in which legislation is interpreted, and places obligations on government that must be fulfilled. In a constitutional legal system, government derives it powers from the Constitution. This involves a limitation on the power of government, allowing government to pass only laws that are consistent with the Constitution (De Waal *et al.*, 2000:6–11). The South African Constitution states that "law or conduct inconsistent with it (its provisions) is invalid, and the obligations imposed by it must be fulfilled" (Section 2).

The Constitution goes further, as Arthur Chaskalson, in the Third Bram Fischer Lecture in 2000, noted: "the Constitution provides that the state must take action to achieve the progressive realisation of socio-economic rights to housing, health care, food, water and social security" (Chaskalson, 2000:29). For example, in the Bill of Rights, the clause on housing states:

(1) Everyone has the right to have access to adequate housing.
(2) The state must take reasonable legislative and other measures, within its available resources, to achieve the progressive realisation of this right.
(3) No-one may be evicted from their home, or have their home demolished, without an order of court made after consideration of the relevant circumstances. No legislation may permit arbitrary evictions. (Section 26 of the Constitution)

In addition the equality clause gives everyone the right to equality before the law. It states that "equality includes the full and equal enjoyment of all rights and freedoms" and ensures that the particular rights set out in the Bill of Rights are generally available [Section 9(2) of the Constitution].

The question that arises is how the Bill of Rights and, particularly, socioeconomic rights such as the right to housing (or water or health care) is to have a real impact on ordinary people's lives. What imperative is there for the government to deliver on these promises? Until recently, the question was largely academic; it had never been tested in the Constitutional Court. During 2000, however, a landmark court case took on this issue, suggesting that the Constitutional Court would like to intervene and even instruct government to take action.

The Constitutional Court was asked to interpret the right to housing clause in the Bill of Rights in a case in which a community was being evicted by a municipal council. The Court held that the housing right creates an obligation on the government to take proactive steps to supply the homeless population with shelter, even if not immediately. At the very least, programs should be put in place and plans made to provide housing (*Grootboom and others v. Government of the Republic of South African and others* CCT 38/00).

Arthur Chaskalson also stated that certain constitutional rights such as the right to dignity, equality, and freedom have little meaning when people are living in extreme poverty and without basic amenities. These rights can be achieved only when socioeconomic conditions are transformed to make realization of these rights possible. He noted that this would take time, but the courts must give effect to the transformative goals of the Constitution and proactively intervene to create an imperative to realize the socioeconomic rights (Chaskalson, 2000).

Aside from the socioeconomic rights mentioned in the Constitution, the Constitution imposes new roles and responsibilities on different spheres of government. These are couched in developmental terms and require the sphere of government in question to deliver social goods. This is particularly the case in respect of local government. For example, the objects of local government are set out in the Constitution, inter alia, as follows:

(a) to ensure the provision of services to communities in a sustainable manner;
(b) to promote social and economic development; . . . [Section 152 (1) of the Constitution]

The next section goes further and states that a municipality must

structure and manage its administration and budgeting and planning processes to give priority to the basic needs of the community, and to promote the social and economic development of the community. . . . (Section 153 of the Constitution)

At a formal level, then, the shift to Constitutionalism has begun the process of ensuring access to Marshall's bundle of civil, political, and social rights.

SLATE NOT WIPED CLEAN

Although much has changed, particularly regarding socioeconomic rights, so much has also stayed the same. The basic nature of capitalism and control over material resources in the country have barely shifted. Although the Constitution heralded a new legal era, many of the old apartheid laws were still in place in 1994 when the first truly inclusive elections were held and in 1996 when the final Constitution was passed. The Constitution began the process of demolishing the nine "bantustans" and unifying the country geographically. However, a number of laws relating to town planning, land tenure, and squatting were not automatically removed. In fact, the Constitution provided that all laws that were in force when the Constitution took effect were to continue, subject to any amendment or repeal and subject to "consistency with the new Constitution" [Section 2 (1) of Schedule 6 of the Constitution].

In the arena of planning law, each of the pre-1994 provinces (Cape Province, Orange Free State, Natal, and Transvaal) had a town planning ordinance regulating land use management and new land development in "white" areas. A number of proclamations also regulated land use, development, and ownership in black areas, the former "non-independent bantus-tans." The nominally "independent bantustans" had their own land use and development law. After 1994, when the boundaries of the nine new provinces came into effect, the administration of these laws was shifted, but the laws continued to be enforced. Only much later, by 2000, did consolidation and repeals create more legal certainty and uniformity, though the process was incomplete. In as many as five provinces there are still four or more laws that regulate land use, development, tenure, and planning, most having been passed during the apartheid years.

What is the significance of this? For one thing, it creates cumbersome systems of planning and development that favor the well-organized, finan-cially strong developers and create difficulties for the poor and marginalized. For example, the laws relating to land development are outside the control of rural black peasants who want to transform their agricultural systems and modernize. Land control has barely changed.

Despite this, by 2000 the legal terrain had become far more favorable to supporting development and urbanization. Yet, changes in the law in this area have been gradual and sometimes nonexistent. Formal rights are there, but real changes that give material substance to them, particularly social rights, have yet to materialize.

The remainder of this chapter will use the inner city area of Johannesburg to illustrate how the former racially divisive legal system has shifted toward a legal system that is more open and facilitatory. Despite the shifts in the for-mal arena of law, these changes have not been sufficient to alleviate urban poverty and struggle. In the inner city, perhaps more than anywhere, the lim-itations of looking to the law as a solution to urban poverty can be seen.

INNER CITY LIVING IN JOHANNESBURG AFTER APARTHEID

If the law reflects material conditions in a society, then the law has changed to reflect the broader political and social transformation of South Africa. Civil, political, and certain social rights (such as the right to a free educa-tion and the right to a pension) have been extended to all. But the nature of the urban form in Johannesburg has not fundamentally altered since 1994. The poor, who are mainly black, still live in ghettoes and shack settlements on the peripheries of the city and the rich live in the leafy suburbs.

The inner city looks different than it did in the 1970s, but these changes have not altered the social reality. All that has happened is that as the area has become overcrowded and poorer and the wealthier white population has left. Until the mid-1970s, the inner city was predominantly inhabited by white middle-class people. Entry to the inner city by blacks was disallowed by the Group Areas Act and the shifts in inner city demographics began only in the late 1970s and accelerated in the 1980s (Dugard, 1978:79).

Johannesburg's inner city is made up of a central business district (CBD) surrounded by a number of high-rise flatland areas. Up to the 1970s the flatland areas—Hillbrow, Berea, and Joubert Park being the best known—were predominantly inhabited by whites (Cloete, 1991:94). The area has changed since the late 1970s. There has been an increase in the overall percentage of blacks coming into the area, many poorer than the whites that used to live there. Now, an estimated 95% of the population of the Johannesburg inner city is black. The inner city has therefore undergone very substantial racial desegregation.

The increased numbers of poorer and mostly black people living in high-density flats have changed the demography of the inner city and increased social problems. For instance, there has been a rise in landlord–tenant conflict, overcrowding of buildings, building abandonment, a decrease in money being spent on maintaining building standards, and an increase in slum conditions. (The decline of Hillbrow is well documented in Morris, 1997: 153–175.) There is also evidence of high levels of unemployment or underemployment. This has led to an increase of marginal people eking out a living by begging on the streets, assisting drivers to park, picking up garbage, and selling fruit.

The CBD used to be the shopping, retail, and commercial center for upper and middle-class whites of Johannesburg. Capital flight away from the city has been profound and the nature of economic activity in the CBD has altered. The retail sector now serves a predominantly black and poorer market and the businesses are smaller, black owned, and probably more marginal than the businesses that were dominant up until the end of the 1970s.

Street trading was never permitted during the apartheid days. The Group Areas Act controlled where blacks could practice business, the Influx Control laws controlled entry of blacks into the cities, and municipal by-laws strictly controlled all informal trading. But toward the end of the 1970s there was a move toward deregulation. As mentioned, the Group Areas Act and Influx Control laws were repealed and this enabled black and poorer people to come freely into the cities. By the late 1980s, much of the pavement in the CBD and in most inner city areas was used by a new breed of entrepreneurs (i.e., informal traders) to ply their wares. The effect this had was profound. Neat sidewalks became bustling areas of economic activity, yet also untidy, congested, and in some instances dangerous places, particularly for pedestrians. The municipality of Johannesburg lost control of the sidewalks and was unable to keep the public spaces clean and well managed. All this contributed to massive capital flight.

Another change that took place from the mid-1970s onward was the change in the major transport system for black commuters from rail and bus to minibus taxis. Hundreds if not thousands of taxis began to rank at various places on all sides of the CBD, initially in an entirely unregulated manner. Many major entry routes into the city became roads on which taxis formed long queues, picking up and dropping off thousands of commuters. It has taken the city and the provincial government almost three decades to control the taxis adequately and only now have real management measures for taxis, such as proper ranking areas and public facilities for commuters, been put in place.

For people living in the inner city areas, the congestion caused by taxis and street trading and the lack of municipal control of the public spaces have meant a deterioration in the living environment. For instance, in discussions with residents in the Joubert Park area, a flatland area in the inner city of Johannesburg, most residents (lower middle-class blacks) complained that when they walk out of their buildings there is litter all over the sidewalk, taxis lined up on their streets, high crime because of the shifting population, and generally hazardous conditions. Even the poorer inner city residents are struggling to live in the unmanaged inner city environment.

PROGRESSIVE LAW IN THE INNER CITY

Despite broad deterioration in conditions in the inner city, since 1994 the situation has gradually stabilized. For the new residents and tenants there have been improvements. The people living in the flatland areas have gained greater security and have a more certain future. Even homeless squatters have benefited by certain changes in the law.

Security of tenure has become more real both through the broad reform of landlord–tenant legislation, which has improved the situation of lessees and the introduction of more secure tenure forms. The aim of the legislators has been to create more certainty in the law governing private lease arrangements. In the past, the common law alone prevailed. This allowed unscrupulous land owners to lease their land or buildings to uneducated and desperate lessees and to put clauses into lease agreements that were unfairly prejudicial to the latter.

Both national and provincial laws regulating the landlord–tenant relationship have been passed, specifically the Rental Housing Act 50 of 1999 and the Gauteng Residential Landlord and Tenant Act 3 of 1997. The Gauteng law, in particular, includes detailed regulations that set out what should be contained in a lease contract and what forms of contract or behavior on the part of lessors are unfair and disallowed practices. The Act contains a section headed "unfair practices," which sets out common problems that arise between landlords and tenants, and regulations that define these practices more precisely. They include matters relating to changing of locks, deposits, eviction, house rules, and nuisances [Section 9 (3) of the Residential Landlord and Tenant Act]. Disputes concerning these practices can be brought to a special dispute resolution board.

Increased security of tenure for blacks in inner city areas, like Johannesburg, has been encouraged by the introduction in the 1990s of the possibility of ownership of flats by occupants by means of sectional title. Sectional title is a secure form of tenure that is governed by statute and gives to owners full ownership rights of the unit that he or she occupies and shared ownership of the common property such as the passages, lifts, lobbies, and gardens. It is as strong a form of tenure as freehold ownership; it is registered in the Deeds Registry and is mortgageable via the Sectional Titles Act 95 of 1986. Over the past decade, many buildings in the inner city of Johannesburg have been converted from rental buildings to sectional title buildings.

Cooperative ownership has also been developed and is being tested in the inner city. For the past decade, an NGO called COPE Affordable Housing has been developing a cooperative housing model to implement in a number of inner city buildings. The model is based on the premise that the people who live in the building should be co-owners of the building. When they leave, they should be given some equity, if there is any. For the model to be implemented successfully, a change in legislation has been necessary and there is currently work being done to amend the Cooperatives Act 91 of 1981, which has traditionally been a law regulating agricultural cooperatives.

People living in flats who have entered into lease agreements with landlords (whether written or oral) are liable for eviction under the common law for breaches in the lease agreement, such as a failure to pay rent. The new landlord–tenant laws have codified that situation and allow for a tenant to be evicted for a breach of a lease agreement if proper notice is given.

Those who occupy land or buildings unlawfully and who are not "lessees" in a relationship with a landlord are covered by other laws. For example, the Prevention of Illegal Squatting (PISA) Act 52 of 1951 had been used in urban areas to remove people from land that they were occupying illegally. The Constitution, though, made PISA unconstitutional, as some of its provisions allowed for evictions without due process and an order of court. The Constitution states that "(3) No one may be evicted from their home, or have their home demolished without an order of court made after considering all the relevant circumstances. No legislation may permit arbitrary evictions" (Section 26 of the Constitution). PISA was later repealed and replaced by the Prevention of Illegal Eviction From and Unlawful Occupation of Land Act 19 of 1998.

Prior to the passage of the Prevention of Illegal Eviction Act, homeless people in the inner city areas of Johannesburg were evicted by the local municipality using PISA. It allowed for evictions of people from land to take place more easily than via current law. The types of evictions that took place were usually of people who had erected shacks on street pavements in busy commercial areas or who had set up shack settlements under highways in the CBD. Also, there were cases of homeless people having put up shacks in parks.

The Prevention of Illegal Eviction Act does not disallow evictions, it merely makes the procedures administratively fairer. For example, notice must be served on an unlawful occupier prior to a court case for his or her eviction [Section 4 (2) of the Prevention of Illegal Eviction Act]. Evictions cannot take place unless there has been an order of the court and, in certain cases, alternative accommodation must be provided. In a few exceptional cases, an eviction can take place without following all the processes laid out in the Act but pending the outcome of a final court order.

In a recent case the Maritzburg High Court considered whether the protection offered by the Prevention of Illegal Eviction Act applies only to squatters who are on someone's property illegally from the start or whether it extends to people who were once legal tenants. Advocates argued that the Prevention of Illegal Eviction Act should be applied to tenants who have lease arrangements with landlords as well as to other "illegal" occupiers. If this judgment had been upheld, it would have given far greater protection to tenants facing eviction than they currently have. For instance, evictions would be prohibited unless consideration was first made regarding the personal circumstances of tenants and the availability of alternative accommodation and other issues such as whether the tenant is elderly or disabled. These kinds of questions have to be considered when using the Prevention of Illegal Eviction Act. However, the argument was not adopted (*The Sunday Times*, 4 February 2001).

INTEGRATED DEVELOPMENT PLANNING

In addition, new laws have consciously aimed to break down apartheid urban forms by introducing a new planning paradigm. The Development Facilitation Act 67 of 1995 set forth development principles that were to inform all land development. One of the Act's principles is that there should be an integration of the former poorer marginalized black areas with the established (former white) urban areas (Section 2). The Development Facilitation Act also has a chapter that requires local authorities to draw up "land development objectives," i.e., plans that demonstrate, among other things, the manner in which such integration is to take place (Section 28). The concept of integration was taken further when the Local Government Transition Act 209 of 1993 was amended in 1996 to include the concept of integrated development plans. These plans must be drawn up by local authorities and used to integrate development and management of the area of a municipality with regard to the principles of the Development Facilitation Act.

The significance of integrated development planning is twofold. On the one hand it puts integration at center stage of planning. Not only spatial integration is involved, which is obviously a prerequisite given the spatial layout of apartheid towns and cities, but also social integration, which implies filling in the spaces between townships and white suburbs. Social integration means that development should occur in such a manner that people's social and economic needs are taken into account, for instance, by ensuring that clinics, schools, and other amenities are available, that transportation is accessible, that the environment is sound, and that people's overall well-being is considered.

The second important aspect of integrated development planning is that it must be driven and implemented by local authorities. This is a significant change from the past when there was much less emphasis in law on the role of local authorities in respect of development and when public authorities played a less immediate role. The centrality of the role given to local government, moreover, is not limited to planning. The Housing Act 107 of 1997 defines new roles and functions for all the different spheres of government and, significantly, gives the role of housing delivery to local government.

Legal imperatives are now in place that will force city planners to consider the inner city areas holistically and require them to plan for the growth and development of these areas. In response, city planners are already introducing renewal strategies in neighborhoods such as Hillbrow and Joubert Park. These strategies aim to stabilize the residential areas and manage the taxis, street trading, and other public environment issues so that the areas work in a more functionally sound manner (Emdon et al., 1999).

THE LAW STILL DOES NOT PROTECT THE URBAN POOR

It has clearly become easier for certain people to live in the inner city, particularly employed people who are willing and able to live by the rules. Contrary to popular myths about the new black population being an "underclass" of indigent households, most black households in the inner city are employed. In fact, many households have more than one full-time breadwinner (Oelofse, 1997). They can secure accommodation and, if they pay their rent, can remain in such accommodation without being evicted. If they are lessees, they must pay their rent and adhere to the lease agreement. If they live in housing cooperatives, they have a stake in the building and will gain some equity upon leaving. And if they are owners of flats in sectional title schemes, their situation is secure if they and their fellow occupants pay the levies to maintain the buildings.

However, a small minority of households cannot afford inner city housing and live in overcrowded poorly maintained buildings. (Approximately 12% of inner city households have no full-time breadwinner [Oelofse, 1997].) For such people, the conditions needed to maintain stability are not always there. Some of these people form part of a broader homeless population estimated in 1999 to be in the region of 5,000 people (Emdon *et al.*, 1999). The homeless population live under bridges, in derelict buildings, and in parks and are in constant conflict with landowners and the council over their situation.

Over the years, a number of buildings in the inner city have become rundown and derelict and have attracted poorer people. Some of these buildings have been abandoned by their owners because they have been unable to cover their operating costs over the years and have run up huge arrears with the municipality for electricity, water, and rates. Owners of these buildings are often reluctant to sell them, because a precondition, in law, of a sale going through is that all moneys owed to the local municipality must be paid.

Some sectional title apartment blocks are failing miserably as well. Owners have stopped paying the monthly levies to the body corporate, which is meant to maintain the buildings, or owners have sublet the flats they own to tenants and then do not pay the levy. The buildings become run-down and owners cannot sell their flats and stop making mortgage payments to banks. The poor living in overcrowded buildings, the homeless, the people who have lost their savings because of failed sectional title buildings, and other marginal underemployed people live under very difficult conditions in the inner city. For them, the new dispensation does not really help. The example of street trading illustrates rather starkly how little protection a certain group of poor people in the city can expect from the law.

THE CASE OF STREET TRADERS

Street trading in the inner city mushroomed in the 1980s. At least part of the reason that the municipality lost control so completely was a result of encouragement by influential lawyers in the early 1980s who were funded and promoted by the Free Market Foundation. The basic argument forwarded by these groups was that the remaining legal prohibitions on trading—i.e., Council licencing laws—should be repealed and there should be minimal regulation. The Businesses Act 71 of 1991 heralded this permissive approach and repealed most existing licensing legislation.

The broad effect of the Businesses Act was felt only later. What had been a quaint revival of street life in the 1980s, became a massive problem of unrestricted and uncontrolled street trading in the 1990s. As time passed, it became apparent to most city users and to local authorities that street trading had to be managed. In 1999, regulations were passed by the Johannesburg municipality restricting street trading to certain areas and also setting conditions relating to hygiene and cleanliness.

The Johannesburg municipality has also embarked on a program aimed at creating fixed market areas where street traders may ply their trade, removing them from the pavements in highly commercial areas. The program is proceeding slowly as traders have resisted the notion that they should be told to trade only in specific places. Consequently, a struggle has ensued between the city officials and the poorer trading community over rights to trade, location of trading, and related issues.

The law on street trading, then, responded to the period of deregulation of street trading in the 1990s by becoming more control oriented. This highlights, perhaps most starkly, the fact that even with the liberalization of the law in South Africa, the extension of rights to all, and the new democratic dispensation, the prevailing interests of capital are stronger than the interests and concerns of the poor and marginalized. Street traders who feel aggrieved by the city's attempts to manage them are often people who make a meager living through their trading and are generally poor. The city, on the other hand, wants to stem the flow of white capital and black consumers out of the city and is prepared to act against the traders in the interests of big capital. Thus the potential job creation among marginalized traders is being dampened by those in control of state resources, especially the law.

LOCAL COUNCIL TAKES CONTROL

For the marginalized poor living and working in the inner city, the struggles with the council and property owners are not very different from those that were waged in the apartheid years. These people are constantly threatened with eviction, face electricity and water cut-offs, have no security of tenure, and cannot make a living free of intervention.

The city managers see the problem differently. They are concerned about capital flight from the CBD and view their role as reviving the inner city, creating more stability, and managing the urban poor and marginalized in these areas so that order is re-established. Recently, for example, the council decided to demolish twelve inner city buildings that are mostly business premises and that have been occupied by homeless people. Most of these buildings have become health hazards and, in many, makeshift corrugated structures have been erected. The city council is using health by-laws to obtain the court orders to demolish. Unlike in the apartheid past however, they cannot simply gain such court orders and evict the inhabitants. They have to find alternative accommodation for the people living there and obtain separate court orders to evict (*The Star*, 9 October 2000).

The law is meant to make things fairer for both sides. Certainly, the reforms made to landlord–tenant legislation create a more equitable balance between the needs of tenants vis-à-vis the needs of landlords. And the broader protections offered in the Constitution and by the more favorable legal environment should provide the urban poor with more leverage and security than they had before. Despite this, the poor still inevitably lose.

So, although at the level of law there may have been progressive changes and shifts, the social and economic conditions that underlie the law have not been dramatically and fundamentally altered for marginalized people in inner city areas. The new ruling elite have not taken up the needs and interests of the urban poor.

The solution is not clear-cut either. Those that run the city want to improve the public environment—the streets, pavements, parks, taxi areas etc.—to attract business and want to renew Johannesburg to reverse its overall decline. The marginalized poor want cheap accommodation, jobs, and better welfare support. They do not want to move to the peripheries of Johannesburg and into sprawling shack settlements that are far from economic opportunities. They want to stay in the city center. In the end they may be forced out, not by apartheid laws but by the new democratically elected local council that sees itself as promoting a clean, tidy, well-run city able to compete globally with other big cities. The council is unlikely to be very tolerant toward people who cannot contribute and who constitute obstacles to its vision of the city. That capital also sees them as an obstacle is a further limitation on the law and legally granted social rights.

REFERENCES

Bernstein, A. 1991. "The Challenge of the Cities," in M. Swilling, R. Humphries, and K. Shubane, editors, *Apartheid in Transition*. Cape Town: Oxford University Press, pp. 322–333.

Chaskalon, A. 2000. "Human Dignity as a Foundational Value of Our Constitutional Order." *South African Journal of Human Rights* 16.

Cloete, F. 1991. "Greying and Free Settlement," in M. Swilling, R. Humphries, and K. Shubane, editors, *Apartheid in Transition*. Cape Town: Oxford University Press, pp. 91–107.

De Waal, J., I. Currie, and G. Erasmus. 2000. *The Bill of Rights Handbook*. Kenwyn: Juta & Co.

Dugard, J. 1978. *Human Rights and the South African Legal Order*. Princeton, NJ: Princeton University Press.

Du Plessis, L. 1999. *An Introduction to Law*. Cape Town: Juta & Co.

Emdon, E. 1993. "Privatisation of State Housing with Special Focus on the Greater Soweto Area." *Urban Forum* 4(2):1–13.

Emdon, E., M. Stewart, I. Mkhabela, M. Nell, R. Gordon, M. Makhetha, and O. Holicki. 1999. *Greater Johannesburg Metropolitan Council Housing Strategy*. Prepared for the Greater Johannesburg Metropolitan Council.

Hindson, D. 1987. "Orderly Urbanisation and Influx Control: From Territorial Apartheid to Regional Spatial Ordering in South Africa," in R. Tomlinson and M. Addleson, editors, *Regional Restructuring Under Apartheid: Urban and Regional Policies in Contemporary South Africa*. Johannesburg: Ravan Press, pp. 74–105.

Lloyd, D. 1964. *The Idea of Law*. Hammandswork, UK: Penguin.

Marshall, T. H. 1964. "Citizenship and Social Class," in T. H. Marshall, editor, *Class, Citizenship and Social Development*. New York: Doubleday, pp. 67–124.

Morris, A. 1997. "Physical Decline in an Inner City Neighbourhood." *Urban Forum* 8(2):153–175.

Oelofse, M. 1997. "Inner city Residents Housing Usage and Attitudes Survey." Inner City Housing Upgrading Trust, Johannesburg.

STATUTES

Black (Urban Areas) Consolidation Act 25 of 1945
Black Communities Development Act 4 of 1984
Businesses Act 71 of 1991
Constitution Act 108 of 1996
Cooperatives Act 91 of 1981
Development Facilitation Act 67 of 1995
Gauteng Residential Landlord and Tenant Act 3 of 1997
Group Areas Act 41 of 1950
Housing Act 107 of 1997
Local Government Transition Act 209 of 1993
Native Land Act 27 of 1913
Native Trust and Land Act 18 of 1936
Prevention of Illegal Eviction from and Unlawful Occupation of Land Act 19 of 1998
Prevention of Illegal Squatting Act 52 of 1951
Rental Housing Act 50 of 1999
Sectional Title Act 95 of 1986
Upgrading of Land Tenure Rights Act 112 of 1991

13

Johannesburg Art Gallery and the Urban Future

JILLIAN CARMAN

The Johannesburg Art Gallery (Figure 1) was established in 1910 as part of an attempt to create a cultural infrastructure that would encourage suitable family settlers from England. Its foundation collection, funded by mining magnates, promoted British cultural values to the exclusion of a South African content. Today its collection is predominantly South African with many black artists represented. The history of the gallery is one of interaction with its local environment both in the way its collection has developed and through its location in Joubert Park.[1]

Johannesburg and its art gallery have always had a love–hate relationship. Public and city council attitudes have been characterized by praise, damnation, and sheer indifference in equal measure and often simultaneously. The negative attitudes have sometimes been unfair, provoked more by unfortunate circumstances and ignorance than real wrongs. For example, typical of the gallery's bad timing, it was unable to be the venue for an exhibition during the Urban Futures 2000 conference because leaks, a recurring and worsening problem, were being fixed.[2]

The gallery's location has also been to its disadvantage: on the southern boundary of Joubert Park next to a railway cutting and busy taxi rank. The cutting was originally meant to be covered over, a plan devised by the gallery's architect, Edwin Lutyens. The gallery has always perched on the edge with its back to the park, instead of being in the center of a grand garden extending over a concealed railway into the Union Grounds to the south. In 1986, the Meyer Pienaar extensions to the original Lutyens building attempted to turn the entrance around, with the main access from the park to the north, but this never really worked. Finding the entrance by way of the

Figure 1

Aerial View of Johannesburg Art Gallery-2000. Photo courtesy of Jillian Carman.

park was unclear and often unsafe, and the main entrance has reverted to the south.

Joubert Park today is not a safe place for visitors. Even in its heyday, when there was a resident horticulturist, bands performed, and visitors flocked to see the aquarium in the hot-house,[3] the gallery was often on the margins both physically and metaphorically. In 1936, for example, when Johannesburg celebrated its golden jubilee with major exhibitions at the old show grounds (the present-day west campus of the University of the Witwatersrand) and illuminations and arches festooned the center of the city, with even the gallery façade being floodlit,[4] the souvenir guide curtly stated that unlike the library, the art gallery played a very inactive part in the life of the city.[5] The gallery had the misfortune at that time of still being rather small (two new pavilions eventually opened in 1940) and it did not have a competent curator. The post had recently been made a full-time one, but it still did not carry permanent-staff privileges and no suitable applicant had been found to fill it.[6] So when Johannesburg was celebrating its fiftieth birthday in 1936 with the most important loan exhibitions it had ever experienced, the Johannesburg Art Gallery was neither an exhibiting venue nor was its curator on the Fine Arts or Hanging committees of the large art section of the exhibitions.[7] It was, indeed, playing an inactive part, largely because it was the victim of circumstances and negative perceptions.

The Johannesburg Art Gallery is again in crisis, a familiar condition, but one that is particularly acute 90 years on. Joubert Park has altered dramatically during the past ten to fifteen years with the influx of low-income flat-dwellers, street vendors, and minibus taxis. It has acquired the dubious reputation of being the prime crime spot of central Johannesburg.[8] The management of the gallery during this time has been increasingly ineffectual in response to these circumstances. Visitors have fallen by two-thirds,[9] municipal funding has been progressively restricted,[10] and the concept of collection-based western-type museums is not so much challenged and debated by the wider public, as it is elsewhere in postcolonial societies,[11] as treated with monumental indifference. The Johannesburg Art Gallery would be in danger of dying through sheer apathy were it not for a recent development that, ironically, has emerged from the very environment that has been aggravating the gallery's survival: the Joubert Park Project.

The Joubert Park Project is "a collaborative public art initiative that aims to promote and activate the social and cultural climate of the Joubert Park precinct" and "to specifically engage with the conditions, the changes and possibilities of inner city living in Johannesburg—particularly in view of the inner city plans for urban renewal and residential rehabilitation."[12] Various partners are involved, principally Jack Mensink of the Artificial Shelter Foundation who initiated the "So where To" Dutch/South African public art project in 1996 out of which the Joubert Park Project grew. Other partners include the Western Joubert Park Precinct Pilot Project, the Joubert Park Neighbourhood Centre, the Market Photography Workshop, the Ziyabuya Children's Art Festival, the Fine Arts Department of the University of the Witwatersrand, and the Johannesburg Art Gallery. A group of artists and cultural practitioners coordinates the Joubert Park Project, funding comes from the Department of Arts, Culture, Science and Technology and the National Arts Council, and the gallery is used as a facility for various community projects. The first major project was a series of workshops, coordinated by professional photographers and artists, for the park's freelance photographers and youth from the neighboring communities. The resulting exhibition opened in the gallery on the Joubert Park Project's *Open Day* of December 3, 2000. This marked the launch of a major multidisciplinary public art exhibition to be staged in the park precinct and the gallery in late 2001.

The Joubert Park Project is not the first initiative linking the Johannesburg Art Gallery with its neighborhood. But it differs fundamentally from the gallery's previous attempts to engage with its immediate environment; the initiative comes from outside the gallery, is a serious and sustained effort, and is part of a network of inner city renewal programs.[13] It is not being developed in isolation or within the confines of the Johannesburg Art Gallery paradigm. The latter would doom the project to failure, given the ineffectual management of the gallery and the unfortunate public perception of it.[14] The project is being developed through local community involvement with the gallery as one of various participants. By virtue of the gallery being used as a principal venue, it presents the most positive opportunity to date for the gallery's survival in the urban future: the opportunity for the gallery to be reinvented and claimed as a neighborhood resource. This is probably the only way the gallery can regain a positive public presence and thus ensure its continued existence, an opinion expressed on different occasions by both Peter Stark, an international expert on urban renewal, and Makhosi Fikile Dhlamini, a Sangoma who conducted a divination through bone throwing at the gallery just before the launch of the Joubert Park Project.[15]

My interest in the Johannesburg Art Gallery is primarily from the standpoint of an art historian and museologist. The gallery can survive only if it inspires popular ownership, but I would also argue for its survival on the grounds of historical importance. The Johannesburg Art Gallery is a unique witness both to the period of its foundation in 1910 and the cultural growth of Johannesburg from 1910 to 2000. Furthermore, it has the unusual distinction of being the only municipal museum or library in Johannesburg that was open to all races during the years of segregation. It is an essential part of the history of Johannesburg. I shall present my argument for the preservation of the Johannesburg Art Gallery in three parts: the foundation collection and its building, the development of the collection as a record of Johannesburg's cultural history, and the phenomenon of a cultural institution in Johannesburg being accessible to all races during the apartheid years.

THE CONTEXT OF THE FOUNDATION IN 1910

The Johannesburg Art Gallery was established largely through the driving force of Florence Phillips, who was born in 1863, grew up in the Karoo town of Colesberg, and spent the first years of her married life with Lionel Phillips in a tin house in Kimberley. She met Lionel on the diamond fields of Kimberley, to which he had gone from London in 1875 to make his fortune. She moved with him and their children to Johannesburg and the gold fields of the Witwatersrand in 1889, by which time Johannesburg was three years old and the Phillipses fortunes were starting to flourish. There was a temporary setback when Lionel was imprisoned in 1896 for his role in the Jameson Raid,[16] and the family went into a period of exile in England. This hardly curbed their life-style, which was financed through Lionel's gold-mining interests. They maintained lavish residences in the early 1900s in London and in the country at Tylney Hall in Hampshire. By the time Lionel received the ultimate social accolade of a baronetcy in 1912, the Phillipses had relocated their base to South Africa, had built their Johannesburg home Villa Arcadia, and had founded the Johannesburg Art Gallery.

Florence Phillips drove the gallery project and Lionel and his fellow mining magnates provided the funds. But the nature of the collection and the building that houses it are due almost entirely to the Anglo-Irish art dealer Hugh Lane. He was asked to put the collection together when he met Florence Phillips in England in April 1909. Through his influence, the eminent British architect Edwin Lutyens was appointed to design the building.

The Johannesburg collection opened on November 29, 1910 in temporary premises in the South African School of Mines and Technology.[17] The Duke of Connaught performed the opening ceremony during a state visit from Britain, which included the opening of the first Union parliament in Cape Town and the laying of foundation stones for the city hall and Rand Regiments Memorial in Johannesburg and the Union Buildings in Pretoria. The collection had aroused great interest in London where it was shown before being shipped to South Africa. Hugh Lane was well known in England, Ireland, and the United States for his dealings in Old Master paintings and modern British and French art.[18] He had founded the Municipal Gallery of Modern Art in Dublin,[19] which opened in January 1908 and earned him a knighthood, and he was renowned for his ability to seek out and identify old master paintings. Moreover, he owned a famous French Impressionist collection. After his successes in Dublin and Johannesburg, he created the nucleus of Cape Town's Michaelis collection of seventeenth-century Dutch and Flemish paintings and was appointed Director of the National Gallery of Ireland in Dublin. His brilliant career was cut short when he drowned aboard the Lusitania in May 1915. His death precipitated a lengthy battle between Dublin and London over his French Impressionists.

Lane's French collection, of which the most famous piece is probably Renoir's *Umbrellas,* had been on loan to his Dublin art gallery until September 1913 when he withdrew it in protest over the Dublin municipality's rejection of Edwin Lutyens' plans for the Dublin gallery building.[20] He then lent the collection to the National Gallery in London, which at that stage still had the Tate firmly in control. The joint trustees of the National Gallery and Tate quibbled over what to display and Lane, who had altered his will to leave the collection to the National Gallery, was so incensed by the National Gallery's attitude that he appended a codicil to his will leaving the collection to Dublin, which had been his original choice until the fallout of September 1913. This unwitnessed codicil was the center of bitter disputes from the time of Lane's death in 1915 until recently, with Dublin and London both claiming Lane's French Impressionists. In terms of the current settlement, the key pieces in the collection are shared on extended loan by the National Gallery in London and the Hugh Lane Municipal Gallery of Modern Art in Dublin.[21]

The point of this digression into Anglo-Irish hostilities is to show the important position Lane occupied in the art world of the early twentieth century. From our viewpoint of nearly 100 years later, with the National Gallery and various Tates at the apex of the museum hierarchy, and the Johannesburg Art Gallery perched on the edge of a railway cutting somewhere near the bottom, it is difficult to believe that the positions were reversed in the early 1900s and that the Tate was "the laughing stock of intelligent people."[22] The Tate and National Gallery were in the iron grip of conservative trustees who preferred to acquire minor anecdotal works from the Royal Academy rather than a Wilson Steer or a French Impressionist. Lane was the first person, through his loans to the National Gallery, to place French Impressionist paintings in an English public collection.[23] His choice of modern British and French art, in both Dublin and Johannesburg, was highly acclaimed by progressive curators and critics, who lamented that they had to travel beyond the shores of England to see the best of contemporary art-making represented in a museum collection.[24] On the strength of what he had done for Dublin and Johannesburg, it was even suggested that Lane should be appointed a director of the Tate and be given sufficient funds to build a modern foreign gallery in London.[25]

Although funded by Randlord money and reflecting what the Randlords wanted by asserting the superiority of a particular culture, the Johannesburg collection of 1910 was really Lane's creation. In it he perfected what he had done with the Dublin collection: he selected items that displayed contemporary British art and the British and European roots on which it drew. It was regarded in England as probably the finest small collection of its day— and it predated the Tate and National Gallery in displaying French Impressionists and modern British artists.[26]

Lane also played a crucial role in securing Edwin Lutyens as the architect for the building. Lutyens by that time was one of the most prominent of the Edwardian architects, and was engaged in a project in Rome when he agreed to do the Johannesburg Art Gallery. Through Lane's persuasion— and Lord Curzon's[27]—Lutyens was brought to Johannesburg in late 1910, shown the chosen site in Joubert Park, and officially appointed in April 1911. Lutyens nominated a local architect, Robert Howden, to be his collaborator both during the first phase of the building, which opened in an incomplete state in 1915, and again in the 1930s when Lutyens reworked his 1911 design and two of the originally planned four pavilions were added.[28] Lutyens' original design of a building surrounding a central courtyard with pavilions at each corner was never realized, nor were his grandiose plans for a park-like setting. But at least a major part of the building was erected. Lane's promotion of Lutyens as the architect for his Dublin gallery came to nothing, and Lutyens fulfilled no other museum designs.

Johannesburg therefore has the only museum ever built to Edwin Lutyens's plans. It occupies a unique position in both the history of museum building in the early twentieth century and the history of Lutyens's career, which reached new heights when he was appointed architect of New Delhi in 1912 with Herbert Baker.[29] The gallery reinforces the temple-like façade used in art galleries of the nineteenth to twentieth centuries in Britain, North America, and Australia, such as the first Tate of 1897, the Art Gallery of New South Wales in Sydney, and the National Gallery of Art in Washington, D.C.[30] It is the only art gallery of this type of design at this period in South Africa[31] and links the African sub-continent with wider museum developments of the time and, through its designer, Lutyens, with the British imperialism that informed New Delhi. Lutyens's building embodies the imperialist ideals of the gallery's patrons rooted in the context of the town.

The Johannesburg Art Gallery was founded when Johannesburg was only 24 years old and was still recovering from the Second South African War (Anglo-Boer War, 1899–1902). The British may have been victorious over the Boers, but the Boers were not behaving like the vanquished and were mobilizing political strength. The young town was generally regarded as a place of temporary residence and had no state-funded cultural infrastructure of libraries and museums and the mining industry, the reason for Johannesburg's existence, was in a precarious position.[32] Both British culture and the productivity of the gold mines were under threat.

The gallery provided an opportunity for certain powerful British mine owners, the Randlords, to reaffirm the superiority of the British way of life and construct a civil society that would attract suitable immigrant families to service the mines and their community. In the gallery, British superiority is pronounced through the choice of an imperialist British architect for the building and though a bias in the foundation collection toward British artists and subjects. No South African school of painting is in the foundation collection and no Boer leaders are in the proposed "Nucleus of a National Portrait Gallery" of people who are making history in South Africa.[33] The only local artist included in the foundation collection is Anton van Wouw,[34] who is placed in the "Statuary" section without the distinction of a national school, and the only South African imagery appears in seven items out of 130.[35] British culture predominates in the nature of the collection and of the institution itself. In a brand of imperialism peculiar to local conditions, this statement of cultural superiority is appropriated and adapted to further social stability, which at that time increasingly depended on reconciliation between the British and their former foes, the Afrikaner. Of necessity, imperialism was diluted in an attempt to accommodate local South African aspirations. So we have the peculiar situation of a British-biased collection opening to the public in 1910 and moving into its imperialist permanent home in 1915 at the time the core collection was beginning to adapt and change its imperialist bias in response to local circumstances.

THE DEVELOPMENT OF THE GALLERY IN RESPONSE
TO THE ENVIRONMENT

When Florence Phillips met Hugh Lane in April 1909, he persuaded her to collect contemporary art for the new gallery, concentrating on the British school and the European schools that had influenced the development of modern British art. Florence Phillips appears initially to have wanted Dutch old masters or a loan exhibition from England of furniture and other applied arts. She did not consider including South African art because, as she stated in a letter to the press at the time, there was not yet a South African school of painting.[36] Hugh Lane, never having visited South Africa before he brought out the fully formed collection, was not familiar with local art-making nor did he see the necessity to be so. He had certain ideas about modern art that he wished to pursue and Florence Phillips and her co-patrons acceded to them in an astonishingly quick and complete way. This arms-length patronage is a distinctive feature of the Johannesburg foundation collection: almost none of the works came as gifts from personal collections. They were bought on behalf of the donors by an independent dealer who complied with a general aim of promoting British culture. The first purchases were made a few days after Phillips and Lane met: three paintings off Philip Wilson Steer's retrospective exhibition at the Goupil Galleries. Phillips apparently sold her blue diamond ring to finance them. Thereafter she extracted funds from her mining friends to finance Lane's purchases. By the time she set sail for South Africa in November 1909, more than £20,000 had been promised for the new gallery, and Lane had already spent £6,000.[37]

Lane chose to purchase contemporary British art and exemplars of its origins. A number of the British artists, who include Steer, Walter Richard Sickert, and Spencer Gore, were influenced by French Impressionists. Lane thus chose examples to show the general source of inspiration, such as works by Eugène Boudin, Camille Pissarro, Claude Monet, and Alfred Sisley. He also procured a highly important marble bust of Eve Fairfax by Auguste Rodin, the finest of several versions that are in museums worldwide. Among the Victorian precursors of the British modern school he acquired examples by David Wilkie, William Powell Frith, and Edwin Landseer.

Lane identified directions in which the collection should develop, one area being the Pre-Raphaelite School. Shortly after the nucleus had opened in November 1910, Robert Ross and Henry Tonks, who replaced Lane as London-based directors, acquired works in this area using funds from the mining magnate Sigismund Neumann. By this time, the Johannesburg Art Gallery had a part-time temporary Johannesburg-based curator, Albert Gyngell, but he had virtually no powers in the choice of acquisitions. When the Neumann gift was made in 1912, no South African paintings were in the collection. By the time the Neumann gift reached Johannesburg in late 1916, after a lengthy storage in the basement of the Tate, the gallery had finally acquired a South African painting by Gwelo Goodman. This was presented by Lionel Phillips in 1916, undoubtedly in response to local expectations.

There had been great dissatisfaction among Transvaal architects about the choice of an English architect for the building,[38] and local artists were equally affronted by their exclusion from the collection. Goodman himself had written to the London-based director, Robert Ross, in 1913: "I am still suffering under the great disappointment of not being represented at Johannesburg (my native country is South Africa). . . . It seems to me all the harder because 12 other galleries have purchased my work in the last year or two, including Liverpool . . . Ottawa National Gallery, Toronto . . . etc."[39] Three years later, a work of his was in the Johannesburg Art Gallery and realignment of the gallery's focus had begun, a process that would characterize the development of the collection for the rest of the twentieth century. In these early days, though, the shift of emphasis was very limited, not least because the Johannesburg-based curator's post remained part-time and temporary until the 1930s and the gallery did not receive a municipal purchasing budget until the 1920s, when it was described as a "contribution to the Art Gallery Committee."[40] No proper acquisition policy was in place and purchases and the acceptance of gifts seemed to be on the whim of donors and the Art Gallery Committee. A London-based buying committee, consisting of the directors of the Tate and National Gallery and Henry Tonks of the Slade, advised on purchases during the late 1920s to early 1930s,[41] but, according to a later director, Anton Hendriks, in 1947, "the results were a complete failure . . . the agents did not know the Council's collection and the paintings did not fit in, and some were duplicated, and others were so bad that they have never been exhibited."[42]

During this lean period up to the late 1930s there were, however, three highlights: the first was the Howard Pim bequest in 1934 of 551 prints, which included fine examples by Rembrandt and Drer and established the gallery's Print Cabinet as one of the most important in South Africa. The second was the addition of two Lutyens-designed pavilions, which opened in 1940 and dramatically increased the exhibition space. The third was the appointment of Anton Hendriks as director in 1937.[43]

Hendriks was director until his retirement in the early 1960s. He was the first professional Johannesburg-based director to be appointed and he effected major developments in the collection. His policies for the future of the gallery were tabled at council in July 1946 when he declared that

> A modern art gallery . . . has an active function to perform as an educational institution in the life of the city. In order to convert the Johannesburg Art Gallery from a static show place to an institution which will fulfil this function as part of the city life, the existing collections, which are merely the foundations of the more representative collections of the future, must be built up and completed according to a general plan, and new exhibits must be shown from time to time.[44]

In the same report he set out his development plan, dividing the collection into English School, French School, nineteenth-century Dutch School (Netherlands and Belgian), Rest of Europe, and South African painting. All but the last, South African painting, had been present in Hugh Lane's original collection. Of the South African collection Hendriks said:

> It has been the policy of the Art Gallery Committee, for the last few years, to build up a collection showing the best and most significant aspects of South African painting, and a small but fairly representative collection is being formed. The educational value of such a collection displayed next to the best European painting cannot be over-estimated.

Under Hendriks the emerging South African collection grew and consolidated. He was the first director to acquire, in 1940, a work by a black artist, Gerard Sekoto—a unique event, as no further works by black artists were purchased until the early 1970s.[45] He acquired works by an older generation of South African artists such as Pieter Wenning and Strat Caldecott, who should have been in the collection at, or soon after, its inception. And he recognized the importance of contemporary South African artists such as Ruth Everard Haden, Maggie Laubser, Alexis Preller, and Willem de Sanderes Hendrikz.

The South African School was not the only one to benefit from Hendriks' directorship. He expanded the European collections, in particular by establishing the seventeenth-century Dutch collection when he procured the Eduard Houthakker gift in the late 1940s, coincidentally about the time the National Party came to power. This was a concerted move to accommodate the cultural heritage of Dutch-origin South Africans who had felt alienated by the British emphasis of the foundation collection. The French late nineteenth- and early twentieth-century collection also flourished under Hendriks's care. It was temporarily augmented by the loan from the 1940s to 1960s of the privately owned Hague collection, which included paintings by Cézanne, Daumier, Van Gogh, Pissarro, and Renoir.

Contemporary art—local and foreign—fared less well toward the end of Hendriks's directorship. Funds became more scarce and his firm guiding hand became more of a stranglehold. The Lutyens building also fared badly, in fact it nearly ceased to be a gallery. In the 1960s, plans were made to move the collection to new and larger premises in Parktown and the Lutyens building was put up for sale in late 1960. Various proposals for the reuse of the building were discussed: a railway museum, a bus terminus, a crèche, a music school, and an eye research institute.[46] Fortunately nothing came of these plans. Because the Lutyens building was declared a national monument in January 1993, insensitive transformations should now be prohibited.

Uncertainty also surrounded the position of director in the early 1960s. Hendriks wished to retire in July 1962, but he stayed on until 1964 so that a suitable replacement could be found. The search for a replacement continued without success and his assistant, Nel Erasmus, who eventually assumed the full post of director, was appointed acting director in late 1966.[47] The city council's undermining of the status of Nel Erasmus (her height and gender were apparently to her disadvantage), understaffing, and underfunding not surprisingly led to a lack of public confidence in the gallery. "The Johannesburg Art Gallery is static, it lacks vitality, it is nothing but a richly embellished mausoleum" announced The Star, April 30, 1965, saying that its press files were full of criticisms. Among the list of gripes were no constructive policy on the part of the council, no guides, no explanations of the exhibits, no catalogues, and a neglect of South African art. The following year, possibly in response to criticisms, the council increased the gallery's annual buying grant from R10,000 to R30,000. And the gallery clarified its collecting policy:

> The future policy will be to continue to buy only the best South African works and to acquire suitable overseas art as the means allow. . . . The gallery will not specialize in South African art because it is felt that other South African galleries and museums are doing this.[48]

Until her resignation in 1977, Nel Erasmus made her mark on the collection primarily in the area of contemporary international art. Perhaps her most famous acquisition was a Picasso drawing of a harlequin, partly funded by the newly formed Friends of the Johannesburg Art Gallery. It caused a huge controversy and was the best public relations exercise the gallery could have hoped for. The representation of South African art was still considered to be particularly poor however, and the gallery was criticized in progressive quarters for not acknowledging the historical roots of the black majority of Johannesburg citizens.[49] Nel Erasmus also addressed other criticisms. A catalogue list was published in 1968, free pamphlet guides were offered in the display areas, and the first voluntary guides were trained in 1975.

The latter two initiatives were done under the guidance of Nel Erasmus's assistant, Pat Senior, who was the director from 1977 until 1983. Senior strengthened the gallery's research base by initiating an archive system for the gallery collection,[50] trained additional groups of voluntary guides, obtained council approval and funding for major extensions to the Lutyens building, and founded a conservation department. She also continued the contemporary international buying trend, consolidated the gallery's important early twentieth-century British Moderns collection, augmented the South African collection, and initiated a guest artist program whereby local artists were invited to create installations of their works on gallery premises. This was one of the gallery's first steps at engaging the outside communities.

After Pat Senior's death in 1983 Christopher Till was appointed director until 1991 when he became Director of Culture for Johannesburg. With Till, the gallery entered a new phase in which it focused almost exclusively on South African art. He supervised the completion of the Meyer Pienaar and Partners' extensions that had begun under Pat Senior, procured the most important donation since the inception of the gallery (the Anglo-American gift of R6,000,000), and organized a major sculpture competition to coincide with the opening of the new extensions in 1986. Two of the four winning pieces were placed in the park and a third was placed at the road entrance. Through inviting participation in a competition and placing artworks in the areas outside its boundaries, the gallery was moving into the community in a new way.

Under Till, the exhibition and acquisition policies underwent some of the most important developments in the gallery's history and the gallery began to build a reputation for its curated exhibitions, research, and educational publications. In the exhibition area, a series of groundbreaking shows with research catalogues was initiated. The first and probably the one with the most impact was *The Neglected Tradition* (1988) in which the inherited "white" history of South African art was challenged and the contribution of black artists was reassessed and integrated into the mainstream. In the acquisition area, historic works by black artists were actively sought and purchased and contemporary South African art, made by graduates of blue-chip institutions as well as rural dwellers, began to be acquired on an unprecedented scale. Even more important was the institution of a new part of the collection: "traditional" southern African art. This comprises items such as carved sticks, beadwork, headrests, and stools, which, in the past, were considered more appropriate for anthropological than art historical museums. Their incorporation into the Johannesburg Art Gallery, along with a reassessment of exhibitions, indicated a fundamental shift in attitude to the history of art and the old divisions of "fine" and "applied" art. An exhibition and catalogue (*Art and Ambiguity*) in 1991 of a major loan collection of traditional southern African art, the Brenthurst Collection, enabled the gallery to make a forceful statement about its change of direction and to show that it was located in Africa rather than the periphery of a first world empire. Acquisitions in the traditional southern African area have been actively pursued under Till's successor, Rochelle Keene, and this collection today constitutes the major part of the Johannesburg Art Gallery.[51]

A further development in the late 1980s was the establishment of the education department and the opening of the gallery library to the public. Although the gallery for many years had been engaged in educational activities in the way of lectures and tours, assisted by voluntary guides, the formalizing of this role has been particularly significant. Educational outreach has been a principal tool in facilitating access to an institution that was imperialist in origin and intent and that, despite transforming radically, is still viewed by many as elitist and representing an alien culture. There have also been more formal links with art educators through the gallery's association with the Imbali Teacher Training Project (initiated in 1989 to improve art education for teachers), the Art Educators Association (their Art Resource Centre moved from Joubert Park to the gallery's premises in 1996), and cooperation with the Curriculum Development Project for the Creative Arts.[52] The latter resulted in a series of resource books for use by learners and teachers, the first of which was Lesley Spiro Cohen's *Jackson Hlungwani* of 1993.

The consolidation of the gallery's educational role has underscored what is probably the gallery's primary function today: a resource in the service of its local community. It serves as a venue for cultural events, such as those of the Joubert Park Project, and as an educational center where there is a shift from passive display to a facility offering informal skills training, learner-focused library resources, and published material and exhibitions in support of new curricula. The gallery as public facility is reason enough for regarding it an essential part of the urban environment, but particularly so because it has always allowed access, even if grudgingly at times, to visitors of color.

THE GALLERY AND OPEN ACCESS

Unlike other Johannesburg cultural institutions, such as the Johannesburg Public Library (JPL) and the Africana Museum (now MuseuMAfricA), the gallery's doors have been open to all races from the day it opened, through the increasing segregation of the 1920s to 1940s and during the high apartheid years of the 1950s to 1970s. This is one of the most intriguing parts of its history, not least because it is largely unrecorded. At about the time the council resolved in January 1974 to repeal discriminatory resolutions and to allow "Non-Whites" admittance to the Libraries, Museums, Art Gallery, and Zoo,[53] the gallery issued a publicity pamphlet announcing that "The Johannesburg Art Gallery has always been open to all races."[54] There appears to be no record in council minutes to prove or disprove this. By contrast, there are references to the JPL being reserved for Europeans in 1924, to recreational places like parks being proclaimed whites-only areas in 1950, and to the Geological and Africana museums being reserved for whites.[55] One needs to look to more unofficial records than council minutes to find that black visitors, although not always made to feel welcome and even at times prohibited, were generally allowed access. R. M. Gandhi, for example, complained by letter to the Town Clerk in November 1917 that he had been refused admission to the gallery and was advised that the curator was not acting under council instructions in doing so and that "coloured" people were allowed access.[56]

Howard Pim described a visit in the company of a black artist in 1928, evidently an unusual enough event to warrant a written comment to the Town Clerk. Pim informed the Town Clerk in February 1928: "I took Moses Tladi to the Gallery on Saturday afternoon, and think the General Purposes Committee might like to hear from me as to this new departure. There were a fair number of people in the Gallery, but they all took Moses' presence there as quite a matter of course and there was no hint of any difficulty."[57] There seems to have been no official reaction to this "new departure." However, a letter in the *Rand Daily Mail* of November 18, 1929 from M. R. and B. M. Desai complains that they were refused admittance to the gallery because "instructions had been received from the Town Clerk that coloured persons, including Indians, should not be allowed in until it had been decided by the City Council whether or not they were to be admitted." An article in the *Rand Daily Mail* of the next day notes "Some months ago the [art gallery] committee agreed to admit to the gallery a certain native who is studying art, and also other natives . . . so far, no other exception has been made to the rule by which only Europeans may visit the gallery."[58]

Some correspondence exists between the curator of the art gallery and the Town Clerk at this time relating to the matter, the final word seeming to be the Town Clerk's letter of December 12, 1929: "I have to advise you that the General Purposes Committee, at its meeting on 4th instant, agreed to the recommendation of the Art Gallery Committee that educated and cultured Indians and Coloured Persons (exclusive of Natives) be permitted to visit the Art Gallery on Thursday afternoons from 2 p.m. till closing time."[59] This recommendation does not seem to have been reported to council or to have been binding. The prohibited "natives" were allowed access during the 1930s, although, like "Indians and Coloured Persons," they were usually restricted to Thursday afternoons, traditionally the domestic worker's day off.[60] At the Art Gallery Committee meeting of May 17, 1940, which approved the purchase of Gerard Sekoto's *Yellow houses: A street in Sophiatown,* a motion to introduce restrictions on "native visitors" was objected to by the curator, Anton Hendriks, and overruled by the Town Clerk.[61]

A file in the Johannesburg Art Gallery archives titled "Non-European Visitors, 13/8" contains two letters dating from late 1948—the year the National Party came to power—which again suggest that Hendriks encouraged visitors of color. The first is from Hugh Tracey of the African Music Society asking if Africans can visit the gallery, and the second is a copy of Anton Hendriks's reply: "I am not aware of any local regulation in connection with visiting the art gallery by Africans. We have quite a number of natives who are specially interested in painting who come to the gallery" (November 4, 1948). There are no further letters in this file, indicating either sloppy recordkeeping or that this was essentially a nonissue that generated no further correspondence. One must remember that the entrance to the gallery was not through the racially segregated park but from the south boundary, so accessing the gallery was unlikely to be conspicuous enough to cause outside comment.[62]

There are also the occasional references in the gallery's archives that suggest black visitors were admitted. An example is the note prepared for Mr. Shorten for "his publication on Jhb. 80th anniversary" May 26, 1966, in which "non-European professional groups" are listed among the outside bodies who request conducted tours.[63] Personal recollections, however, provide the most compelling evidence. The artist Paul Ramagaga recalls visiting the gallery in 1950 to see the European collection.[64] And the artist and teacher Cecil Skotnes recalls access being granted to students from the Polly Street Art Centre, a recreational venue under the control of the Non-European Affairs Department of the Johannesburg city council.

Polly Street Art Centre, according to David Koloane, was "the first centre for black artists . . . [and is] one of the most significant milestones in the
emergence of a professional class of black artists in South Africa.[65] Skotnes was
appointed cultural recreation officer of the center in 1952 and taught here
until it was closed in 1960. Skotnes recalls offering modeling classes in brick
clay to the students during 1956–1957 and wanting them to study southern
African terracotta figures in the Africana Museum. However, the chief librar-
ian of the JPL, R. F. Kennedy (under whose direction the museum fell),
refused access on the grounds that he supported "totally and absolutely" the
policy of apartheid. Skotnes then made an appointment with the director
of the gallery, Anton Hendriks, to request access for his students to study
sculpture there. His reception was entirely different. Hendriks asserted that
he was against any form of cultural apartheid and that unaccompanied black
visitors were welcome any time or day. He introduced Skotnes to the head
of the gallery security and made his instructions clear that black visitors were
welcome, and urged Skotnes to contact him should there be any problems
regarding access. There never were, recalls Skotnes. His students regularly
visited and were always welcomed.[66]

This recollection is endorsed by Hendriks' assistant at the time, Nel
Erasmus. She recalls there were no specific rules prohibiting black visitors,
but that city councillors were generally so indifferent to the affairs of the
gallery they would probably never have known about its visitor profile.
Erasmus was, however, increasingly anxious during the 1960s about the
unwritten open access policy: she was breaking the law and would lose her
job if found out. A particular fear was that a black visitor might sit on one
of the benches and a white visitor might lodge a complaint, thus leading to
an official enquiry and dismissal.[67] In hindsight this seems ludicrously petty,
but it was quite credible in the context of the 1960s when many aspects of
apartheid legislation were "perfected."

The council's general inattention owes much to three factors: the circumstances of the gallery's foundation, its method of reporting to council,
and its relative distance from council structures. Apart from the deeds of
donation of 1910 and 1913 relating primarily to the artworks and the Art
Gallery Committee structure,[68] there was no tabling of rules and regulations
regarding visitors, such as those tabled when the council took over the JPL
in July 1924. This indicates the council's indifference from the beginning.
According to John Maud, who was commissioned by the council in 1935
to write a history of the local government of Johannesburg, the council took
responsibility for its art gallery in 1910 only because the collection was

offered as a gift, because of the council's "natural anxiety . . . to improve the heritage of the city," and because of the deal proffered by the Transvaal government, by which it would endow the revenue from the market to the council in return for the council accepting an art gallery. Having agreed to this deal, Maud continues, the council then reneged on adequate funding. It did not incur any net expenditure until 1913–1914 and thereafter hardly lavished municipal revenue on its art gallery.[69] I have already referred to this when discussing the nonexistent purchase grant in the early years and the temporary part-time nature of the curator, who was made full-time in 1935, was placed in the same leave position as a permanent member of staff in 1940, and admitted to the municipal pension fund only in 1949.[70]

The council's lack of interest is also reflected in the gallery's reporting methods to council. The Johannesburg Art Gallery and its governing trustees to whom it reported, the Art Gallery Committee, fell initially under the Town Clerk's department. The Town Clerk's department in turn reported to the General Purposes Committee, which then reported to council. In due course, the Art Gallery Committee reported to various other bodies, such as the Art and Culture Committee and the Health and Amenities Committee, but never directly to council. On the other hand, when the JPL was taken over by the council in 1924, it reported directly to council. When the Geological and Africana museums were accommodated in the JPL shortly afterward (1927 and 1933, respectively), they jointly reported as the "Library, Africana and Geological Museums Committee."[71] The JPL and museums were associated from the beginning; the art gallery was often forgotten as a fellow museum, an omission that persists in council and tourist publicists' minds to the present day.[72] Furthermore, the JPL and its museums were physically close to council offices in the years of segregation—on either side of the Library Gardens in the center of town—whereas the gallery was some distance off.[73]

There are three additional reasons why the gallery escaped discriminatory regulations: none of its exhibits could be touched, its premises were never crowded, and officially sanctioned art was considered above politics. At a time when library books could not be shared, separate transport and lifts for blacks were introduced, only whites could sit on park benches or even enter parks, and a councillor tried to prohibit the use of sidewalks by blacks,[74] not to mention the other discriminatory measures that imply a complete tactile phobia on a casual basis (domestic workers excepted), the nontactile display methods and space in the gallery—provided no black visitor sat on a bench—appear to have saved it from scrutiny. In addition, there were unlikely to have been enough black visitors to pose a threat of crowding. If the average black citizen today, after concerted efforts to attract her or him, still finds a temple-like museum, such as the Johannesburg Art Gallery, off-putting and has little desire to enter,[75] the citizen in a racially segregated society is even less likely to have wanted to gain access. Artists and members of professional groups would have been more interested in visiting and were allowed access, as we have seen. But they evidently did not arrive in large enough numbers to cause notice. Anyway, there were not enough white visitors to arouse the kind of bigotry that was manifested in crowded places.[76] From the late 1940s to early 1950s visitors (and we can presume they were mainly white) averaged 62,000 per financial year (July to June), they peaked during the Van Riebeeck celebrations of April–May 1952[77] when over 77,000 visitors were recorded, gradually declined to 39,912 in 1963/1964, then started to pick up again, reaching over 60,000 in 1965/1966.[78] The Africana Museum had nearly twice the number of visitors during the same time.[79]

In support of my third reason, that officially sanctioned art was deemed above the sordid realities of politics, I cite the case of the Pretoria Art Museum. Its foundation stone was laid in a racially exclusive city by the grand master of apartheid, H. F. Verwoerd, in October 1962.[80] Yet when the building opened under the directorship of Albert Werth in May 1964, its policy was "to see the work of black artists as an integral part of South African art and not to exhibit it separately," and it allowed access to all races.[81] Werth recalls that the foundation regulations specified that black visitors could have access only on Thursday, and that this seems to have been usual among South African art galleries at that time.[82] Werth was required to state the Thursday restriction in the signage. After a couple of months he removed the restriction notice from the entrance board and allowed access to black visitors on any day of the week. But very few black people came in the 1960s and anyway the Pretoria city councilors were usually so indifferent to their art museum that they would not have noticed. The few times the Pretoria councilors expressed interest in the museum were when Werth tried to acquire vaguely controversial artworks, such as paintings of nude women or (on one occasion) a sculpture by a black artist, Lucas Sithole.[83]

Neither the Pretoria Art Museum nor the Johannesburg Art Gallery was a vociferous opponent of government policy. They would not have received local government funding if they had been. So the reasons for their open admission lie elsewhere: in the indifference of municipal officials to their art museums, the nontactility of objects, the lack of crowding, the anodyne public collections of that time, the level of cultural compliance of the black visitor who would actually want to enter the buildings, and the patronizing belief among state-employed white museum directors that art is uplifting and is above politics and other sordid matters. In short, officially sanctioned art museums and art-making during the patriarchal years of apartheid may have had a semblance of equity only because they were not invested with any real power. Art and institutions that challenged the status quo and posed a threat to the social order of the day were proscribed.

An interesting rider to this is that one of the few selling outlets for black artists at the time was the Artists' Market Association ("Artists under the Sun") founded in 1962 to cater for "Sunday Painters." It used outdoor venues such as the Zoo Lake, Pieter Roos Park, and Joubert Park and was open to all races. One of its founder members was the Sophiatown and Soweto art teacher, John Koenakeefe Mohl.[84] The art shown here—and still being shown until the present day at Zoo Lake—is marked by conservative compliance, catering to the lowest common denominator. It would hardly have posed a threat to the apartheid regime.

Black visitors to a place like the Johannesburg Art Gallery, although not openly discriminated against, were nevertheless in a position of subservience. As Gavin Younge said in 1979, at a conference that focused on the problems of serious and committed art production in the context of South Africa in the 1970s, "Although blacks are perhaps not openly discriminated against by the art institutions and the art public, there is nonetheless the impression that, proportionately speaking, they have very little say in the dominant modes of cultural expression in this country."[85]

THE GALLERY IN THE URBAN FUTURE

The history of the Johannesburg Art Gallery is not so much about its impact on the Johannesburg environment, but Johannesburg's impact on the gallery. Although I have tried to establish the gallery's international importance, the gallery had little impact on its locality in terms of its collection and building. But its locality, almost immediately, had an impact on the gallery: its building was not completed, its park-like setting bridging the railway cutting was not realized, and its collection began to deviate from Hugh Lane's original policy of collecting European art. In addition, the dynamics of curators, council authority, and public opinion affected the visitor profile, the development of the collection, and the nature of the institution. Visitors of all races have always had access, representation of South African art-making, originally excluded, now predominates, and the gallery has started to transform from temple to resource.

The Johannesburg Art Gallery is both cultural recorder and resource. It is a unique witness to colonial imperialism of the early twentieth century, the developing cultural life of an urban community, the dynamics of community participation, and the role that art and museums played under apartheid and can play today in a democratic South Africa. In adapting to its situation in Joubert Park and to wider concerns—or, more accurately, in the appropriation of the gallery by its environment—the Johannesburg Art Gallery reveals its potential as a major community resource.

The changing social structure of Joubert Park has had a critical negative impact on the gallery in recent years. Ironically, this environment is precisely what is likely to ensure the gallery's survival. The Joubert Park Project is allowing the gallery to be seen from a different and positive perspective and is thereby facilitating the reinvention of the Johannesburg Art Gallery as a cultural institution owned by its community. In the divination of the Sangoma Makhosi Fikile Dhlamini, the gallery is described as the space of the ancestors: "Ancestors of different races, cross culturally, dwell in the Gallery." However, the business of the gallery is not running smoothly or successfully, and the working environment is stressful and conflictual, because the gallery and its ancestors have not been "addressed, remembered and introduced to the people."

> The metaphor of the ancestors and ancestral discontent is a matter of integration. The ancestors when appeased reflect a holistic integration, and generate guidance, protection and blessing. The integration process should involve the ancestors of the Gallery, the artists, management and employees, as well as the surrounding community. The Gallery space must be introduced to and integrated into the environment. Without integration and opening of the space and the energies which comprise that environment, the Gallery will be out of balance and in conflict.[86]

The method proposed for the integration and opening of the gallery space involves animal sacrifices, vigils, and ceremonies with Sangomas from all different cultures. This is a process recommended by prominent spokespeople in the Joubert Park community. It is an unusual solution for a troubled museum, emanating from "the community" that the gallery has so often glibly said it is there to serve. The test will be whether the gallery management has the will and ability to give itself over to its environment. Allowing such a deep and committed community involvement would be a sure way of reinventing the gallery and ensuring its survival in the urban future.

NOTES

1. A version of this chapter, "White Elephant or Essential Service? The Place of the Johannesburg Art Gallery in the Urban Future," was given at the Urban Futures 2000 conference, July 2000. It is based on research for a Ph.D. dissertation on the foundation of the Johannesburg Art Gallery (JAG), registered with the University of the Witwatersrand, Johannesburg, and partly funded by the National Research Foundation. For their help while I was preparing the chapter I wish to thank Claire Wiltshire of the Johannesburg Local Government Library, Jo Burger of the JAG library, and my supervisor, Professor Elizabeth Rankin of the University of Auckland, New Zealand.

2. The R1.8M repairs to the roof and foundations were begun in June and completed in August 2000 (*The Star*, May 29, 2000, *Johannesburg Art Gallery News*, July–August 2000 and September–October 2000). The roof of the original building had required periodic reconditioning since it opened in 1915 (recorded in council minutes), but this has been a manageable problem. The structure of the 1986 extensions, however, had had an unacceptable level of leaking since the extensions opened and, prior to 2000, had not been successfully treated. Unprecedented heavy rains in the summer of 1999–2000 not only exacerbated the problem but raised the water table of the surrounding ground, with the result that leakage also occurred through the walls and floors.

3. For a description of the time of the first resident horticulturist (Richard Adlam, curator of Joubert Park 1893–1903) see "Thomas Richard Adlam: Sunrise and Advancing Morn: Memories of a South African boyhood," in M. Fraser, editor, *Johannesburg Pioneer Journals 1888–1909*. (Cape Town: Van Riebeeck Society, 1986). Bands performed in Joubert Park until the 1950s (recorded in council minutes). The aquarium in the conservatory, a popular attraction, was enlarged in 1947 with ten more tanks (council minutes, November 25, 1947).

4. Council minutes, June 23, 1936.

5. M[orris] S[evitz], "Cultural Amenities of Johannesburg," *The Exhibition Visitors' Social And Business Guide*, Johannesburg, September–October 1936, p. 27.

6. Council minutes April 30, 1935 and August 25, 1936. A new appointment, that of P. Anton Hendriks, was finally made in 1937 (council minutes, April 27, 1937).

7. Members of the JAG's Art Gallery Committee, however, were on the Fine Arts Committee for the Empire Exhibition, as listed in the catalogue of 1936. There was also a Jury for the selection of South African works chaired by Leo François. The art section of the Empire Exhibition styled itself the Art Gallery and had a Director (Mrs. L. Ruxton) and Assistant Director (Mrs. M. Kottler), which, in view of the vacancy at the other permanent art gallery, must have been rather misleading. *Empire Exhibition 1936. Kunsgalery. Art Gallery. Rykstentoonstelling 1936* (illustrated catalogue). Letters sent to P. Anton Hendriks, Pretoria, by the organizers in 1936, JAG archives.

8. The changes in the area and neighboring Hillbrow are recorded in A. Morris, *Bleakness & Light: Inner-City Transition in Hillbrow, Johannesburg* (Johannesburg: Witwatersrand University Press, 1999). The negative impact of the area on the JAG is recorded in the JAG's annual reports and newspaper articles, for example, *The Star*, May 29, 2000, where the director, Rochelle Keene, states: "Where we are situated has negatively affected our attendance figures and the taxi rank is a serious problem because of the perception that the area is unsafe."

9. Visitor figures in the JAG's annual reports indicate a drop from 124,378 in 1988/1989 to 43,447 in 1998/1999. The Chairperson of the Friends of the JAG at their annual general meeting, October 28, 2000, reported a further steady drop in figures.

10. The JAG's annual report of 1997/1998 notes the radical impact on the JAG of the severe financial crisis of the Greater Johannesburg Metropolitan Council: all capital projects were cut, the roof of the temporary exhibition area, which had leaked intermittently since 1986, could not be repaired, the temperature and humidity systems could not be upgraded, no purchase budget was allocated, and critical vacant posts could not be filled. Of the twenty-three curatorial and administrative posts, eight were vacant and, in comparison with ten years before, the security staff complement had been halved.

11. For the debate on a new type of museum and the rejection of the familiar European/North American model, with a comparison of the postcolonial Canadian and South African experiences, see Duncan Ferguson Cameron, "The Death of the Museum and the Emancipation of Cultures," *SAMAB [South African Museums Association Bulletin]* 2000. 24(1):1–11.

12. *Project outline*. Information on the Joubert Park Project was supplied by Bié Venter. Other sources include *Johannesburg Art Gallery News*, November–December 2000; press reports, in particular *Mail & Guardian*, December 1–7, 2000, and *Beeld*, November 30, 2000; text panels and activities at the *Open Day* December 3, 2000; Kathryn Smith, "Joubert Park Renaissance," ARTTHROB, no. 40, December 2000, http://www.artthrob.co.za.

13. The inner city projects are summarized as follows in *Saturday Star*, December 9, 2000: The Park Central (formerly Jack Mincer) taxi rank, Park City taxi rank, Westgate transportation facility,

Metromail, Faraday precinct, Newtown cultural precinct, Hillbrow/Berea regeneration, Constitutional Hill precinct, Rissik Street post office, and Greater Ellis Park precinct. For some unexplained reason the Joubert Park Project is not included in the list, but is discussed in the supporting article.

14. Even the announcement of the Joubert Park Project in the *Mail & Guardian,* December 1–7, 2000, had a barbed reference to "the moribund Johannesburg Art Gallery."

15. Peter Stark of the Centre for Cultural and Policy Research, University of Northumbria, was involved with the urban renewal of Newcastle/Gateshead in the UK. He made his comments about the JAG during the Urban Futures 2000 conference, at which he presented a paper on *The Potential of the Re-invented Cultural Institution to Contribute to Urban Renewal in South Africa.* Makhosi Fikile Dhlamini conducted a bone throwing divination at the gallery, in the presence of the director, November 29, 2000. A further divination was conducted on the open day itself, December 3, 2000. See the end of this chapter for an interpretation of the divination.

16. The Jameson Raid of December 29, 1895 was an attempt to legitimize British seizure of the gold fields of the Witwatersrand, which were under control of the SA (Transvaal) Republic. It was devised and led by Dr. Leander Starr Jameson in secret consultation with the Prime Minister of the Cape Colony, Cecil Rhodes, and with the backing of prominent mining magnates. The plan was for an armed force under Jameson to "rescue" uitlanders (British nationals with no political rights in the SA Republic) rising in Johannesburg against republican injustices. But the Raid left too soon from Pitsane Photloko on the border of the Transvaal Republic—before the fabricated uprising in Johannesburg—and was forced to surrender near Krugersdorp on January 2, 1896. Those directly implicated in the Raid (with the exception of Jameson, who was tried in London) were imprisoned and tried in Pretoria. Four of the leaders, one of whom was Lionel Phillips, were sentenced to death, but their sentences were commuted to prison terms, and subsequently to fines and a pledge not to interfere in politics for three years. Rhodes resigned as Prime Minister. The Raid was a principal cause of the breakdown in English–Afrikaner relations prior to the outbreak of the Second South African War. For further information see *The Jameson Raid: A Centennial Retrospective,* various authors (Johannesburg: Brenthurst Press, 1996).

17. The JAG's temporary home in Eloff Street was the forerunner of the University of the Witwatersrand. The JAG moved to its present Edwin Lutyens building in Joubert Park toward the end of 1915. For background on the collection and donors see T. Gutsche, *No Ordinary Woman. The Life and Times of Florence Phillips* (Cape Town, 1966); M. McTeague, "The Johannesburg Art Gallery: Lutyens, Lane and Lady Phillips," *The International Journal of Museum Management and Curatorship* 1984, 3(2):139–152; and J. Carman, "Acquisition Policy of the Johannesburg Art Gallery with Regard to the South African Collection, 1909–1987," *South African Journal of Cultural and Art History* 1988, 2(3):203–213.

18. For details on Lane see R. O'Byrne, *Hugh Lane 1875–1915* (Dublin: The Lilliput Press, 2000).

19. Renamed "The Hugh Lane Municipal Gallery of Modern Art" in 1975, the centenary of Lane's birth. James White, "The Municipal Gallery, Sir Hugh Lane and the Friends," in *The City's Art— The Original Municipal Collection: Catalogue.* Reprint of a 1908 catalogue by The Friends of the National Collections of Ireland, Dublin: Hugh Lane Municipal Gallery of Modern Art, 1984.

20. Lane lent his "collection of pictures by Continental artists" to the new Dublin gallery "provided that the promised permanent building is erected on a suitable site within the next few years" ("Prefatory notice," *Municipal Gallery of Modern Art,* Dublin 1908). The Dublin Corporation's rejection of Lutyens's plans for a bridge site in September 1913 was the final in a long list of delays and rejections.

21. For an overview of the dispute see *Images and Insights* (Dublin: Hugh Lane Municipal Gallery of Modern Art, 1993), pp. 29–30; O'Byrne, *Hugh Lane,* pp. 228–241.

22. F. Spalding, *The Tate: A History* (London: Tate Gallery Publishing, 1998), p. 26.

23. Under duress, the National Gallery acquired a Boudin in 1906 but only because he was an older generation Impressionist who was dead by that time. The trustees refused to consider a work by Monet, who was still living—Lane bought the Monet they rejected. K. McConkey, *British Impressionism* (London: Phaidon, 1998), p. 142.

24. The irony is that Lane was not in fact collecting what is today considered the best of early twentieth-century European art—the Expressionists and Cubists. For discussions of the conservative situation in England, see Spalding, *The Tate,* pp. 23–35; J. Carman, "Negotiating Museology in the Early Twentieth Century," in *Negotiating Identities.* Proceedings of the SA Association of Art Historians conference, University of South Africa, Pretoria, 1998, pp. 19–29.

25. Recommendation of D. S. MacColl, Keeper at the Tate from 1906 to early 1911, made to the committee set up by the National Gallery Trustees under Lord Curzon in November 1911 to "Enquire into the retention of important pictures in this country, and other matters connected with the national art collections." M. Borland, *D. S. MacColl* (Harpenden: Lennard Publishing, 1995), pp. 189, 194–199.

26. The Lane collection of Impressionist paintings went on display only in London in 1917. Spalding, *The Tate*, p. 39.

27. Lord Curzon, a trustee of the National Gallery in London 1911–1914 (see note 25), was in Rome with Lutyens when Lutyens received a cable asking him to go to South Africa. He apparently persuaded the reluctant Lutyens to accept the commission. Gutsche, *No Ordinary Woman*, p. 255.

28. They appear to have opened in late 1940, and, like the opening in 1915, to have done so without ceremony. McTeague, *The International Journal of Museum Management and Curatorship*, pp. 151–152.

29. For the importance of the JAG in Lutyens's oeuvre see M. Miller, *City Beautiful on the Rand: Sir Edwin Lutyens and the Planning of Johannesburg*. Paper presented to the seventh national conference on American Planning History, Seattle, 1997; C. Hussey, *The Life of Sir Edwin Lutyens* [Woodbridge: Antique Collectors' Club, (1950) 1989].

30. For other examples and a discussion of this type of building see C. Duncan, *Civilizing Rituals: Inside Public Art Museums* (London: Routledge, 1995); G. Waterfield, editor, *Palaces of Art: Art Galleries in Britain 1790–1990* (London: Dulwich Picture Gallery, 1991).

31. No other art gallery in South Africa at this time had an independent, purpose-designed building. Art galleries like the Tatham in Pietermaritzburg and the Durban Art Gallery were accommodated in their town halls. For histories of the different art galleries see R. Becker and R. Keene, editors, *Art Routes. A Guide to South African Art Collections* (Johannesburg: Witwatersrand University Press, 2000).

32. For a discussion of the context of the time and the agendas behind the foundation of the JAG, see J. Carman, "The Foundation of the Johannesburg Art Gallery and Culture in the Service of Empire and Other Things," in *Proceedings*, SA Association of Art Historians conference, University of Natal, Pietermaritzburg, 1999, pp. 51–59.

33. H. Lane, "Prefatory Notice," *Municipal Gallery of Modern Art. Johannesburg. Illustrated Catalogue*. Johannesburg, 1910, p. iii.

34. The Dutch-born Van Wouw (1862–1945) settled in Pretoria in 1889.

35. Anton Van Wouw's five bronzes (one of President Paul Kruger and four of black ethnic "types") and two watercolors of Cape Town scenes by an obscure Australian artist, A. Henry Fullwood.

36. "We may hope that a South African School of Art will grow up, and the study of the masterpieces, a few of which we have been able to secure for the gallery, should be a help as well as an incentive to local artists. And when our South African School of Art begins to produce work worthy to appear side by side with the best examples of other countries, it is surely desirable that the Trustees of the gallery should be in a position to purchase such work, and so not only encourage rising talent, but form from the beginning a great South African collection," *South Africa*, August 13, 1910.

37. Gutsche, *No Ordinary Woman*, p. 236.

38. A protest against the appointment of Lutyens was unanimously approved at a meeting of the Association of Transvaal Architects February 22, 1911. This was followed by a petition of protest signed by 43 practicing architects of the Transvaal and members of the Association of Transvaal Architects, when it appeared that an agreement to appoint Lutyens had been reached without due authority (council minutes, March 29, 1911). It was finally agreed to appoint Lutyens "as the architect of the Art Gallery Building, and that a South African architect (to be nominated by Mr. Lutyens) should be associated with him as collaborator" (council minutes, April 26, 1911). Robert Howden was appointed the supervising architect in Johannesburg.

39. Goodman to Ross May 1, 1913, Ross Papers, JAG archives.

40. Details of the curator's salary and the purchasing budget are in the annual estimates of the municipal council, from 1911/1912 onward. A brief historic overview of the JAG funding is given in council minutes, November 22, 1949.

41. Johannesburg Art Gallery I: correspondence, purchase lists September 1929–July 1930. Tate Gallery archive.

42. Council minutes, March 25, 1947.

43. Hendriks successfully recommended to council that the unprofessional-sounding "curator" be replaced by "director" (council minutes, November 26, 1945). He also at some stage changed the spelling of his name from Hendricks to Hendriks, by which he is more commonly known, and which I use here for the sake of clarity.

44. Council minutes, July 30, 1946.

45. Carman, *South African Journal of Cultural and Art History*, p. 207.

46. A site in the Pieter Roos Park, Parktown, was designated, although sites near the War Museum in Saxonwold and the proposed civic centre in Braamfontein were also considered. *Rand Daily Mail*, November 28, 29, December, 1, 2, 7, 8, 12, 22, 24, 1960; *The Star*, January 27, 1962.

47. Council minutes, December 6, 1966.

48. *The Star,* April 28, 1966.
49. Joyce Ozynski, "Art: Peanuts for the Gallery," *Snarl* November 1974, 1(2):10–11.
50. For a record of archive and other research initiatives see J. Carman, "Research Strategies at the Johannesburg Art Gallery," *SAMAB* 1989, 18(part 8):275–290.
51. The collecting policy of the 1990s and an overview of policies in the past are set out in *Johannesburg Art Gallery: Collecting Policy,* JAG 1994.
52. These developments are recorded in the JAG's annual reports, in particular for 1992/1993, 1994/1995, and 1996/1997.
53. Council minutes, January 30, 1974.
54. The pamphlet is essentially a reissue of the JAG's diamond jubilee (60th anniversary) pamphlet of 1970/1971, with "The Johannesburg Art Gallery has always been open to all races" added as item 8 of a list headed "Do you Know?" (and with the program of jubilee events for 1971 omitted). It seems the amended pamphlet may have been issued in reaction to the council resolution, however, Nel Erasmus, director of the JAG at that time, does not recall this as a reason for the reissue. She believes the revised pamphlet was associated with the launch of the Friends of the JAG (in conversation with the author, March 22, 2001).
55. Council minutes, July 1, 1924, and September 19 and October 17, 1950. The Geological Museum became a part of the JPL in 1927; the Africana Museum was part of the JPL 1933–1935, after which it was a separate municipal department within the JPL building, subject to JPL regulations [R. F. Kennedy, *The Heart of a City. A History of the Johannesburg Public Library* (Cape Town: Juta, 1970), pp. 548–578]. At a council meeting in mid-1975, the Library and Africana Museum gave an overview of the first year of use of their facilities by "non-white" visitors. Council minutes, July 29, 1975.
56. General Purposes Committee, January 18, 1918. R. M. Gandhi is not to be confused with Mohandas (Mahatma) Gandhi, who left South Africa in 1914.
57. Pim to the Johannesburg Town Clerk, D. B. Pattison, February 20, 1928, Pim Papers Cc 1905–1934, University of the Witwatersrand. Quoted in A. Caccia, "Moses Tladi (1906–1959): South Africa's First Black Landscape Painter?" *de arte* September 1993, 48:14.
58. Cuttings in Town Clerk's file 10/34, 1928–1931, JAG archives.
59. Correspondence in Town Clerk's file 10/134, 1928–1931, JAG archives.
60. Black employees usually filled domestic worker posts. In a letter to S. Rosen of February 3, 1937, the curator A. A. Eisenhofer states: "Our Gallery is open to Europeans every day from 10 am to 6 pm and Sundays Wednesdays and Public Holidays from 8 pm to 10.15 pm. Indian and coloured people are permitted every Thursday. We also encourage the natives." Six months earlier, in a letter of August 12, 1936, Eisenhofer had informed Rev. S. Carter of the Community of the Resurrection: "We always welcome Natives who wish to visit the Gallery and they are always admitted every Thursday between the hours of 2–5 p.m." Other letters and reports during the 1930s mention visits of "Native," "coloured," and Indian school pupils, days not always specified, while a letter from Howard Pim of December 2, 1932 talks about arrangements for a group of "20 Native students" in the charge of Mr. I. R. Rathebe to visit on a Saturday. The Thursday ruling seems to have been flexible. Letters and reports in Town Clerk's file 10/134, 1928–1931; letters file A–D 1933–1936; letters file P–Z 1930–1937. JAG archives.
61. L. Spiro, *Gerard Sekoto: Unsevered ties,* JAG 1989, p. 69, note 212.
62. A bylaw to restrict access to parks, gardens, and open spaces—which were to be reserved for the recreational use of "Europeans only"—was promulgated by the city council October 17, 1950.
63. The typed and hand-written notes were evidently compiled by Nel Erasmus, identified by her handwriting. JAG archives.
64. Interview with Lesley Spiro, *Gerard Sekoto,* pp. 69–70, note 212.
65. D. Koloane, "The Polly Street Art Scene," in A. Nettleton and D. Hammond-Tooke, editors, *African Art in Southern Africa: From Tradition to Township* (Johannesburg: Ad. Donker, 1989), p. 229.
66. Cecil Skotnes in conversation with the author, March 7, 2001.
67. Nel Erasmus in conversation with the author, March 22, 2001.
68. A draft deed of donation was tabled at council October 25, 1910, and an official *Deed of Donatio Inter Vivos* was signed in front of a notary by Lionel Phillips January 21, 1913 and by the mayor W. R. Boustred January 22, 1913.
69. J. P. R. Maud, *City Government: The Johannesburg Experiment* (Oxford: Clarendon Press, 1938), p. 147, Appendix I. Maud was appointed to write the book at a council meeting of March 26, 1935.
70. Council minutes, April 30, 1939, November 19, 1940, and November 22, 1949.
71. See note 55 for the inclusion in JPL structures of the Geological and Africana museums.

72. Recent examples of an art gallery not being considered a museum range from the Greater Johannesburg Metropolitan Council's telephone directory entry for 2000/2001, where the JAG is not listed under "museums," to the Cape Metropolitan Tourism's full-page color advertisement in the *Sunday Times,* July 2, 2000, advertising Cape Town's museum route. It does not include the South African National Gallery or the Michaelis Collection, or any references to the painting collections in the Castle or Groot Constantia, both of which are otherwise described in terms of their historic artifacts, furniture, and porcelain. Paintings, in the popular mind it seems, belong to galleries, and galleries are not museums.

73. The council offices, which used to be housed in the city hall, have since moved to Braamfontein and the Africana Museum (MuseuMAfricA), formerly in the JPL, to Newtown.

74. Separate seating in combined public transport was enforced some time before separate services were introduced, such as the tram service (Council minutes, April 26, 1938). The installation of "Non-European" lifts became compulsory April 28, 1942 (Council minutes). The sidewalk motion was negatived (Council minutes, December 15, 1925).

75. Observation made by Christine Jikelo, University of Fort Hare Museum, in a paper, *If Museum Buildings Can't Move, Collections Can,* 64th annual conference of the South African Museums Association, Robben Island, May 29–June 2, 2000.

76. Apart from the segregation of parks, which people of color evidently frequented in too many numbers for white comfort, a prime example is that of the zoo. Even after the restrictions on Libraries, Museums, and the Zoo were removed in January 1974, it was resolved to maintain separate visiting days for white and "non-white" school groups. Council minutes, January 30, 1974.

77. Unlike the previous major celebrations of 1936 (see above), this time the director was on the organizing committee and wrote the Preface to the Van Riebeeck Festival Exhibition catalogue, and the JAG was an exhibition venue.

78. Figures from a table dated January 1967 in the JAG archives giving "Annual attendance of visitors to the art gallery" from July 1945 to June 1966. There were no records kept from July 1953 to June 1960. The average of 65,000 to 70,000 "over the past few years" given in the notes for Mr. Shorten of May 26, 1966 is misleading (see note 63).

79. *The Star,* April 30, 1965.

80. Verwoerd and the mayor of Pretoria each laid a cornerstone, and Hendriks was one of the advisers for the new museum (*Die Transvaler,* October 20, 1962). A rather indifferent collection had been in existence for some years prior to the 1960s.

81. Albert Werth, "Introduction," in R. de Villiers (compiler), *Looking at Our Own: Africa in the Art of Southern Africa* (English and Afrikaans) (Pretoria Art Museum, 1990), p. 4. The policy of allowing access to all races since 1964 was confirmed by Dirkie Offringa of the Pretoria Art Museum July 7, 2000.

82. Albert Werth in conversation with the author, March 23 and 29, 2001. Werth recalls that the Transvaal Museum also had Thursday openings. Nel Erasmus recalls (conversation of March 22, 2001) that it was usual for other SA art galleries at that time to allow Thursday access and that the JAG was unique in allowing access on all opening days. The "Thursday access" provision challenges the generally held view that black visitors were denied any form of access to museums and art galleries during the apartheid years.

83. Werth's observations (in conversation, March 23 and 29, 2001). The Sithole piece was eventually acquired in the 1960s, despite objections from two councillors because the work was by a black person. This incident is corroborated by Frieda Harmsen (in conversation March 23, 2001), who was on the Pretoria Art Museum committee at that time.

84. S. Sack, *The Neglected Tradition,* JAG, 1988, p. 116; D. Koloane, in Nettleton and Hammond-Tooke, editors, *African Art in Southern Africa,* p. 226.

85. G. Younge, "Dead in One's Own Lifetime—The Contours of Art under Apartheid," *The State of Art in South Africa,* conference proceedings, University of Cape Town, July 1979, p. 37.

86. Semantic interpretation by Johan van der Westhuizen of Makhosi Fikile Dhlamini's divination through bone throwing for the art gallery, November 29, 2000. The divination was conducted in Zulu and is available as a tape and video recording.

Section IV

REREPRESENTING

R obinson and Mangcu end the volume by reflecting on the tyranny of representations. Robinson situates Johannesburg in the discourse of urban studies, specifically in the tendency to think of cities as either global or developing, with the latter including megacities. Neither perspective captures the current situation of postapartheid Johannesburg. She advises scholars and policymakers to think of the city as "ordinary." This is not meant to be demeaning but to focus attention on the needs of its residents and on how a city functions internally. Johannesburg has to be imagined as a city in a specific place and at a specific point in history, not as a member of a category.

Mangcu takes a different tack. His concern is with the connection between political culture and government policy. For him, the postapartheid era has brought an increased emphasis on technocratic decision making based on an integrationist and nonracial ideology. He calls for greater recognition of the historical importance of the black consciousness movement to the apartheid struggles, and its continued salience to the nation. In this way, he returns us to the multiple identities that inhabit Johannesburg.

The book thus concludes on the issue of race. Even Robinson's chapter encompasses this theme: global cities are mainly cities of the (European) north and megacities those of the (non-European) south. Postapartheid Johannesburg is unlike its apartheid predecessor, but race still operates to shape the daily lives of its residents, the opportunities it offers them, and the institutions that enable it to function.

14

Johannesburg's Futures

BEYOND DEVELOPMENTALISM AND GLOBAL SUCCESS[1]

JENNIFER ROBINSON

The ending of apartheid offered a moment when South Africa's urban future was relieved of the long burden of separate development and the threat of international isolation. New kinds of geographies were imagined, and new kinds of futures opened up for South African cities. For Johannesburg, these questions of potential futures carry a particular poignancy. Many times residents of the city have compared her towering skyscrapers and dynamic economy to some of the world's largest and most powerful cities. On a continent where observers found it hard to identify successful or even big cities, Johannesburg has stood out. The racial complexities of these ambitions to world status in earlier times are less of a hindrance (although not completely absent) in postapartheid South Africa, but the challenge of achieving this potential remains.

Apartheid's demise has not ended the experiences of segregation and inequality that have shaped the lives of most people living in Johannesburg. New developments seem as likely to reinforce old patterns as transform them, despite many hopes of initiating a new, integrated, and compact city form across the country. Meeting the needs of the poorest citizens of Johannesburg draws planners and politicians to the complexities of improving basic services, addressing housing shortages, and upgrading existing environments. Practitioners and advisers from around the world have brought Johannesburg into a network of development thinking that previously had little impact on urban management.

Johannesburg's future, then, is entwined in two of the most powerful discourses and sets of practices shaping cities. The idea of being a global city and the urgent requirement to improve living conditions for the poor are respectively the parameters of the world city hypothesis and developmentalist understandings. Historically, these discourses have been associated with quite different groups of cities. The world city hypothesis (and its near neighbor, the global city hypothesis) draws on a framework that highlights the achievements and form of cities in advanced economies, usually Western ones. Developmentalist approaches bring into view cities in some of the poorest countries of the world. Contemporary urban studies are substantially divided into theoretical accounts that stress Western cities and those that deal with the difficulties of developing or improving cities in the Third World.

This has two consequences. First, non-Western cities are either considered to be in need of major interventions to catch up with the normative Western city form or they are considered comparatively along criteria laid down by Western cities (see Dick and Rimmer, 1998, for a recent example). They are seldom turned to for theoretical insight or to reframe what it is that cities or cityness are considered to be. At a moment when much social theory has been strongly inflected with a postcolonial critique eager to dislodge old imperialist assumptions about where and how knowledge is produced, urban theory continues to express its ideas on the basis of a very restricted range of cities from the wealthiest countries of the world. Second, this division sets up a hierarchy of cities that deploys a regulating fiction about what makes a successful city, encouraging cities of all kinds to undertake costly and often destructive investments to attain the highest status as a global or world city.

This chapter explores the case of Johannesburg as an antidote to this divisive tradition in urban studies and as a practical example of how cities can be imagined outside of the global/developmentalist division. Drawing on an emerging set of approaches that prefers to insist that all cities are ordinary cities, it explores the diverse but ordinary (although no less challenging) future of Johannesburg.

JOHANNESBURG (NOT) IN VIEW, OR FALLING BETWEEN CATEGORIES
NOT QUITE GLOBAL

At different stages in her history, Johannesburg has had great hopes of a global position. Keith Beavon told the story like this in a local newspaper:

> Notwithstanding the evils of segregation, Johannesburg in the past, and particularly in the '30s, was recognised as a "world city", largely because it had just built its CBD. At the end of 1932 South Africa came off the gold stan-

dard and foreign capital poured into South Africa via the city and its institutions. Between 1933 and 1938 capital inflow was equal to some 66% of the total capital inflow over the first 40 years of the city's existence. In the CBD, skyscrapers popped up like champagne corks and thousands of new apartments (for whites only) were built in the inner-city residential zone. It was also the era of the large and grand department stores. They and a plethora of high order speciality shops helped make the CBD, and particularly Eloff Street, the shopping mecca not only of South Africa but of the continent. (Beavon, 2001 in *Sunday Times,* January 7, 2001)

Today these ambitions seem to have the potential to be realized. They are reinforced by an urban literature that casts Johannesburg as a city with globalizing characteristics and the potential to play an internationally significant role.

John Friedmann initiated almost two decades of urban research on the basis of his "world city hypothesis" (Friedmann and Goetz, 1982; Friedmann, 1986). Concerned to explore how the changing dynamics of the world economy affects cities, one of the by-products of the approach has been attempts to rank the major cities of the world according to the functions they perform within, and their integration into, the global economy. As Friedmann [1986 (1995):319] explains, "Key cities throughout the world are used by global capital as 'basing points' in the spatial organisation and articulation of production and markets. The resulting linkages make it possible to arrange world cities into a complex spatial hierarchy." Primary world cities are associated with economic activities of the broadest global scope; the importance of secondary cities, especially in peripheral countries, "depends very much on the strength and vitality of the national economy which these cities articulate" [Friedman, 1986 (1995):319]. In a framework where most cities in poor countries are completely "off the map" of world city theorists (Robinson, 2002), Johannesburg is the one city in Africa that consistently appears as a "world city," second(ary) class [Friedmann, 1986 (1995):321].

Updating this earlier work, Friedmann returns to some of the cities that he had placed on the lower rungs of the world city hierarchy, including Johannesburg:

In my original formulation Johannesburg was the only world city in Africa. But this was before the international boycott of South Africa and prior to the current political struggle of the black majority for political control of the country. This struggle is likely to continue and to create large uncertainties, which will make it difficult for Johannesburg to recapture its world city position. (Friedmann, 1995:39)

This statement is alarming. First, every other city in Africa is excluded from the world economy (see Simon, 1995 on alternative views). Second, the identification of a hierarchy has slid very quickly into a normative ambition—to become a world city seems to be something that cities such as Johannesburg naturally want to do.

Why are large and dynamic cities being excluded from the top range of what world cities are? What is it that they are being enticed to pursue to rise to this status, a status conferred by a small group of Western-based urban theorists?

Although status within the world city hierarchy is based on a range of criteria, including national standing, location of state and interstate agencies, and cultural functions, the primary determination of status is economic. As Friedmann [1986 (1995):317] notes, "The economic variable is likely to be decisive for all attempts at explanation." This has been taken to an extreme by writers who focus on the top rank of world cities, labeled, "global cities." Here the explanation for the dominance of a few cities at the top of the global urban hierarchy rests on the concentration in these cities of capacities for control and coordination of a globally dispersed economy (Sassen, 1991). Colocation, far from being undermined by technological innovations in communication and transport, is being encouraged. One case in which this has been observed is in the apparent value of proximity for global corporate headquarters, financial sector firms, and producer services enterprises that support and enable the complex products that global management and global financial activity require. Economic clusters for the production of management and coordination of the global economy are found in the three top ranked world cities (New York, London, and Tokyo) identified by Friedmann and others. Success in the ranks of cities is associated with participating in the most global of economic activities and with the most prized aspects of it, coordination, management, and finance.

The global city hypothesis defines the status of the top tier of cities on the basis of their participation in a small sector of the economy that has claim to "global" reach. To compete, cities are enticed into an emphasis on managing transnational investments, coordinating information flows, and encouraging producer services industries (Jessop and Sum, 2000; Morshidi, 2000; Olds and Yeung, 2002). If these are not realistic options, cities are drawn, sometimes at great cost, to access global flows of some other kind (e.g., tourism, footloose manufacturing) (Robins and Askoy, 1996; Kelly, 2000). For those cities that do not make the grade, a subsidiary category of cities with "global city functions" can be identified, including places such as

Miami and Sydney, which coordinate transnational flows and investments in a regional context (Sassen, 1994).

Postapartheid Johannesburg falls easily into this category of cities. As businesses invest or even move parts of their activities offshore, and more large companies engage in or facilitate business links with other sub-Saharan African countries, Johannesburg can be viewed as a city with global city functions. The predictions of the global city hypothesis are that these activities become more connected to the global economy and to other global cities, and more disconnected from other national economic activities, and especially from sectors of society and economy that are peripheral to the global economy. The global city theorists' analytic hierarchization of economic activity invites cities eager to succeed to promote globalizing sectors, perhaps at the expense of other more local or peripheral activities.

From the point of view of a theory determined to rank and to categorize cities, Johannesburg is billed as a city on the cusp of global status. The implicit assumption is that cities can (and should) exploit their capacity to link into the global economy and any advantages of location that they offer to a range of firms whose specialist needs can be met there. To compound economic reductionism, the status of cities is determined in relation to the geographic reach of only certain economic activities. The logical consequence of such approaches is to reinforce a competitive aspiration to rise to the "top of the heap" and to encourage a distorted emphasis within cities on certain kinds of transnational activities. This approach rests on a quintessentially neoimperialist and global-scopic vision, a view from apparently nowhere in particular, able to assess and label all cities everywhere. In reality, this vision is located in the centers of power and privilege that valorize and prioritize the activities of the most powerful in a few (old and new) imperial centers.

Anthony King (1990) proposes that we consider all cities to be world cities. To do this, we need to be mindful of the range of processes that shapes cities and to ensure that we do not privilege the "global" as a site of operation, or only fixate on certain phenomena with an apparently "global" reach. King's suggestion draws us away from any enthusiasm to categorize cities. The categories of Western and Third World cities assumed substantial differences in the international division of labor. Global and world city hierarchies have replaced this older division without displacing many of its assumptions. Cities that are not part of the successful components of the global economy are peripheralized, marginalized, described as catatonic (Knox, 1995), and seemingly are doomed to remain so. They are portrayed as defective, lacking in dynamism, and their only hope is to follow the example of successful cities and hook into global flows. That way they can find their way onto the radar of urban theory and be counted successful.

If the viewpoint is changed from that of the global urban theorist to the located observer, commentator, or urban manager trying to come to grips with the challenges of worldly cities (global or not), then these limitations of categorizing, hierarchizing, and global-reductionist perspectives are exposed. On this basis, urban theorists need to follow a more cosmopolitan path to theory production, one built up from a diversity of ordinary cities rather than from the categorizing imperative of the imperial/imperious observer.

Johannesburg, then, falls out of the view of world and global city theorists, except for the small element of its economy that is linked to some so-called "global city functions." As a "not-quite" global city, it receives more attention than any other city on the African continent. The "not quite" description, though, invites a form of postcolonial critique (Bhabha, 1994) whereby the standards on which cities such as Johannesburg are being judged—not global, such as New York—are exposed as developed in and in response to places of power. Moreover, like other African cities Johannesburg also falls from view when considered through another popular lens of urban theorists: the idea of the megacity and its close association with developmentalist theories.

NOT QUITE BIG ENOUGH OR BAD ENOUGH?

Cities that are large in population but do not qualify on the global theorists' criteria as connected to the global economy are usually relegated to the category of "megacity." As Sassen (1994:51) observes, "In the developing world we see trends toward the continuing growth of mega cities and primacy as well as the emergence of new growth poles resulting from the internationalisation of production and the development of tourism."

Writing in response to a brief to consider world cities and megacities in Latin America, Alan Gilbert (1998:179) notes that "Latin America probably contains several mega-cities . . . But do any of these cities warrant inclusion as world cities?" Half-jokingly, he suggests that the only cities worth including as world cities are Cali and Medellin, "the centres of Latin America's only major transnational corporations, the drug cartels" (Gilbert, 1998:180). This is an important point (not only a joke). It directs us to the significance of a more diverse range of interconnections among cities than those defined by hegemonic capitalism.

Megacity is a residual category. Conventionally, it includes cities that do not qualify for being significant economically but for some inexplicable reason keep growing. This should remind us that they must be important to the people who move there (McGee, 1995). Megacity is also a rather precarious category, relying on some arbitrary cut-off point, which has been set at different times at 4, 8, and 10 million inhabitants (Gilbert, 1998). This suggests a lack of clarity as to exactly what a megacity is. It is also hostage to the fortunes of history, as cities do or do not grow as fast or as slowly as predicted and as people invent new ways of being in cities and defining their urban experiences. The growth of megacities does seem to be slowing. In both Latin America and Southeast Asia, there is evidence that the expansion of secondary cities is now replacing large city growth. A number of population projections made by international agencies for megacities are having to be revised downward (Lopez de Souza, 2001).

Even though megacity may be a residual category, and one that is changing over time, in some parts of the world it is also of little relevance. Rakodi (1997) observes that there are no truly megacities in Africa. She notes though that Cairo and Lagos are rapidly becoming of that order and that, on some definitions, Greater Johannesburg just qualifies. (She puts the population of Johannesburg at just over 8 million, although other estimates are lower, around 6.3 million.[2]) Of course, other cities on the continent are known to be large and possibly heading toward megacity status, but, given the paucity of data, it is hard to know quite how large they are. Rakodi mentions Kinshasa, Nairobi, and Abidjan.

Megacity may be used as a catch-all phrase for big but not powerful cities, but in different parts of the world megacities assume different forms, grow for different reasons, and provoke quite different policy responses (McGee, 1995; Douglass, 1998). In some areas the role of these cities in fostering new kinds of economic growth on the urban periphery has been encouraged and new urban–rural interactions viewed favorably (McGee, 1989). In others, the problems associated with megacities have led to policies of containment and the focus on size and form has provoked responses that emphasize their physical fabric. Policy interventions have ranged from efforts at deconcentration to attempts to stimulate differentiation of urban space (Yeung, 1989).

By itself, size is neither a very helpful way of describing or defining cities nor does it indicate appropriate policy responses. Although concerns with the size of cities have been expressed in many different parts of the world and at different times in history, the contemporary conceptualization of the megacity also invites theorists to draw on interpretations that place them in the category of "Third World Cities." Lacking an economic base, these cities continue to grow and are seen to follow a distinctive (deviant?) path in terms of economic activity, political organization, and spatial form. In sum, they lack many of the defining features of cityness, according to Western-based understandings of urbanism. This is not to disavow the developmentalist urbanists' concerns, for they deal with urgent matters of life and death for the inhabitants of these (and other) cities: water, sewerage, diseases, health, education, and livelihoods Pugh (1995) offers a relatively recent review). But the megacity concept limits theoretical and policy attention in these cities to elements of city life that feed an alarmist assessment of their deficiencies.

In the same way that global and world city approaches ascribe the characteristics of only parts of cities to the whole city, megacity approaches extend to the entire city the imagination of those parts that are lacking. Where the global city approach generalizes the successful locales of high finance and corporate city design, the megacity approach builds toward a vision of poor cities as comprised of relatively unproductive and informal built environments. Many other aspects of city life in these places are obscured, especially questions of culture, innovation, and the creative production of diverse forms of urbanism—all valuable resources in the quest for improving urban life (Askew and Logan, 1994; Hansen, 1997; Simone, 1998). Envisioning city futures on the basis of these partial accounts is limiting. From the point of view of urban theory, these megacity experiences do not contribute to expanding the definition of cityness: rather they signify its obverse, what cities are not.

The megacity concept can be linked to developmentalist approaches to cities that have a longer history and have been very influential in shaping state and NGO interventions in cities in poor countries. The apparently distinctive features of the Third World City (high rates of migration, burgeoning informal sector, and poor infrastructure) influenced practitioners to focus on the improvement of housing in sprawling informal settlements and the provision of basic services and infrastructure. Initially these were provided through government subsidies, as shacks were removed and formal housing provided. The substantial cost implications of these initiatives coincided with the emergence of a new policy consensus in the mid-1970s that took a more positive view of squatter settlements. Site and service and self-help housing on the periphery in the form of home ownership and secure land tenure supported an understanding, particularly in Latin American cities, that peripheral settlements were the end point of the rural in-migrants cycle, initially seeking accommodation in inner city slums (Burgess *et al.*, 1997:Chapter 7).

These policies reinforced peripheral urban sprawl and a continued understanding that developing "Third World Cities" meant focusing on those areas that seemed to distinguish them (unfavorably) from cities in the West. In the imagination of the urban developer, the city had shrunk to be synonymous with the peripheral areas of poorly housed and serviced popu-lations. In the late 1980s, the urban policy consensus of the mid-1970s began to be dislodged by neoliberal policies focused on urban productivity and urban poverty (World Bank, 1991). Emphasizing enabling and partnership strategies for housing provision and the importance of infrastructure and city-wide managerial capacity, the urban imagination of development agen-cies began to stretch.

If poor cities have escaped the lens of global urban theorists and mega-cities encompass only the largest of these, a developmentalist literature has at least had cities such as Johannesburg, and smaller settlements, in view. But only selectively. As Nigel Harris (1992) has observed, the city-wide economy has been neglected in favor of the needs of specific areas, notably those poorly serviced and on the periphery. Together with international agencies such as the World Bank, Harris and others are building toward a dynamic sense of the city as a diverse whole. The potential for economic growth and development in any given city (small or large, successful or not) can be realized more effectively if the entirety of the city is considered and priorities determined across its different areas and sectors. For the World Bank, this means balancing four identified needs:

> If cities and towns are to promote the welfare of their residents and of the nation's citizens, they must be sustainable, and functional, in four respects. First and foremost, they must be **livable**—ensuring a decent quality of life and equitable opportunity for all residents, including the poorest. To achieve that goal, they must also be **competitive, well governed** and managed, and financially sustainable, or **bankable.** The strategy proposes an agenda for helping cities develop along these four interrelated dimensions—a compre-hensive development framework for the urban arena. . . . The urban policy agenda outlines some broadly common goals for all cities and local govern-ments. But it would be implemented very differently in different places, with the pace, priorities, and operational instruments depending on the political commitment and capacities of the local and central government and other key stakeholders. (World Bank, 2000:8)

Beyond global competition and developmentalist alarm about mega-urban futures, are South Africa's city managers able to adopt a distinctive approach to imagining the future of their cities?

JOHANNESBURG'S FUTURES: GOING GLOBAL OR DOING DEVELOPMENT
GLOBALIZATION AND DEVELOPMENTAL LOCAL GOVERNMENT

Urban theory's dominant paradigms have had a powerful effect in the world. John Friedmann tells of a trip he took:

> A few years ago, I was invited by the government of Singapore to speak on world cities. In private conversations with senior government officials it became clear to me what the government really wanted. Singapore was embarking on "the next lap" (Government of Singapore, 1991), and officials hoped to hear from me how their city state might rise to the rank of a "world city." The golden phrase had become a badge of status, just as "growth poles" had been in an earlier incarnation. There was little I could say that the government did not already know. But to me this question pointed to an ongoing competitive struggle for position in the global network of capitalist cities and the inherent instability of this system. (Friedmann, 1995:36)

The concept of "world city" was traveling beyond the author's control. Friedmann suggests we look to capitalism for the reason why the world city idea has had such purchase in the frequently destructive striving for increasing status among cities. But perhaps the author is too modest. The concept of the world city has built into it the idea of hierarchy and status—hierarchies that replay older divisions between cities that are seen as successful and whose achievements other lesser cities are invited to pursue. In a previous era, this division (not entirely defunct) was couched within the rubric of Western and Third World cities.

This divisive form of urban theory can severely limit perspectives on the possible futures of particular cities. Following the megacity/global city dualism, each perspective invokes a future that emphasizes either those areas or sectors with globalizing potential or those that fail to conform to some or other norm of city living. Between the hype/hope of the global city and the horror of megacity urban failures, urban managers have to look elsewhere for creative solutions to their problems. Globalization and developmentalism seem to be equally awkward alternatives. Although urban theory is located in Western imaginations, urban managers and thinkers around the world are confronted with the diverse realities in their own contexts. A more cosmopolitan urban theory, consciously locating itself in places other than the West or the all-knowing global observer, might, like the urban managers and theorists in Johannesburg, enable us to think about cities differently.

Like the government of Singapore, Johannesburg's urban planners and managers are well informed about global and world city approaches, and have engaged with the conclusions of this literature in a way that attempts to build proactively on its analysis, as the following extract from a recent planning document illustrates:

> Globalisation of the economy and the shift towards the service sector, high-technology industries, flexible labour markets and global procurement places Gauteng and Greater Johannesburg in a position of needing to ascertain the role that location and government can play in this international economy. A multidimensional economic approach is required that addresses the overall objectives of the city, that maximises a range of opportunities and above all will ensure an economic base that can withstand large shifts and changes in the global market. [Johannesburg Metro Spatial Development Framework (GJMC, 1998)]

The Local Government White Paper of 1998 notes that globalization, encouraged by the government's overarching program for Growth Employment And Reconstruction (GEAR), could potentially lead to considerable restructuring in South Africa's cities, including a measure of intermunicipal competition to attract industries (RSA, 1998:31). Pursuing the globalizing city logic, Johannesburg's planners observe with some concern, would reinforce existing lines of privilege, entrenching the location of employment in the north of the city at a great distance from most low-income housing in the south. It could also privilege a jobless growth scenario as high tech and global management functions offer little to the largely unskilled poor population of the city (GJMC, 1998). All sorts of other competing demands face the Johannesburg authorities. Apart from a severe fiscal crisis, the challenges of poverty, environmental issues, maintaining and improving an apartheid-designed infrastructure, and coping with declining employment in manufacturing all press on decision makers (Rogerson, 2000).

Another rubric for local development initiatives in Johannesburg competes with the globalizing city vision. This stresses the urgent need for service delivery and addressing poverty across the city. It involves an emphasis on quite different areas than those brought into view by the global city approach. Here, the former African areas and peripheral informal housing developments, far from the dynamic center of Sandton (northern business area) and the ailing central business district (CBD), are more closely in view.

Substantial efforts within the city (and through national policies) promote a broadly developmentalist agenda to deliver much-needed services to Johannesburg's poorer residents. Upgrading, poverty reduction, and basic service delivery have all become the focus of major research and policy initiatives (Parnell and Pieterse, 1998; Beall *et al.,* 2000).

National urban policies provide a framework for a developmentalist view of local government, but the motivations for this and influences on it are multiple, including the direct challenge of overcoming the damages of apartheid to the social and physical structure of the city. Antiapartheid traditions of local organizing remain alive, too, in the hopes of building consultative forums in which planning and spending priorities will be developed. But international developmentalist and donor ideas (such as building social capital and overcoming exclusion) also shape the government's local agenda.

Within the same city, quite different policy agendas and an imagined future circulate. City managers have to grapple with these divergent elements and devise responses to the multiplicity of a city's economies and social networks. On the one hand, there is the chimera of becoming a global city (Bremner, 2000)—so often promised in the literature! The route to this— to be connected to the economic networks that matter—has driven some policymakers to identify a desire to become a "smart city" (like Singapore and Kuala Lumpur). Rogerson (1999:97) writes of "strategic initiatives to reconfigure Gauteng [the province in which Johannesburg is located] as South Africa's 'smart province' with its emphasis upon building up strengths in export-oriented niche activities and in high technology production . . . (and which propose that) . . . high technology production should be the base towards which long term manufacturing growth is supported." On the other hand, the challenges of redressing the inequalities of apartheid and meeting the needs of the city's poorest residents remain. How can the city imagine its future creatively across this diversity?

STRATEGIES FOR CITY DEVELOPMENT

Paralleling each other, Johannesburg's frameworks for development and the World Bank's city development strategy see the need to find a way to hold together the city's diversity in imagining its future. The four principles of the World Bank's city development strategy spelled out earlier need to be balanced against one another. Bankability and competitiveness need to be traded off against livability, especially for the poor—although as the UNCHS (2001:xxxiii) Report suggests, "the connection between the logic of the market and the logic of liveability is anything but automatic." Moreover, effective forms of governance involve substantial interface with communities and civil society groups, which is crucial, if difficult to sustain. By aiming to draw together people from different constituencies to assess the specific potential for future growth, the City Development Strategy (CDS) represents a significant new phase in urban development policy. As one of the key World Bank advisors notes, "To help civic leaders articulate a shared vision for the city's future, CDSs aim to set out community visions, priorities and actions and help guide the allocation of resources" (Campbell, 1999:19). Of course the politics of the negotiations among the different local organizations are likely to be profound—popular movements are likely to contest fiscal prudence in service provision or perhaps argue against local government choices to invest in low employment generating but globally competitive forms of industrialization.

In 1998 the Spatial Development Framework for the metropolitan area drew together the following "interrelated strategic objectives—or strategic thrusts: Good Governance, Economic Growth, Equity and Integration, and Quality-of-Life . . . as part of a co-ordinated programme for the future direction of Johannesburg." Financial crisis necessitated a more thoroughgoing review of the city's future organizational and developmental structure, which has been undertaken under the rubric of *iGoli*[3] *2002*, an immediate and short-term urgent restructuring of the city's financial and organizational base. Although short term, some of the decisions taken within this framework lay a longer term path of potentially private sector service delivery for the city. This is linked to a longer term visioning process, *iGoli 2010*. Although drawing on broader cost-recovery and business principles associated with neoliberalism, the council's restructuring under *iGoli 2002* was, according to Beall *et al.*, (2000:x), urgently needing to address the racially based administrative fragmentation and chaos inherited from the apartheid era. As they note, "There is a very clear sense across the city that, without success in the technical transformation of local government, the pro-poor goals of reconstruction will not be achieved and this is the fundamental premise of iGoli 2002." Some commentators, as well as unions and antiprivatization bodies, contest this claim, and suggest that poverty and inequality have fallen from view in this process of organizational change (Tomlinson, 1999c).

In the *iGoli 2010* process the city has initiated a "partnership" of key stakeholders: government, labor, community, and business, and is calling on "international experts experienced in transforming major cities around the world into world class, globally competitive cities" (GJMC, 2000:10). The organizers describe the ambitions of *iGoli 2010*:

> The primary object of this long-term plan is to deal with the ongoing development paradox that the city finds itself in. This refers to achieving a balance between addressing basic needs and service backlogs on the one hand and ensuring economic growth and competitiveness on the other. The council's approach addresses both elements simultaneously since improvements to one element contributes to the benefit of the other element. (iGoli Online, *iGoli 2010* Partnership: http://wwww.igoli.gov.za/)

With the slogan "Building an African World Class City" and aiming for global competitiveness, the challenges to the coalition to find innovative ways to make this vision a reality are substantial. To ensure that activities with a global reach find synergies with more locally based (even survivalist) activities and that the transnational links that are supported and prioritized by the local state reflect the diversity of flows and influences from various parts of the world and the continent demand a more creative imagination than urban theory can offer in either its globalist or developmentalist versions.

ORDINARY CITIES, ORDINARY FUTURES
ORDINARY CITIES: EMERGING APPROACHES

In place of the global and world city approaches that focus on a small range of economic and political activities within the restrictive frame of the "global" or megacity approaches, a number of writers are offering more generalized accounts of cities. Michael Storper (1997) has focused on the economic creativity of urban agglomerations in his description of the "reflexive city." He generalizes the need for "proximity" in economic interactions to cement relations of trust among complex organizations and between individuals and organizations. Storper sees the city as providing a key context for these reflexivities, so crucial to the "untradeable" and "tacit" elements of economic life. However, rather than being limited to a focus on the workings of single industry production complexes or to production chains, or filieres, reflexivity is a generalized possibility in city life. Storper (1997:245) suggests then that we think of "the economies of big cities . . . as sets of partially overlapping spheres of reflexive economic action . . . (including) their conventional and relational structures of coordination and coherence." Cities, then, remain attractive locations for business activity across a range of sectors and offer an environment that enables economic production and innovation.

Amin and Graham (1997) concur, suggesting that (at least to some extent) cities generally foster creativity. In Western policy circles, they argue, there is a rediscovery of "the powers of agglomeration" and an excitement about cities as creative centers. Agreeing that many accounts of cities highlight only certain elements of the city (finance services, information flows) or certain parts of the city—both leading to a problem of synecdoche—they rather describe (all) cities as "the co-presence of multiple spaces, multiple times and multiple webs of relations, tying local sites, subjects and fragments into globalizing networks of economic, social and cultural change . . . as a set of spaces where diverse ranges of relational webs coalesce, interconnect and fragment" (Amin and Graham, 1997:417–418). Within this spatialized imaginary of cities as sites of overlapping networks of relations in which people, resources, and ideas are brought together in a wide variety of different combinations and within complex geographies of internal differentiation and dis/ordering, the futures of cities are uncertain, to be made—and limited—by the historical circumstances of that city (Allen *et al.*, 1999; Pile *et al.*, 1999). Power relations are, of course, not absent.

These approaches stress the importance of acknowledging overlapping networks of interaction that stretch beyond the physical form of the city and connect it to other places in the world. The range of potential international or transnational connections is substantial: cultural, political, urban design, urban planning, informal trading, religious influences, financial, institutional, intergovernmental, and so on (Allen, 1999). These various webs of connections are of differing reaches, some to the immediate hinterlands of cities, some stretching across the wider region, or to other continents. An important point is that the scale of the "global" is an imagined space made up of transnational links of all kinds that follow a diversity of quite specific trajectories that seldom actually cross the whole globe. By invoking the "global," urban theorists and practitioners have lost sight of the specific and defined reaches of different transnational flows. Even global finance, the most mobile and extensive of economic activities, does not stretch everywhere, and involves specialist activities with different reaches and fields of expertise (Hirst and Thompson, 1996).

To the extent that the regulating fiction of the global city is sustained by a twin form of reductionism—economic (and only a small segment of economic activity) and global (economic activities with a purportedly global reach are privileged)—this alternative spatialized account of the multiple webs of social relations that produce ordinary cities could help to displace some of the hierarchizing and excluding effects of this approach. Within this framework, cities can be understood to shape their futures in and through these varying networks and their intersections within the space of the city (Pryke, 1999). There is no need to become a "global" city as all cities are part of networks with varying reaches. To be "globally" connected to other parts of the world is an "ordinary" feature of cities and can be as character-istic of informal sector traders as of Wall Street. Cities that have not made it onto the map of global theorists (see Robinson, 2002) are nonetheless shaped by a range of transnational circuits, all important in imagining their possible futures. International NGOs in Nairobi (Simon, 1995), for exam-ple, or the cluster of activities organizing international labor migration in Manila (Tyner, 2000), West African traders finding a space for networking in the inner city of Johannesburg (Simone, 1998), or Zairian-Parisian traders fostering networks through Kinshasa (McGaffey and Bazenguissa-Ganga, 2000)—the variety of economic, social, and political networks needs to be appreciated and understood in imagining the future of a city, whether it be London or Lusaka.

ORDINARY FUTURES FOR JOHANNESBURG'S INNER CITY?

The inner city of Johannesburg brings into view the diverse networks of varying reaches that make up (ordinary) cities. It also challenges the habit within urban studies of working with categories. Policymakers and com-mentators have grappled with Johannesburg's long-term decline and its changing characteristics, struggling to find inspiration from urban theory and policy to guide their interpretations. From being the heart of Johannes-burg's international and national company headquarters and finance and invest-ment companies, the inner city has been changing its social and economic character (at least since the early 1970s). Richard Tomlinson (1999a, b) argues that it can no longer be seen as synonymous with the central business dis-trict or home to most of South Africa's company headquarters and insur-ance, banking, and financial sectors, although some of these activities remain. The skyline of the inner city retains its distinctive, modern image and dom-inance of the urban landscape, but by all accounts the central area is now just another large business area among many, albeit one whose distinctive cen-trality assures it an important role in a range of different enterprises.

Many of the major companies have decentralized to the northern sub-urbs, but some have stayed. In addition, the center of the city has acquired a distinctly African identity. Informal traders dominate the streets, African migrants from across the continent come to the area to trade and to live, and there has been a substantial change in the kinds of tenants in office and residential space. Routine banking and office jobs remain in the area, as it is accessible to large African townships such as Soweto where many of their skilled and semiskilled workers live. But a substantial proportion of the commercial, insurance, and banking activities previously attracted to the inner city's centrality has been locating in the northern suburbs, with more congenial surroundings and more up-to-date office buildings (Beavon, 1998; Tomlinson, 1999b). New kinds of activities have taken up the vacated office space, including small and medium enterprises such as clothing, black professionals, NGOs, and government (Tomlinson and Rogerson, 1999).

In the face of these changes, Johannesburg's residents and urbanists have felt the need to chart a different future for the inner city, drawing on new images of cityness. Tomlinson, for example, asks whether the future of Johannesburg can any longer be imagined through a Euro-American urban development lens. One alternative is to imagine the future of Johannesburg as an African city: "American conceptions of inner city decline and appropriate policies may blind policy-makers to local opportunities not found in American inner cities" (Tomlinson, 1999a:1). Among these are substantial retail trading opportunities with migrant traders from South Africa's northern neighbors; a supportive environment for small and medium enterprises, especially clothing; many informal trading opportunities; diverse cultural events; and the continuing advantages to some large companies of centrality combined with relatively cheap top quality office accommodation.

A number of similar economic activities find an opportunity to "cluster" in the inner city area and benefit from proximity to one another and to larger producers, suppliers, and markets. Clothing production is an emergent cluster, with new small and microenterprises that are largely African or immigrant owned and linked to older and larger factories in Fordsburg, Lenasia, and Mayfair. Availability of low rent and ample space, relatively low crime rates (compared to townships), and proximity to other similar enterprises have made the inner city an attractive location for these new firms (Tomlinson and Rogerson, 1999). Other more established clusters, such as finance and jewelry production, together with plans for a major cultural quarter (partly in place) make the task of promoting the inner city's economy much less difficult than the publicity surrounding its decline might suggest.

However, the question as to what kind of city the inner city area of Johannesburg could become weighs on the minds of policymakers. Tomlinson (1999a:12), for example, suggests that "We do not fully understand what this overlay of African and Western cultural, economic, infrastructure, management, social and value systems means for the inner city, but we do know that the inner city has irrevocably changed. . . . There needs to be a paradigm shift in our presentation of the city." Together with policymakers, councillors, and other academics, Tomlinson has been exploring the possibility of thinking of Johannesburg as an African city, leaving behind its Euro-American focus of the apartheid years (see also Bremner, 2000). Now, the slogan of the postapartheid council is that Johannesburg is (or should become) an "African World Class City," speaking to both the apparent need to compete globally and the value of taking her place in Africa.

This is a good example of new approaches to cities and their futures that do not limit future possibilities by categorizing or labeling cities but that instead bring into view a distinctive, diverse history to each city. We need to pay attention to the long experiences that cities such as Johannesburg have of different kinds of links and interactions across the region, continent, and to different parts of the globe, and acknowledge that on the basis of these the city has the opportunity to imagine a particular type of future for itself, and an identity not limited by preordained categorizations or the trajectories of other cities.

The Inner City Office in Johannesburg is grappling with these diverse uses and histories gathered together in the former CBD. One initiative has been to demarcate distinctive precincts within the city, encouraging different kinds of developments (clusters) across the area (GAPP Architects and Urban Designers, 1999). This acknowledges the multiplicity of uses in the city and attempts to find a way to manage their coexistence. Different kinds of flows into and beyond the inner city are channelled into specialist areas where the opportunities for synergies in services and the built environment could possibly be realized. Different kinds of activities are far from harmonious, though. Different constituencies benefit from different conditions, and have different capacities to ensure they are realized. Hawkers who benefit from passing trade on pavements are a hindrance to big businesses keen to maintain their corporate image and property values; and South African hawkers find the presence of foreign informal traders a threat. Homeless children, petty thieves, and people involved in illegal activities might find a congenial home on crowded city streets or deserted parts of the inner city, but for commuters and residents they make the city seem dangerous. Negotiating diverse needs through partnerships and consultation is not an easy task. But most benefit the city as a whole, and their presence draws in shoppers, traders, workers, and investment from all over the city, the rest of the continent, and beyond. To privilege only businesses whose global reach follows traditional paths to the West and its associated economies or whose activities involve formal associations with businesses across the continent is to miss the dynamism of existing activities that stretch, for example, to rural hinterlands in the country (as traditional markets do) or draw in cultural inputs from other South African cities and neighboring countries.

The inner city has always been a place of flows through which literally hundreds of thousands of people pass each day and from which investments and businesses across the country are managed. These flows may change in content and direction (Mumbai or Lusaka rather than London), but the potential for the inner city to remain a vibrant context for facilitating interactions and networking within and across these enterprises remains. Attempting to follow the chimera of Western global competitiveness, to become a global city, would undermine the dynamic potential of inner city Johannesburg. Within a developmentalist frame, the excellent infrastructure, relatively high-income residents, and solid, if poorly maintained and overcrowded accommodation would make this area a very low priority for council attention. Between global city status and developmentalism, the inner city has found ways to imagine its own present and future as part of an ordinary city—just like all other cities—in its own distinctive idiom.

CONCLUSION

The consequences of these reflections on the future of Johannesburg for urban theory and urban development are quite substantial. Moving from a divided approach to cities, where some are interpreted as structurally irrelevant and others as globally powerful, to one where all (ordinary) cities are understood as complex, diverse, and contested environments for living, necessitates changes in how urban theory and urban policymaking are done. The old tracks, from centers of power to peripheries of irrelevance, need to be redirected.

Johannesburg offers us a metaphor for the changes that are needed. It is re-imagining its place in the world, hankering less after Chicago and New York, as it might have in the 1930s, and instead directing its enquiring gaze across cities nearby and far away to discover links, connections, and lessons for future planning. The broader but still cosmopolitan trajectories of Johannesburg's urban imagination suggest paths for theorizing. These avoid the all-seeing neoimperialist global theorist, mapping and ranking cities around the world. They also avoid the parochialism of the categories that divide cities from one another or set up impossible ambitions.

Just as Johannesburg's citizens and managers must grapple with that city's complexity, and devise creative ways of thinking about its future, and negotiating present dilemmas, so urban theorists need to move beyond globalization and developmentalism, and embrace the ordinary, but dynamic, complexity of urban life.

NOTES

1. I should like to thank people who attended various seminars—at Birmingham, Loughborough, and Kings College London Geography Departments; at the OU-Bristol-Durham Cities Seminars, the 2000 AAG, and the Urban Futures Conference in Johannesburg—where early versions of this chapter were presented. All the discussions made me think harder about these points, as did the individual comments of Richard Tomlinson, Steve Pile, and Robert Beauregard.
2. Thanks to Richard Tomlinson for this information.
3. "Igoli" is the vernacular name for Johannesburg—"place of gold."

REFERENCES

Allen, J. 1999. "Cities of Power and Influence: Settled Formations," in J. Allen, D. Massey, and M. Pryke, editors, *Unsettling Cities*. London: Routledge, pp. 181–228.

Allen, J., D. Massey, and M. Pryke, editors. 1999. *Unsettling Cities*. London: Routledge.

Amin, A., and S. Graham. 1997. "The Ordinary City." *Transactions of the Institute of British Geographers* 22:411–429.

Askew, M., and W. Logan. 1994. *Cultural Identity and Urban Change in Southeast Asia: Interpretative Essays*. Victoria: Deakin University Press.

Beall, J., O. Crankshaw, and S. Parnell. 2000. "Local Government, Poverty Reduction and Inequality in Johannesburg." *Environment and Urbanisation* 11:107–122.

Beavon, K. 1998. "Johannesburg: Coming to Grips with Globalization from an Abnormal Base," in F-C. Lo and Y-M. Yeung, editors, *Globalisation and the World of Large Cities*. Tokyo: United Nations University Press, Chapter 13.

———. 2001. "The City That Slipped." *Sunday Times* January 7, pp. 6–7.

Bhabha, H. 1994. *The Location of Culture*. London: Routledge.

Bremner, L. 2000. "Reinventing the Johannesburg Inner City." *Cities* 17:185–193.

Burgess, R., M. Carmona, and T. Kolstee, editors. 1997. *The Challenge of Sustainable Cities: Neo-liberalism and Urban Strategies in Developing Countries*. London: Zed Books.

Campbell, T. 1999. "The Changing Prospects for Cities in Development—The Case of Vietnam," in *Business Briefing: World Urban Economic Development*. Official briefing for World Competitive Cities Congress. Washington, D.C.: World Bank, pp. 16–19.

Dick, H. W., and P. J. Rimmer. 1998. "Beyond the Third World City: The New Urban Geography of South-east Asia." *Urban Studies* 35(12):2303–2321.

Douglass, M. 1998. "World City Formation in the Asia Pacific Rim: Poverty, 'Everyday' Forms of Civil Society and Environmental Management," in M. Douglass and J. Friedmann, editors, *Cities for Citizens*. Chichester: John Wiley, pp. 107–137.

Friedmann, J. 1986. "The World City Hypothesis." *Development and Change* 17:69–84. Reproduced in Knox, P. L., and P. J. Taylor, editors. 1995. *World Cities in a World-System*. Cambridge: Cambridge University Press, pp. 317–331.

———. 1995. "Where We Stand Now: A Decade of World City Research," in P. L. Knox and P. J. Taylor, editors, *World Cities in a World-System*. Cambridge: Cambridge University Press, pp. 21–48.

Friedmann, J., and W. Goetz. 1982. "World City Formation. An Agenda for Research and Action." *International Journal of Urban and Regional Research* 6:309–344.

GAPP Architects and Urban Designers. 1999. *Aide Memoire 1. Toward a Spatial Framework for the Greater Johannesburg Inner City*. Johannesburg: GJMC Inner City Office.

Gilbert, A. 1998. "World Cities and the Urban Future: The View from Latin America," in F. Lo and Y. Yeung, editors, *Globalisation and the World of Large Cities*. Tokyo: United Nations University Press, Chapter 8.

Greater Johannesburg Metropolitan Council (GJMC). 1998. *Spatial Development Framework*, Johannesburg.

———. 2000. *Igoli 2002: Making the City Work*. Johannesburg.

Hansen, K. 1997. *Keeping House in Lusaka*. New York: Columbia University Press.

Harris, N. 1992. *Cities in the 1990s*. London: UCL Press.

Hirst, P., and G. Thompson. 1996. *Globalization in Question*. Cambridge: Polity.

Jessop, B., and N. Sum. 2000. "An Entrepreneurial City in Action: Hong Kong's Emerging Strategies in and for (Inter) Urban Competition." *Urban Studies* 37(12):2287–2313.

Kelly, P. F. 2000. *Landscapes of Globalization. Human Geographies of Economic Change in the Philippines.* London: Routledge.

King, A. 1990. *Urbanism, Colonialism and the World-Economy*. London: Routledge.

Knox, P. L. 1995. "World Cities in a World System," in P. L. Knox and P. J. Taylor, editors, *World Cities in a World-System*. Cambridge: Cambridge University Press, pp. 3–20.

Knox, P. L., and P. J. Taylor, editors. 1995. *World Cities in a World-System*. Cambridge: Cambridge University Press.

Lopez de Souza, M. 2001. "Metropolitan Deconcentration, Socio-political Fragmentation and Extended Suburbanisation: On Brazilian Urbanisation in the 1980s and 1990s." *Geoforum* 32:437–447.

McGaffey, J., and R. Bazenguissa-Ganga. 2000. *Congo-Paris. Transnational Traders on the Margins of the Law.* Oxford: James Currey.

McGee, T. 1989. "Urbanisasi or kotadesasi? Evolving Patterns of Urbanisation in Asia," in F. Costa, A. Dutt, L. Ma, and A. Noble, editors, *Urbanisation in Asia. Spatial Dimensions and Policy Issues.* Honolulu: University of Hawaii Press.

———. 1995. "Metrofitting the Emerging Mega-Urban Regions of ASEAN: An Overview," in T. McGee and I. Robinson, editors, *The Mega-Urban Regions of Southeast Asia.* Vancouver: University of British Columbia Press, Chapter 1.

Morshidi, S. 2000. "Globalising Kuala Lumpur and the Strategic Role of the Producer Services Sector." *Urban Studies* 37(12):2217–2240.

Olds, K., and H. Yeung. 2002. "From the Global City to Globalising Cities: Views from a Developmental City-State in Pacific Asia." *International Journal of Urban and Regional Research*.

Parnell, S., and E. Pieterse. 1998. *Developmental Local Government.* Isandla Working Paper.

Pile, S., C. Brook, and C. Mooney, editors. 1999. *Unruly Cities?* London: Routledge.

Pryke, M. 1999. "City Rhythms: Neo-liberalism and the Developing World," in J. Allen, D. Massey, and M. Pryke, editors, *Unsettling Cities.* London: Routledge.

Pugh, C. 1995. "Urbanisation in Developing Countries: An Overview of the Economic and Policy Issues in the 1990s." *Cities* 12(6):381–398.

Rakodi, C., editor. 1997. *The Urban Challenge in Africa: Growth and Management of Its Large Cities.* Tokyo: United Nations University Press.

Republic of South Africa (RSA). 1998. The White Paper on Local Government, *Government Gazette*, March 13, 1998.

Robins, K., and A. Askoy. 1996. "Istanbul between Civilisation and Discontent." *City* 5–6:6–33.

Robinson, J. 2002. "Global and World Cities: A View from Off the Map." *International Journal of Urban and Regional Research*.

Rogerson, C. M. 1999. "Industrial Change in a Developing Metropolis: The Witwatersrand 1980–1994." *Geoforum* 30:85–99.

———. 2000. "Local Economic Development in an Era of Globalization: The Case of South African Cities." *Tijdschrift voor Economische en Sociale Geografie* 91:397–411.

Sassen, Saskia. 1991. *The Global City: New York, London, Tokyo.* Princeton, NJ: Princeton University Press.

———. 1994. *Cities in a World Economy.* Thousand Oaks, CA: Pine Forge Press.

Simon, D. 1995. "The World City Hypothesis: Reflections from the Periphery," in P. L. Knox and P. J. Taylor, editors, *World Cities in a World-System.* Cambridge: Cambridge University Press, pp. 132–155.

Simone, A. 1998. "Globalization and the Identity of African Urban Practices," in H. Judin and I. Vladislavic, editors, *blank_____: Architecture, Apartheid and After.* Rotterdam: NAI, p. D8.

Storper, M. 1997. *The Regional World.* London and New York: Guilford.

Tomlinson, R. 1999a. *An Economic Strategy for the Johannesburg Central City.* Draft Manuscript. Graduate School of Public and Development Management, University of the Witwatersrand.

———. 1999b. "From Exclusion to Inclusion: Rethinking Johannesburg's Central City." *Environment and Planning A* 31:1655–1678.

———. 1999c. "Ten Years in the Making: A History of Metropolitan Government in Johannesburg." *Urban Forum* 10(1):1–40.

Tomlinson, R., and C. Rogerson. 1999. "An Economic Development Strategy for the Johannesburg Inner City." The strategy was prepared as part of the United Nations Development Program's Urban Management Program City Consultation Process, on behalf of the Greater Johannesburg Metropolitan Council's Inner City Section 59 Committee.

Tyner, J. A. 2000. "Global Cities and Circuits of Global Labour: The Case of Manila, Philippines." *Professional Geographer* 52:61–74.

UNCHS. 2001. *Cities in a Globalising World: Global Report on Human Settlements 2001*. London: Earthscan (UNCHS).

World Bank. 1991. *Urban Policy and Economic Development: An Agenda for the 1990s: A World Bank Policy Paper*. Washington, D.C.: World Bank.

———. 2000. *Cites In Transition*. Washington, D.C.: World Bank.

Yeung, Y-M. 1989. "Bursting at the Seams: Strategies for Controlling Metropolitan Growth in Asia," in F. Costa, A. Dutt, L. Ma, and A. Noble, editors, *Urbanisation in Asia. Spatial Dimensions and Policy Issues*. Honolulu: University of Hawaii Press.

15

Johannesburg in Flight from Itself

POLITICAL CULTURE SHAPES URBAN DISCOURSE

XOLELA MANGCU

This chapter is about the prefigurative thrust of social movements and its implications for contemporary urban discourse in Johannesburg.[1] It is about the way that the incorporation of symbols and the demands of protest movements determine the direction of subsequent institutional relations.[2] Public policy priorities, it claims, are shaped by the political culture of societies.

On a concrete level, the understanding of South African cities as sites of globalization, economic growth, and service delivery is best understood by locating it within the modernist discourse of nonracialism that has been associated mostly with the ruling African National Congress (ANC). This latter modernism has its antecedents in a longer history of enlightenment universalism. Properly understood, modernism celebrates the superiority of scientific and technical methods as the basis for rationally organizing society. Modernism is thus closely associated with the triumph of reason over emotion, of objectivity over subjectivity, and of universalism over particularism.[3]

Peller has identified the relationship between modernism and integrationist nonracialism: "integrationist beliefs are organized around the familiar enlightenment story of progress as consisting of the movement from mere belief and superstition to knowledge and reason, from the particular and therefore parochial to the universal and therefore enlightened." Because race has no scientific basis, then, nonracialism becomes its symmetric opposite.[4] According to this logic, all manner of race or cultural consciousness is of necessity retrogressive, irrespective of its content.

This symmetrical logic has been the foundation of the ANC's politics and has led to the displacement of culture in urban policies. The dominant political culture of nonracialism that reached its crescendo in the 1980s resulted in a technocratic urban discourse with virtually no room for the role of cultural identities in the formation of a new urban citizenship. Scientific technique replaced politics and the universal language of global economic progress displaced local innovation. Peller argues that the meaning of race has been grafted onto other central cultural privileges of progress so that the transition from segregation to integration and from race consciousness to race neutrality mirrors movements from myth to enlightenment, ignorance to knowledge, superstition to reason, and the primitive to the civilized. This inability to conceptualize a liberating and progressive conception of race identity has "distorted our understanding of the politics of race in the past and obscures the ways that we might contribute to a meaningful transformation of race relations in the future."[5] Yet "race" is a cultural phenomenon around which people, especially oppressed people, have constructed entire cultural histories. As West puts it: "people, especially poor and degraded people, are also hungry for meaning, dignity and self-worth."[6]

For example, the township of Soweto just outside of Johannesburg is more than a mere racial location, even though its apartheid designers intended it as such. Soweto is above all a cultural space in which black people have constructed a particular social history. That conception is missing in Johannesburg's modernist urban discourse. The city's modernist path was not an inevitable outcome of broader ideological struggles within the liberation movement that ended with the nonracial triumphalism of the ANC and the demise of the politics of culture. Racial identity has not been dispelled, nor should it be.

CULTURE AND DEVELOPMENT IN THE TOWNSHIPS

In a recent paper on social capital in African townships, Bozzoli argues that before the advent of apartheid in 1948 the townships consisted of a private world of social networks. Although they were characterized by poverty, townships also developed "a rich and fairly deeply institutionalized cultural life which found expression in shebeens, schools, gangs, families, sports and many other forms."[7] Thus, despite the harshness of the political and economic conditions, people carved out spaces of cultural creativity and material survival through a process of syncretic adaptation.

This syncretic adaptation found its expression through, inter alia, the creolization of European languages, the adoption of American music (especially jazz), and the development activities of organizations such as the Red

Cross and the YWCA. According to Bozzoli "nothing entered townships without being given local meaning."[8]

Bozzoli's analysis is important. White writers about townships have too often presented a picture of townships as nothing more than sites of poverty and oppression. Townships are for the most part described as zones of danger, despair, and pathology. The accurate physical description of townships as rows and rows of monotonous and drab matchbox houses also hides from outsiders the cultural activities that happen within them. This conflation of physical monotony and cultural monotony can be found in this description from one of South Africa's respected urban planners, David Dewar:

> Rather than being Arcadian retreats, the townships have very little growing in them. The public spaces are inhospitable, dangerous, and frequently serve as dumping grounds for rubbish. Atmospheric pollution is severe. Large amounts of residual land, awaiting new or expanded facilities that in all probability will never materialize, destroy all sense of human scale, and *there is no tradition of making positive urban spaces.* Finally because of the high degree of planning and control, *the settlements are invariably sterile, monotonous, and boring,* despite attempts to provide variety through design techniques such as convoluted road configurations.[9]

Elsewhere, Dewar notes that "social and commercial facilities and other vibrant urban activities are notable mainly by their absence."[10]

Despite physical deterioration, numerous political and cultural spaces did exist, though these efforts were always contested. Soon after the National Party was elected in 1948 on a radical ticket of apartheid, it sought to close the spaces by criminalizing political protest. Laws such as the Public Safety Act of 1953 enabled it to declare a state of emergency when it deemed necessary and the Criminal Law Amendment Act increased penalties for acts of defiance. The National Party outlawed the African National Congress and the Pan Africanist Congress (PAC) in 1960 and sent their leaders to long prison terms. By the early 1960s the grandmaster of apartheid, Hendrik Verwoerd, and his Minister of Justice, B. J. Vorster, had managed to build what O'Meara has described as "walls of granite" around apartheid: "Under the new Minister of Justice, B. J. Vorster, South Africa was rapidly turned into a grubby (and increasingly corrupt) police state. . . . By 1964 the underground networks of the ANC and the PAC had been wiped out, the various rural revolts brutally suppressed, and the member unions of the South African Congress of Trade Unions bled white."[11]

The birth of the black consciousness movement in 1968 was an attempt to reclaim these political and cultural spaces. Drawing inspiration from liberation movements in the United States and elsewhere on the African continent, black consciousness activists sought much more than political rights. The father of the movement, Steve Biko, put it this way: "We don't want to be mere political Africans. . . . We want to be social Africans."[12] The new movement sought to restore values of self-respect, self-reliance, and dignity to the black community and to transfer those values to the envisaged democratic society. As Sam Nolutshungu put it: "the movement presents itself as a secession from a world that rejects or frustrates or cannot comprehend its concerns and needs, but the secession precedes the shattering of that world and the creation of another."[13] For the student leaders of the movement this meant embracing the antecedent values and social networks that Bozzoli described. Barney Pityana, one of the movement's founders, encapsulated the reigning sentiment: "it is essential for the black students to strive to elevate the level of consciousness of the black community by promoting awareness, pride and capabilities."[14] The students also realized their limitations as a social force and established a national political organization that would, for the first time since the banning of the ANC and the PAC, give explicit political voice to the aspirations of black people.

More importantly for understanding the relation of political culture and racial identity was the formation of the Black Community Programmes (BCP) in 1970. The BCP emerged out of the South African Council of Churches and Christian Institute. It built schools, day-care centers, and clinics throughout the country. It established home-based industries and cooperatives in remote rural settings and townships and published community newspapers and journals such as the *Black Voice, Black Review, Black Perspective,* and *Creativity in Development.* The BCP also opened community-based research institutes such as the Institute for Black Research.[15] The period has been generally referred to as the Black Renaissance because of the revival of cultural, literary, and political activity within black communities throughout the country.

A 1975 issue of *Black Review* argued that in evaluating community projects it was necessary to look beyond the structures created to the level of consciousness attained by the people involved. The civic outcomes include the development of a generation of leaders, e.g., Mamphela Ramphele, Barney Pityana, and Malusi Mpumluana among others, who have come to play key roles in present-day South Africa, despite the murder of movement leaders such as Mapetla Mohapi and Steve Biko in 1976 and 1977, respectively. As had been the case after the 1960 bannings of the ANC and the PAC, the 1977 bannings of black consciousness organizations were followed by an upsurge of a new leadership cadre in the 1980s. The development economist Albert Hirschman succinctly describes these cycles of renewal: "the social energies that are aroused in the course of a social movement do not disappear when that movement does but are kept in storage and become available to fuel later and sometimes different social movements. In a real sense, the original movement must therefore be credited with whatever advances or successes were achieved by those subsequent movements: no longer can it be considered a failure."[16]

CIVIC MOVEMENTS AND URBAN REFORM IN JOHANNESBURG

The subsequent social movements were qualitatively different in their outlook. The black consciousness movement had concentrated on inward-looking cultural strategies of community development as a basis for the outward political struggle. The new civic movements of the 1980s concentrated almost exclusively on the political challenge to the apartheid state. Impatient with the steady pace and organization building of the 1970s, the "young lions" of the 1980s brought a dizzying pace to township struggles. Seeking to make the country ungovernable they substituted mass mobilization for organization building. The main targets of mobilization were the black local authorities introduced by the apartheid government in 1982 as part of its reform strategy. The then Prime Minister P. W. Botha had established these local authorities to coopt urban blacks into the political system without extending the vote at the national level. These bodies were in turn supervised by the notorious Joint Management Committees of the State Security Council, a body comprised of senior political and military leaders.

Civic movements in townships throughout the country declared consumer, rent, and bus boycotts to force the local councilors to resign. Townships were transformed into literal war zones as local youths battled the police. As an example, Alexandra township just outside Johannesburg was caught up in what came to be known as the Six Day War between February 14 and February 20, 1985. The funeral of the victims of that war drew more than 40,000 people. This was community mobilization on a scale never seen before. Upward of 25,000 government troops were deployed in the townships and more than 26,000 people were detained without trial over a period of a year starting in 1986.[17]

The government restored short-term political order by declaring successive states of emergency in 1985 and 1986. Consistent with P. W. Botha's carrot and stick approach to politics, the government adopted "a winning the hearts and minds approach" by offering millions of rands to upgrade township infrastructure in places such as Alexandra. However, the government sought to cover the costs through increased rental and service charges. This provoked the ire of the civic movement and the campaigns to destroy the black local authorities intensified. Rent boycotts were accompanied by consumer boycotts and work stayaways. Increasingly local white councilors and business people entered into negotiations with civic leaders through joint forums without the discredited black councilors. In lieu of local authority structures, civic organizations became a de facto component of local government.

To illustrate, the Soweto People's Delegation led by prominent community leaders such as Cyril Ramaphosa and Frank Chikane met with representatives of the provincial authorities and the three black local authorities in Soweto to demand the scrapping of any rent arrears that had accrued over the duration of the boycotts. They also argued that local residents had, over decades, paid off the value of Soweto's rental housing and should now own them. They called for the upgrading of services tied to the creation of one tax base for an integrated Johannesburg. "One city, one tax base" campaigns took place all over the country.

The provincial authorities ultimately agreed to write off the debt while the Soweto People's Delegation encouraged people to pay an interim service charge toward bulk service provision. Discussions also led to the formation of a metropolitan chamber to begin a longer term debate about democratic local government for Johannesburg. The chamber consisted of almost every urban player: local councilors, civic and residents associations, white ratepayers, and political parties. Business and worker groups were given observer status. Swilling describes the chamber "as a glorious experiment in participatory governance."[18]

CIVICS AND TECHNOCRATIC CREEP

The collapse of local authorities also meant that pressure was put on civic leaders to deal with the practical problems of everyday life in the township. Increasingly, the civic organizations relied on technical assistance from newly established service agencies comprised mainly of white academics at major universities such as the University of Witwatersrand. These agencies specialized mainly in housing, service delivery, and local economic development.

The most prominent of these agencies was Planact based in Johannesburg. Mark Swilling, himself a Planact staff member, explains the role of these agencies: "they wanted to use their skills to empower communities, by providing them, rather than the state or business, with information and specialist analysis."[19] Alan Mabin, one of the founding members of Planact, wrote that "a significant body of urban planners tried actively to assist the most disadvantaged sections of society." But ultimately these planners and activists represented what Mabin describes as "the left-modernist idea of the state as an instrument for reconstruction."[20]

Just as apartheid ideologues had imagined that they could use the state to shape society in their image, these planners thought they could use the state to reshape society in a more progressive manner. As Scott puts it: "many of the great state-sponsored calamities of the twentieth century have been the work of rulers with grandiose and utopian plans for their society. One can identify a high modernist utopianism of the right, of which Nazism is surely the diagnostic example. The massive social engineering under apartheid in South Africa, the modernization plans of the Shah of Iran, villagization in Vietnam, and huge late-colonial development schemes (for example, the Gezira scheme in the Sudan) could be considered under this rubric. And yet there is no denying that much of the massive, state-enforced social engineering of the twentieth century has been the work of progressive, often revolutionary elites."[21]

Within a short period of time, the central urban actors were those with the technical expertise to participate in the planning forums that were sprouting all over the place. Oligarchy had set in and, as Michels pointed out early in the last century, these oligarchical tendencies were a "matter of technical and practical necessity."[22] The long-term consequences were that the urban policy process increasingly occurred away from public view. Moreover, a number of participants in the Johannesburg Metropolitan Chamber disagreed with Swilling's descriptions of the process as a "glorious experiment in participatory governance." Some critics argue that white intellectuals such as Mark Swilling and William Cobbett and others "were like little boys playing computer games . . . building their own empires. They did nothing to empower the civic leaders." An ominous division of labor had begun to emerge: investors, business leaders, urban managers, and consultants became the real and active citizenry while communities were reduced to nothing more than fictitious citizens or what Partha Chattarjee calls "empirical objects of government policy, not citizens who participate in the sovereignty of the state."[23] Although the technical details of projects are important and need to be correct, this particular activity of urban management is far from sufficient for the creation of an urban citizenship. Urban politics had permanently shifted from the life world of social movements to the systems world of bureaucrats and technical experts, all in the name of empowerment and reconstruction.

The discourse of modernist reconstruction in South Africa found its highest expression in the ANC's Reconstruction and Development Programme (RDP). The RDP would be the instrument by which the new government would reconfigure the South African physical and socioeconomic landscape. Central to reconstruction was the idea that human needs were best met through the rational and efficient delivery of service. Service delivery, especially of housing and water taps, became the watchword of development. Site and service schemes, far from places of work, dotted the South African landscape. In some communities, old historical buildings were destroyed in the name of development.

FROM LEFT-WING TO RIGHT-WING MODERNISM

The RDP Ministry office was closed in 1996, but it was not replaced by a less modernist program. In its place came a market-based approach to urban development within an increasingly conservative national economic framework. Soon after, the national government had adopted a new policy framework—Growth, Employment and Redistribution (GEAR), at the center of which were policies such as privatization. Thus the local government of Johannesburg embarked on a privatization of municipal services that set it on a collision course with the South African Allied Municipal Workers Union (SAMWU). The focus was exclusively on balancing the city's budget. The old leftist-modernists in organizations such as Planact were replaced by new rightist-modernists in organizations such as the Centre for Development and Enterprise Development. The latter became an important and influential lobbyist for the idea of cities as sites of global competition, mobilized around the generative metaphor of world class cities.[24] The global market had replaced the state as the engine of modernization. Local development needs were now seen as dependent on being world class. Being world class meant clearing the city underbrush of street hawkers and informal taxis so that the real forest of formal world class businesses could flourish.

Many aspects of both kinds of modernism are politically anathema to the cultural politics of the black consciousness movement. One obvious problem is the leadership role played by white experts in development policy. It is quite possible to argue, as Swilling does, that the planners were simply responding to the exigencies of community leaders, who were in turn responding to the needs of desperate communities. In other words, white urban professionals could not really be expected to carry the banner of African cultural themes and values if the communities were not raising them. But I can also see Steve Biko, were he alive today, challenging the communities and their leaders to develop their own capabilities. I can see him reacting strongly to the incipient "begging bowl" model of development. I can see him challenging communities to reframe their demands and expand them to include issues of pride and capabilities.

In their book *India: Economic Development and Social Opportunity*, economists Dreze and Sen argue that poverty is not just a matter of income deprivation: "Poverty of a life . . . lies not merely in the impoverished state in which the person actually lives, but also in the lack of real opportunity . . . to choose other types of living. Even the relevance of low incomes, meager possessions, and other aspects of what are standardly seen as economic poverty relates ultimately to their role in curtailing capabilities (that is their role in severely restricting the choices people have to lead valuable and valued lives). Poverty is thus ultimately a matter of capability deprivation."[25] The authors conclude that the success of development programs must ultimately be judged in terms of their capacity to expand people's choices in the long term, or what Young describes as groups engaged in "politicized self-help."[26]

CONCLUSION: BRINGING CULTURE BACK IN

The history of social movements in South Africa is the basis for an indigenous planning. At present, too much emphasis is being given to outside and Western technical expertise. Social movement experiences have been abandoned for scientism. A phalanx of American consultants have come to dominate the policymaking process in South Africa, bringing in sophisticated models of how to make Johannesburg globally competitive, while ignoring its own assets. These consultants do not bring in progressive models that use local assets as the basis for global competitiveness.[27]

Prominent writer Njabulo Ndebele[28] has argued that the townships could be sites of fruitful urban research about how people are making and remaking their lives where they are, not in response to a hypothesized global economy. He provides an example that might be instructive for modernist policymakers who want to remove the underbrush of informal taxis from the streets of Johannesburg and replace them with formal sector and larger buses. This is an example about urban innovation in his hometown of Duduza.

As Duduza grew in size, a local taxi service began. Now you can see the small Mazda 232 cars, called *amaphela,* cars running up and down the streets of Duduza. At first, the established taxi services that use minibuses sought to flex their muscles and run the new service out of business. The community rallied behind the new service. They had recognized that it considerably enhanced internal mobility, facilitating easier communication. The economic potential for this development and the impact it could have on other aspects of township life have not been fully realized. A civic culture has yet to evolve that is able to capitalize fully on emerging opportunities and enhance the sense of communal autonomy. But what we do have are the makings of an internal economy that ought to be the focus of policy.

The challenge facing Johannesburg is to think of urban futures that are as attentive to the development of human values, capabilities, and social networks in the real cities where people live as to globalization. The urban futures of our global cities are inescapably dependent on the present health and capabilities of local places. One hopes that current President Thabo Mbeki's idea of the African Renaissance will lead to a more postmodernist political culture in which people's identities can be appreciated and utilized as the basis of vibrant cultural industries.

NOTES

1. Boggs, C., *Social Movements and Political Power* (Philadelphia: Temple University Press, 1986).
2. For a discussion of the environmental movement as the example of the institutionalization of the language of social movements, see Castells, Manuel, *The Power of Identity* (Oxford: Blackwell Publishers, 1997).
3. For descriptions of the modernist city see Harvey, David, *The Condition of Postmodernity* (Oxford: Blackwell Publishers, 1990) and Scott, James, *Seeing Like A State* (New Haven: Yale University Press, 1998).
4. Peller, G., "Race Consciousness," in D. Danielsen and K. Engle editors, *After Identity* (New York: Routledge, 1995), pp. 69–76. Quote on p. 74.
5. Ibid., p. 74.
6. West, Cornel, *Race Matters* (New York: Vintage Books, 1993), p. 20.
7. Bozzoli, B., "The Difference of Social Capital and the Mobilizing and Demobilizing Powers of Nationalism: The South African Case," unpublished manuscript, 2000 p. 3. Cornel West makes a similar observation for the American case. See West, p. 21.
8. Ibid., p. 4.
9. Dewar, David, "A Manifesto for Change," in David M. Smith, editor, *The Apartheid City and Beyond* (Johannesburg: Wits University Press, 1992), p. 246. Emphasis added.
10. Dewar, David, "Settlements, Change and Planning in South Africa Since 1994," in Hilton Judin and Ivan Vladislavic, editors, *blank_Architecture, Apartheid and After* (Rotterdam: NAI Publishers, 1998), p. G6.
11. O'Meara, Dan, *Forty Lost Years: The Apartheid and the Politics of the National Party, 1948–1994* (Johannesburg: Ravan Press, 1996), pp. 109–110.
12. Biko, Steve, *I Write What I Like* (Oxford: Heinemann Educational Publishers, 1978), p. 131.
13. Nolutshungu, Sam, *Changing South Africa: Political Considerations* (Manchester: Manchester, University Press, 1982), p. 152.
14. Quoted in Fatton, Robert, *Black Consciousness in South Africa: The Dialectics of Ideological Resistance to White Supremacy* (Albany: State University of New York, 1986), p. 68.

15. The first director of the BCP was Ben Khoapa and Steve Biko was hired with responsibility for youth activities.
16. Hirschman, Albert O., *Getting Ahead Collectively and Grassroots Expansion in Latin America* (New York: Pentagon Press, 1984), pp. 55–56.
17. Mabin, Alan, "Reconstruction and the Making of Urban Planning in 20th Century South Africa," in Hilton Judin and Ivan Vladislavic, editors, *blank_Architecture, Apartheid and After* (Rotterdam: NAI Publishers, 1998), p. 66.
18. Swilling, Mark, "Rival Futures: Struggle Visions, Post-Apartheid Choices," in Hilton Judin and Ivan Vladislavic, editors, p. E8.
19. Ibid., p. 66.
20. Mabin, p. 66.
21. Scott, p. 89.
22. Michels, Robert, *Political Parties* (Glencoe, IL: Free Press, 1915), p. 39.
23. Chattarjee, Partha, *Wages of Freedom: Fifty Years of the Indian Nation State* (Oxford: Oxford University Press, 1998), p. 16.
24. For a discussion of the role that metaphors play in generating dominant meanings about the city, see Mier, Robert, *Social Justice and Local Development Policy* (Newbury Park, CA: Sage Publications, 1993).
25. Dreze, Jean, and Amartya Sen, *India: Economic Development and Social Opportunity,* (Oxford: Clarendon Press, 1995), p. 11.
26. Young, Iris, *Justice and Politics of Difference* (Princeton, NJ: Princeton University Press, 1991), p. 85.
27. For a discussion of progressive city planning models in the United States, see Clavel, Pierre, *The Progressive City* (New Brunswick, NJ: Rutgers University Press, 1986) and Clavel, Pierre, *Washington and the Neighborhood* (New Brunswick, NJ: Rutgers University Press, 1991).
28. These examples come from two speeches that Professor Ndebele gave: the Steve Biko Memorial Lecture on September 12, 2000 at the University of Cape Town and the Values in Education Conference in Cape Town on February 24, 2001. Ndebele is currently the Vice Chancellor of the University of Cape Town.

About the Editors

Robert A. Beauregard (Ph.D.) is Professor at the New School University in the Milano Graduate School of Management and Urban Policy in New York City. He is the author of *Voices of Decline: The Postwar Fate of U.S. Cities,* 2nd ed. (Routledge, 2003) and co-editor of *The Urban Moment: Cosmopolitan Essays on the Late Twentieth Century City* (1999). His research focuses on urban economic development and redevelopment, urban theory and planning theory, and the postwar history of industrial cities both in the United States and South Africa.

Lindsay Bremner is Chair of Architecture at the University of Witwatersrand in Johannesburg. She was formally a Councillor on the Greater Johannesburg Metropolitan Council for the African National Congress where she played a key role in urban policy development immediately after the first democratic local government elections. She chaired the Organizing Committee of the Urban Futures Conference held in Johannesburg in July 2000. She has published extensively on architecture and urbanism in Johannesburg, including a five-part series—Contemporary Johannesburg: Spaces, Cultures, Identities—in the *Sunday Times* for which she received the first Bessie Head Fellowship Award.

Xolela Mangcu (Ph.D.) is founding Director of the Steve Biko Foundation as well as Associate Editor and Columnist for *The Sunday Independent* (Johannesburg). Prior to joining the Foundation, Mangcu was a Senior Analyst at the Centre for Policy Studies and before that a Visiting Scholar at Harvard University doing research on community development and democracy. He obtained his Ph.D. from Cornell University in 1997. Later, he became a Warren Weaver Fellow at the Rockefeller Foundation in New York City and taught in the Department of Urban Planning at the University of Maryland (College Park). He also obtained a master's degree in Development Planning from the University of Witwatersrand in 1988 and worked at the Development Bank of Southern Africa.

Richard Tomlinson (Ph.D.) serves as a Visiting Professor at the Graduate School of Public and Development Management of the University of Witwatersrand and as a consultant in urban economic development, housing, and infrastructure. His books include *Urban Development Planning: Lessons for the Reconstruction of South Africa's Cities* (Wits University Press, 1994) and *Urbanization in Post-Apartheid South Africa* (Unwin Hyman, 1990). He has also published numerous articles critiquing urban and regional policies under apartheid and exploring the shape of South Africa's cities in the democratic era. As a consultant, he has contributed to and managed teams that have prepared a number of South Africa's major urban policies. His current research concerns urban policy processes and the pervasive influence of U.S. literature, policy, and institutions in shaping how people in the developing world view cities.

Contributing Authors

Jürgen Bähr is a Full Professor of Geography at the University of Kiel (Germany). His research focuses on population and urban geography in Latin America and southern Africa. He has more than 200 publications, including several textbooks on population geography. He is a member of several national and international organizations and is vice-chancellor of the University of Kiel.

Jo Beall is a Reader in the Institute of Development Studies (DESTIN) at the London School of Economics and works on urban social disadvantage and urban governance, with a regional specialization in southern Africa and south Asia.

Jillian Carman is an art historian and museologist with an interest in the history of museums and public art collections in South Africa. Her research focus is the early history of the Johannesburg Art Gallery, the topic of her Ph.D. through the University of Witwatersrand. She was a curator at the Gallery from 1977 to 1998 and has curated exhibitions and published on aspects of its collection and South African museology. She serves on various museum and educational committees, including the Wits University Forum, which monitors institutional transformation.

Owen Crankshaw (Ph.D) is a Senior Lecturer in the Sociology Department of the University of Cape Town. Prior to this appointment, he was a Senior Researcher at the Human Services Research Council and the Centre for Policy Studies in Johannesburg. He has also lectured at the University of Natal, University of Witwatersrand, and London School of Economics. His research focuses on changing patterns of social inequality in South Africa. Crankshaw has published on racial inequality in the labor market, urbanization, squatting, and neighborhood change. His *Race,*

Class and the Changing Division of Labour Under Apartheid was recently published by Routledge and he is currently writing a book on class formation and settlement among urban Africans during the apartheid period.

André P. Czeglédy (Ph.D.) is Senior Lecturer in Organizational and Business Anthropology at the University of Witwatersrand (Johannesburg). He holds degrees from the University of Toronto, Cambridge University, and London School of Economics and Political Science. His research and publications have focused on nationalism, corporate restructuring, organizational learning, and the culture of international business in central Europe. More recently, his research has examined architecture and urban development in Hungary and South Africa.

Erica Emdon is a housing and development lawyer currently working at the National Urban Reconstruction and Housing Agency (Nurcha). She has worked extensively on legal and development issues in the inner city of Johannesburg. The focus of her work over the past decade has been on using the law to facilitate transformation and urban regeneration.

Martin Gnad is a Ph.D. student in the Department of Geography at the University of Kiel (Germany). His thesis is on postapartheid development in Johannesburg with case studies of Yeoville and Robertsham.

Soraya Goga is completing her Ph.D. in Urban Planning at Rutgers University, the State University of New Jersey (United States). Her Ph.D. dissertation focuses on reasons for the decentralization of office development in Johannesburg and her main research interests lie in examining economic development strategies for local government. She has previously worked at the Durban Metropolitan Council and is currently employed by The World Bank.

Graeme Gotz is an independent researcher/consultant specializing in governance and urban and rural development. Until recently, he served as Manager of the Local Government Programme, a unit within the Graduate School of Public and Development Management (P&DM), University of Witwatersrand. His research focuses on new techniques of government needed to manage an ever more complex urban environment.

Patrick Heller is an associate professor of sociology at Brown University. He has conducted extensive research on development and democracy. His book, *The Labor of Development* (1999), explores the politics of redistributive development in the Indian State of Kerala. He was recently a visiting researcher at the Centre for Policy Studies (Johannesburg) where he researched the relationship between civil society and local government.

Ulrich Jürgens is a Lecturer in Geography at the University of Kiel (Germany) focusing on urban and economic geography in Germany and South Africa. He has nearly fifty publications including a textbook on South Africa (in press) co-written with Jürgen Bähr.

Anna Kesper holds a Masters in International Business Studies (University of Paderborn, Germany, and University of Alcala d. H., Spain) and did three years of Development Administration (UNISA, South Africa). Her M.A. thesis focuses on the informal sector in the Johannesburg central business district while her Ph.D. dissertation elaborates on success factors of formal small manufacturing enterprises in Greater Johannesburg. Kesper is currently based at the School of Geography, Archaeology, and Environmental Science, University of Witwatersrand and works as a consultant on SMME policies.

Pauline Larsen has a background in economic analysis and journalism and has spent the past six years working in property and urban research. In addition to being a director and property economist at Viruly Consulting (Johannesburg), she is currently researching the emergence and development of the Sandton business direct for a Masters degree at the University of Pretoria. Larsen sits on the executive committees of the African Real Estate Society and Women's Property Network, chairs the Sapoa Publications Committee, and lectures on urban and property economics at the University of Witwatersrand.

Ingrid Palmary is a researcher at the Centre for the Study of Violence and Reconciliation in Johannesburg. Her current research focuses on social crime prevention initiatives being undertaken in South Africa's major urban centers. She has published articles on children's experiences of violence in schools. Prior to that she has conducted research that critically examines models of counseling used to reduce family violence. She has a Masters degree in Research Psychology from the University of Natal.

Susan Parnell is an associate professor in the Department of Environmental and Geographical Sciences at the University of Cape Town. Her areas of expertise include urban poverty, segregation, and local government.

Janine Rauch is a senior consultant at the Centre for the Study of Violence and Reconciliation in Johannesburg. She has postgraduate degrees in criminology from Cambridge University and the University of Cape Town. She was employed as an advisor to the Minister of Safety and Security after the democratic election of 1994 where one of her responsibilities was to coordinate the policy process that resulted in the government's 1996 National Crime Prevention Strategy. She has published widely on police reform and crime prevention in South Africa and has advised numerous cities (including Johannesburg) on public safety strategies.

Jennifer Robinson has worked in Natal (South Africa) as well as the London School of Economics and is now at the Open University (England). Her research interests are primarily focused in southern Africa with particular attention to the relations between space and power in the construction of apartheid cities and in the emergence of the postapartheid urban form. Much of this worked appears in *The Power of Apartheid* (1996). More recent research explores urban development thinking in some cities of Africa and aims to craft a postcolonial critique of urban theory. Other interests include feminist political theory and the writings of Julia Kristeva. She is a joint editor of *Geoforum*.

AbdouMaliq Simone is presently teaching international affairs at New School University (United States). His work centers on urban practices and institutions. Simone has held academic positions at the University of Khartoum, University of Ghana, University of Western Cape, the City University of New York, and the University of Witwatersrand as well as working for several African NGOs and regional institutions. Key publications include *In Whose Image? Political Islam and Urban Practices in Sudan* (University of Chicago Press, 1994); "Urban Social Fields in South Africa," *Social Text* 56 (1998); and "Straddling the Divide: Remaking Associational Life," *Informal African City* 25 (2000).

Graeme Simpson is the Executive Director of the Centre for the Study of Violence and Reconciliation in Johannesburg. He has an LLB and a Masters degree in history from the University of Witwatersrand. Simpson is an internationally respected commentator on processes of truth recovery and reconciliation and has worked in Cambodia, Sierra Leone, Indonesia, Guatemala, Northern Ireland, and Bosnia. He was a civilian member of the team that drafted the National Crime Prevention Strategy in 1996 and the 1998 White Paper on Safety and Security.

Elizabeth Thomas is a specialist scientist at the Health and Development Group of the South Africa Medical Research Council, with prior experience in local and provincial planning and informal settlement development. She has a Masters in Town and Regional Planning (University of Cape Town) and is presently completing her Ph.D. at South Bank University (London). Her dissertation is on social capital and is titled "Vulnerability of Women in Informal Settlements in South Africa."

Index

A

Abu Dhabi, 50

Acts (legislation)

Black Communities Development Act 25 of 1984, 218

Black (Urban Areas) Consolidation Act 25 of 194, 217, 218

Businesses Act 71 of 1991, 228

Constitution Act 108 of 1996, 218

Cooperatives Act 91 of 1981, 224

Criminal Law Amendment Act 8 of 1953, 283

Development Facilitation Act 67 of 1995, 226

Free Settlement Act 102 of 1988, 25

Gauteng Residential Landlord and Tenant Act 3 of 1997, 224

Group Areas Act 41 of 1950, 23, 60, 61, 217, 222

Housing Act 107 of 1997, 226

Land Act 19 of 1998, 225

Local Government Transition Act 209 of 1993, 9, 158, 170, 226

Native Land Act 27 of 1913, 216

Native Trust and Land Act 18 of 1936, 216

Native Urban Areas Act of 1927 repealed), 5

Native (Urban Areas) Amendment Act of 1955 (repealed), 23

Natives Resettlement Act of 1954 (repealed), 23, 199

Population Registration Act 30 of 1950, 23

Prevention of Illegal Eviction from an Unlawful Occupation of Land Act 19 of 1998, 225

Prevention of Illegal Squatting Act (PISA) 52 of 1951, 224, 225

Public Safety Act of 1953, 283

Rental Housing Act 95 of 1995, 223

Sectional Titles Act 95 of 1986, 224

Unlawful Occupation of Land Act 19 of 1998, 225

Upgrading of Land Tenure Rights Act 112 of 1991, 218

Urban Areas Act of 1924 (repealed), 5

African Museum (now MuseuMAfricA), 245, 249

African National Congress (ANC) xiii, 9, 153, 155–157, 160, 161, 164, 167, 168, 170, 171, 173–181, 183, 202, 210–212, 281, 283, 285, 288

Alexandra1, 3, 40, 44, 112, 118, 129, 141, 164, 165, 182, 183, 204, 285, 286

Alexandra Civic Association (see civic)

America, 28, 283

American West, 35, 41

Americanization, 44

North America, 25, 27, 58, 238, 282

United States, xii, xiv, 19, 57, 58, 236,

apartheid, 23, 53, 108, 115, 116, 120, 153, 154, 199, 204, 221, 259, 285, 287

anti-apartheid activism / struggle, 159, 209, 215, 270

post-apartheid, xi, 22, 39, 173, 209, 257, 262,

B

Baker, Herbert31, 32, 237

becoming, 83, 124, 125, 126, 127, 135, 144–146

Belgravia, 40

Bellevue, 30

belonging, 36, 83, 124–126, 129, 131–134, 144, 145

Berea, 20, 27, 30, 61, 113, 128, 131, 222, 253

Bertrams, 113, 128

Biko, Steve, 8, 284, 289

Black Consciousness Movement, 8, 257, 284

black economic empowerment, 86, 87
bonds, 165, 168, 172, 176
   boycott, 167
Botha, P.W., 8, 285
Botswana, 187
boycott (see bonds, rents and services)
Braamfontein, 25, 27, 30, 40
Britain (see England)
Bryanston, 30
Burundi, 113

C
Cameroon, 125
Cape Town, 3, 37, 39, 40, 182, 236
central business district (see inner city)
Central Witwatersrand Metropolitan Chamber
   (Chamber) (CWM), 9, 10, 158, 166,
   169, 286, 287
Chaskalson, Arthur, 219
citizen / citizenship, 123, 153, 175, 176, 215,
   281, 287
city / cities, xi, xiv, 4, 8, 57, 130, 151, 229,
   251, 257, 259, 262, 266, 272, 273
   African, 126–128, 262, 264, 275, 276
   African world class, 4, 18, 272, 276
   apartheid / post apartheid, 22, 24, 101
   building, ix, x
   developmentalist approach, 260, 277
   edge city, 13, 28, 67
   global, xii, xiv, 257, 259, 262, 270, 274, 277
   global city function, 262, 263, 264
   global city hypothesis, 260
   hierarchy, 260, 262
   lesser, 268
   megacity, 257, 264, 268
   not-Western, 260
   not-quite global, 260
   off the map, 261
   ordinary, 257, 272, 274, 277
   reflexive, 272
   secondary, 264
   smart, 270
   subterranean, 126
   Third World, xiv, 263, 266–268
   Western, 260, 263
   white, 56
   world, xii, 3, 260, 263, 264, 266, 268, 269
   world city hypothesis, 260, 262, 263
   world class, 29, 192
Cities used for thinking about Johannesburg
   Abidjan, 265
   Cairo, 265
   Chicago, 4, 277
   in America / United States, xii, xiv, 19, 29,
      57
   in Brazil, xiv
   Kinshasa, 265, 274
   Kuala Lumpur, 270
   Lagos, 265

   London, 274, 277
   Lusaka, 277
   Manila, 274
   Mumbai, 277
   Nairobi, 265, 274
   New York, 4, 264, 277
   London, 76, 274, 277
   Saint Louis, 4
   Singapore, 268, 269
City of Johannesburg, vii, 10, 19
civic, 156–159, 161, 162, 166, 167, 172, 175,
   177, 178, 181, 210, 211, 290
   activist, 162, 166
   association / organization, 153, 158, 170
   Alexandra Civic Organisation, 161, 180, 182
   local, 157, 159, 161, 162
   movement, 8, 9, 155–157, 159, 166, 171,
      173, 181, 285
   National Association of Residents and Civic
      Organisations (NARCO), 161, 180
   Soweto Civic Association, 9, 161, 205
   structures, 156, 162, 165
civil society, 155, 156, 160, 161, 163, 166,
   173, 177, 210
Community Agency for Social Enquiry
   (CASE), 161
community based organizations (CBOs) /
   associations, 126, 163, 210
Congo (see Democratic Republic of the
   Congo)
Constitution, 169, 194, 215, 218–220, 229
   Bill of Rights, 219
   Constitutional Court, 219
   Constitutional democracy, 218
   Constitutional sovereignty, 218
Congress of South African Trade Unions
   (COSATU), 178, 283
Cote d'Ivoire1, 25
crime, xii, 27, 28, 62, 66, 93, 97, 98, 101, 104,
   106, 112–121, 172, 190, 207, 233
   and grime, 44, 53, 136
   acquisitive, 115
   capital, 101
   dangerous, 114
   economic, 113
   level / rate, 87, 102, 114, 275
   ordinary, 113
   petty, 172
   prevention / fighting, 104, 121
   property, 106, 107
   sexual, 107, 108
   substance abuse, 119
   trends, 121
   urban, 104
   victims, 118, 119
   violent (i.e. assault, murder and rape), 63,
      83, 102, 102, 103, 106, 114,
      118–120,
Country View, 60, 65, 68

D

Decentralization / decentralized (see suburbs),
44, 72, 78, 79, 81, 88
capital flight, 222, 228, 229
white flight, 62
Democratic Republic of the Congo /
Congo / Congolese, 14, 131, 132
Department of Education, 172
Department of Housing, 193, 194
Department of Provincial and Local Govern-
ment, 10, 193
Department of Transport, 142
Devland, 16
Diepkloof, 13, 16, 115, 164, 182
Diepmeadow, 204
Diepsloot, 162
domestic servants / workers, 6, 23, 246, 248
Doornfontein, 30, 40
Douglasdale, 60
Durban, 37, 39, 87
Durban Roodepoort Deep Goldmine, 209

E

East Rand (see Ekhuruleni)
Eastern Cape, 161, 179, 181
Ekhuruleni, 6, 174, 178, 179
Eldorado Park, 6
elections
democratic, 86, 209
local, 159, 166, 209
England (English / Britain / British), 3, 31, 59,
231, 237, 238
Ennerdale, 6
Ethiopia, 114
ethnic, 94, 126
enclaves, 59, 67
ethnicity, 93
identity, 162, 165
separation, 199
segregation, 56
Europe / European, xii, 4, 23, 25, 27, 29, 56,
61, 73, 257, 282
Eurocentric, 29

F

Finetown, 172, 182
foreign / foreigners, 40, 48, 94, 95, 101, 111,
112, 113, 114
foreign African, 128, 131, 132
Fordsburg, 5, 50, 275
Forestdale, 59, 60, 61, 65, 66, 68
fortification (see gated communities)

G

Gabon, 125
gated communities, 2, 60, 65, 67, 68, 108
gated compound, 32

enclosed neighborhood, 57
fortification, 35, 36
fortified citadel, 68
gated compound, 32
security village, 57, 66
walled communities, 64, 66
walled estates, 13, 47
walled settlement, 66
Gauteng, 30, 50, 88, 89, 93, 117, 161, 163,
169, 171, 176, 179, 180, 182, 188,
193, 205
Gauteng Hawkers Association (GHA),
139–141,
Gauteng Provincial Department of Transport
and Public Works, 142
Ghana, 129
Ghetto / ghettoization, 57, 58, 59, 60, 67, 101,
215
gold, 3, 4, 23, 27, 29, 39, 44, 79, 260
government, 10, 123, 127, 131, 134–136, 138,
168, 185, 191, 194, 211
black local authorities, 8, 9, 157, 285
boycott black local authorities, 8, 9
local (incl. authorities), xiv, 8, 154, 169,
170, 181, 185, 186, 193, 205–208,
211, 220, 268, 286
governance, 123, 127, 136, 128, 153, 177
governmentality, 123, 127, 134, 137, 146
gray areas, 25, 61, 128
graying, 13, 61
Greater Johannesburg Hawkers Association,
112
Greater Johannesburg Metropolitan Counci,
vii, 9, 10, 29, 39, 54, 87, 135, 141,
191, 205,206, 208, 256
group areas, 23, 24, 60
Growth, Employment and Redistribution
Strategy (GEAR), 269, 288
Guguletu, 182

H

Habitat Agenda / best practic, ex, 142
Hawkers (see informal sector)
Highgate, 48
Hillbrow, 13, 27, 30, 40, 61, 63, 128, 133,
136, 141, 222, 226, 252, 253
HIV / AIDS, x, xiv, xv, 3, 16, 17, 109, 153,
185–195, 209
AIDS (acquired immunodeficiency
syndrome), 185
AIDS deaths, 17, 189, 209
AIDS orphans, 189, 194
HIV (human immunodeficiency virus), 185
HIV affected, 186, 189, 195
HIV among women, 17
HIV infection / infected, 19, 186, 188, 189,
195
HIV-positive, 187, 194
HIV prevalence, 187, 188

impact, 186, 190, 192
  Metro AIDS Council, 192
  program, 191, 192
hostel / hostel dwellers, 102, 105, 198, 204
Houghton Estate, 30
housing, x, xiii, 4, 14, 18, 62–64, 153, 159,
    161, 164, 176, 186, 193–195, 197,
    199, 200, 207, 219, 259
  backyard dwellings / shacks, 198, 200–202,
    207, 211, 213, 229
  family, 198, 199
  freehold ownership, 203, 218
  markets, 56, 60
  owners / ownership, 129, 165, 198, 199,
    203, 206–208
  projects, 10, 14
  slumlording, 63
  squatters, 102, 165, 201
Hyde Park, 27, 28

I

Identity / identities, 36–38, 124, 125, 134,
    144, 146, 162, 165, 174, 281
  African, 274
  cultural identity, xi, 282
  reidentification, 141
iGoli
  2002, 10, 17, 18, 29, 40, 169, 271
  2010, 18, 29, 40, 186, 191, 192, 271, 272
  2030, 18, 19
income, 2, 13, 62, 76, 106, 197, 207
  household income, 190, 192, 207
India / Indian, 89, 112, 181, 289
informal economy / sector (see small, medium
    and micro-enterprise), 16, 47, 50, 54,
    64, 89, 128, 138, 161, 175
  business / enterprise, 1, 6, 95
  traders / trading (see street trading), 1, 14,
    16, 53, 222, 274–276
informal settlements / areas / peripheral /
    squatter, 14, 156, 159, 161, 165, 172,
    177, 182, 188, 267
  Finetown, 172, 182
  First, Ruth, 175, 182
  Harry Gwala, 165
  Winnie Mandela Park, 164, 172, 174, 176,
    182
Inner City Fashion District, 87
Inner City Office (ICO), 87, 135–139,
    141–147, 276
Inner City Section 59 Committee, 87
Inkatha Freedom Party (IFP), 205, 214

J

Jeppe, 5
Johannesburg Art Gallery, xiv, 153, 255
    231–242, 244–246, 248–251, 254,
    256

Erasmus, Nel, 242, 243, 247, 255
Hendricks, Anton240, 241, 242, 246, 247,
    252
Lane, Hugh, 235, 237, 239, 241, 250, 253,
    254
Lutyens, Edward, 231, 235–237, 240, 242,
    243, 254
National Gallery (London), 236, 237, 240,
Phillips, Lady Florence and Sir Lionel, 235,
    239
Tate, 236, 237, 240
Johannesburg International Airport, 43, 50
Johannesburg Stock Exchange, 28, 65, 74, 81,
    95,
Joubert Park, 128, 222, 223, 226, 231, 233,
    237, 244, 251, 253
Joubert Park Project, 233, 234, 253

K

Killarney, 28
Klipspruit, 5, 204, 207
Kruger, Paul (President of the Transvaal
    Republic), 23, 38
KwaZulu-Natal188. 204, 207, 214

L

land development objectives, 170, 171, 206,
    226
legislation (see Acts)
Lenasia, 6, 275
life expectancy, 17, 189, 192

M

M1 (freeway), 14, 16
Malls / shopping centers, 28, 44, 4–49, 54, 57,
    112
  Cresta, 46
  Dobsonville Shopping Centre, 49
  Eastgate44, 46, 48, 49
  Melrose Arch, 52
  Sandton City, 44, 46
  Southgate, 44, 46, 48
  Westgate, 46, 49
Mandela, Nelson, ix, x, 182, 187, 209, 210
Marshall, T.H., 215, 220
Masondo, Amos, 177, 182, 191
Mayekiso, Mzwanele, 160, 161, 163, 182
Mayfair, 275
Mbeki, Thabo (President), 185–187, 290
Meadowlands6, 163, 197–206, 208–213
Metropolitan Trading Company (MTC),
    138–142
Midrand, 40, 60, 71, 74, 83
Migrant / migration / immigrant, 14, 61, 64,
    102, 111, 112, 114, 125, 129, 131,
    132, 139, 202, 274
  labor / workers, 103, 204, 274

traders, 44, 50
    urban, 116, 198, 200
Morngingside, 34
Moroccan, 112
Mozambique / Mozambican, 43, 50, 112,

N
National Party, 6, 8, 46, 241, 283
nation building, ix, x, 39
neo-liberal, x, 168, 267
Newtown, 148
Nigeria / Nigerian, 14, 93, 94, 112, 125, 132,
    133, 136
non-government organizations (NGO), 95,
    106, 168, 209, 211, 224, 266, 275
northern suburbs (see suburbs)

O
"one city, one tax base", 9, 168, 285
Orange Farm, 13
Orlando, 6
Orlando East, 6
Orange Farm, 13

P
Pan African Congress, 283, 285
Pageview, 6, 24,
Parkhurst, 31
Parktown, 27, 30–32, 40, 80, 242
Parktown North, 31
Parkwood, 36
Phillips, Lady Florence and Sir Lionel, 30, 31
    (see Johannesburg Art Gallery)
place, 125
    bound, 124, 125, 131
    making, 124
Planact, 162, 286
post-apartheid (see apartheid)
Pretoria (see Tshwane)
property, 77, 78, 80, 176
    assets, 73
    disinvestment, 87
    finance companies / institutions, 64, 76, 80
    funding, 78
    index, 77
    industry, 77
    investment / investors, 74, 87
    management, 77, 79, 81
    owners, 64, 75, 76, 123
    pension funds, 72, 73
property companies / firms, 77
    AFC Holdings, 80
    Ampros, 75, 79
    Liberty Life, 79
    Old Mutual Properties , 74
    Rand Merchant Bank, 75, 79
    Sanlam, 75

R
Randburg, 13, 19, 30, 40, 46
Randlord, 237, 238
rates, 168
real estate, 71
    prices, 64, 65
    investment, 71, 72,
    investors, 75, 64, 65, 71, 72, 75
Reconstruction and Development Programme
    (RDP), 158, 159, 182, 207, 288
Reid, Graeme, 136, 137, 141, 143
rent boycott, 205, 286
    Soweto Rent Boycott, 9, 205, 214
Roodepoort, 206
Rosebank, 27, 28,
Rockey Street, 136, 138, 139
Rockey Street Market, 53, 140, 141, 144–146
Roodepoort, 206
Rosebank, 27, 28

S
Saint Louis, 4
Sandton, xiii, 1, 16, 28, 40, 46, 48, 71, 74, 79,
    269
    Council, 46
    KwaSandton, 148, 149
Santa Cruz, 59, 65, 67, 68
Saxonwold, 30
segregation, xiv, 24, 56–59, 259
Senegalese, 112, 129
Services (e.g. water and sanitation), 18, 169,
    176, 186, 195, 201, 219, 266, 286
    basic services, 189, 190, 197
    bills / billing, 165, 172, 208, 209, 210
    boycott, 8, 9
    backlogs, 18
    delivery, 153, 186, 193
    tariffs, 176
Sharpeville, 4
Santa Cruz, 59, 60, 66
Shoppers
    black, 43, 53
    experience, 48
    passing, 137
    tourist, 52
    weekend, 47
    white, 46, 53, 54
Sierra Leon, 113
small, medium and micro enterprise (SMME)
    (see informal sector and street traders),
    83, 86, 89, 91, 92, 93, 95, 275
    clothing competitors in the inner city
        (China, India, Pakistan), 89
    clothing / garment, 86, 89–98, 275
    furniture, 90, 92
social movements, 156, 167, 281, 285, 289
Sophiatown, 6, 24, 199, 210, 213, 246, 250
South African Communist Party, 112, 178,
    182,

South African Municipal Workers Union
(SAMWU), 288
South African National Civic Organization
(SANCO), 112, 156, 156, 158–176,
178–180, 182, 183, 210, 211
South (management district), 13
Soweto; 5, 13, 16, 19, 43, 44, 47–49, 65, 105,
108, 110, 115, 129, 141, 153, 162,
170, 177, 188, 189, 199, 200–206,
250, 275
Greater Soweto Accord, 9, 205,
Geater Soweto, 197–206,
Peoples' Delegation, 9, 286
Rent Boycott, 9, 205
student uprising, 217
Street traders / trading / hawkers, 16, 43, 44,
53, 83, 102, 128, 136, 137, 139–141,
222, 223, 227, 228,
Suburbs, 6, 87, 109,
development, 28
growth, 27
northern, xii, xiii, 1, 6, 21, 22, 25, 28–39,
40, 44, 46, 48, 89,
residential suburbanization, 71
southern, 23
Sudan, 287
Swilling, Mark, 286–288

T
Taxi, 50, 64, 128, 142, 223, 226, 290
association, 136, 141–145
Diepmeadow Taxi Association, 143
Management Agreement, 143, 144
parking–Jack Mincer, Metromall, Park Cen-
tral, 53, 141, 142, 145, 146
rank, 53, 146, 223
Users' Committee, 143, 144
violence, 128, 141
warfare / wars, 102, 141
Tembisa Residents Association, 182
Tleane, Ali, 161, 163, 164
Tomlinson, Richard, 274–276
transformation, 155, 168, 169
Transformation Lekgotla, 10, 169, 181
transition, 104, 111, 155–157
Transvaal Provincial Administration, 9, 205
Treatment Action Campaign, 187
Triomf, 199
Tshwane, 16
Pretoria, 6, 23, 38, 60, 161, 176

U
United Nations Commission for Human
Settlements, 270
United States (see America)
University of the Witwatersrand (Wits), vii, 27,
233, 234, 253, 286

V
Verwoerd, Hendrik, 283
Vietnam, 287
Violence, 102–105, 107, 108, 112–115, 120,
137, 211
against women, 101, 111
among migrants, 112
crime (see crime / violence)
"culture of violence", 107, 120
domestic, 103, 110
familial, 120
gun, 117, 118
political / socially functional (against
apartheid), 105, 115, 116
sexual, 110
Sexual Violence Forum, 108
structural, 103, 120
taxi, 128, 141
urban, 105
victims / victimization, 106, 118
youth, 101, 114–116
Vorster, B.J., 283

W
Walled suburbs (see gated communities)
Wattville, 164, 165, 183
Westdene36
Western Cape 88, 161, 181, 188
Western Metropolitan Local Council
(WLMC)205, 206. 208
Western Native Township199
White Paper on Local Government 1998, 169,
269
White Paper on National Strategy for the
Development and Promotion of Small
Business 1995
White Paper on Safety and Security 1998, 104
Women, 102, 108–111, 165
at risk, 83
homeless, 109, 110, 113, 120
violence against, 101, 111–113,
workers, 94, 96

World Bank, x, 103, 169, 190, 267, 270, 271
    City Development Strategy, 270, 271
World Trade Organization, 15

X
Xenophobia / xenophobic, 14, 50, 90,
    111–113, 131, 162

Y
Yeoville, 27, 30, 53, 59, 61, 62, 64–68, 93,
    127, 131, 138–140
Yeoville Traders Association, 139, 140

Z
Zambia1, 25
Zimbabwe, 43, 51, 112, 187